Studies in Applied Philosophy, Epistemology and Rational Ethics 3

For further volumes:
http://www.springer.com/series/10087

Bart Custers, Toon Calders, Bart Schermer,
and Tal Zarsky (Eds.)

Discrimination and Privacy in the Information Society

Data Mining and Profiling in Large Databases

Springer

Editors
Dr. Bart Custers (lead editor)
Faculty of Law
Leiden University
Leiden
The Netherlands

Dr. Bart Schermer
Faculty of Law
Leiden University
Leiden
The Netherlands

Dr. Toon Calders
Faculty of Math and Computer Science
Eindhoven University of Technology
Eindhoven
The Netherlands

Dr. Tal Zarsky
Faculty of Law
Haifa University
Haifa
Israel

ISSN 2192-6255
e-ISSN 2192-6263
ISBN 978-3-642-30486-6
e-ISBN 978-3-642-30487-3
DOI 10.1007/978-3-642-30487-3
Springer Heidelberg New York Dordrecht London

Library of Congress Control Number: 2012938581

Printed on acid-free paper

Springer is part of Springer Science+Business Media (www.springer.com)

Acknowledgements

As the editors of this book, we would like to thank the Netherlands Organization for Scientific Research (NWO) for funding the research project on which a large part of this book was based.

Furthermore, we would like to thank all the reviewers that contributed to this book. In alphabetical order, these are: Michael Birnhack, Paul De Hert, Arno Knobbe, Alan Miller, Rutger Rienks, Jan Kees Schakel, Elisabeth Thole, Susan Van Den Braak, Bart Van Der Sloot, Sicco Verwer, Jilles Vreeken, Ran Wolff, Albrecht Zimmerman, Indre Zliobaite, Frederik Zuiderveen Borgesius and Gerrit-Jan Zwenne.

Also many thanks to the members of our valorization panel, who provided us with welcome comments and enthusiastic support: Ron Boelsma, Sunil Choenni, Tineke Cleiren, Paul De Bra, Renzo Ghianni, Heike Goudriaan, Jeannette Quast, Rutger Rienks, Aernout Schmidt, Mine Temürhan and Claartje Thijs.

Finally, we would like to thank Lorenzo Magnani, editor of the book series of which this book is part, and Leontina Di Cecco from Springer Verlag for their support in getting our work published.

Contents

Authors

Toon Calders graduated in 1999 from the University of Antwerp with a MSc in Mathematics. He received his PhD in Computer Science from the same university in May 2003, in the database research group ADReM. From May 2003 until September 2006, he continued working in the ADReM group as a post-doctoral researcher. Since October 2006, he is an assistant professor in the Information Systems group at the Eindhoven Technical University. His main area of expertise is data mining. Toon Calders published over 70 papers on data mining in conference proceedings and journals, was conference chair of the BNAIC 2009 and EDM 2011 conferences, and is a member of the editorial board of the Springer Data Mining journal and Area Editor for the Information Systems journal.

Sunil Choenni holds a PhD in database technology from the University of Twente and a MSc in theoretical computer science from Delft University of Technology. Currently, he is heading the department of Statistical Information Management and Policy Analysis of the Research and Documentation Centre (WODC) of the Dutch Ministry of Security and Justice and is a professor of human centered ICT at the School for Communication, Media and Information Technology, Rotterdam University of Applied Sciences in Rotterdam. His research interests include data warehouses and data mining, databases, e-government, and human centered design. He published several papers in these fields.

Bart Custers PhD MSc LLB is research manager at eLaw, the Centre for Law in the Information Society at Leiden University, the Netherlands. With a background in both law and physics, his research is focused on discrimination and privacy issues of new technologies, particularly data mining and profiling.

In 2004 dr. Custers published the book "The Power of Knowledge" on the technological, legal and ethical effects of data mining and risk profiling. On a regular basis he gives lectures on profiling and privacy issues of new technological developments. He presented his work at international conferences in the United States, China, Japan, the United Kingdom, Portugal, Lithuania and Malaysia. He has published his work, over 60 publications, in both scientific journals and newspapers.

Dr. Custers is also project leader Technology in Policing at the Ministry of Security and Justice in the Netherlands. His work is focused on technologies that may contribute to law enforcement, criminal investigation and prosecution. Examples of such technologies are Automatic Number Plate Recognition (ANPR), wiretapping, fingerprints, forensic DNA research, database coupling, data mining and profiling, camera surveillance and network analyses. Currently, his team is investigating the effectiveness of new technologies in policing and performing privacy impact assessments for these applications.

As a former consultant dr. Custers worked for banks (tracing terrorist funds), insurance companies (tracing fraud), for the Dutch Customs (profiling transports and detecting suspicious cargo), for the Dutch Ministry of the Interior (privacy in the criminal law chain) and local governments.

Paul De Hert is an international fundamental rights expert, with work on human rights and criminal law, constitutionalism and the impact of technology on law. He is interested both in legal practice and more fundamental reflections about law.

At the Vrije Universiteit Brussel (VUB), Paul De Hert holds the chair of 'International, European and Belgian Criminal Law' and 'The History of Constitutionalism'. In the past he has taught 'Human Rights', 'Legal theory' and 'Constitutional criminal law'. He is Director of the Research Group on Fundamental Rights and Constitutionalism (FRC), Director of the Department of Interdisciplinary Studies of Law (Metajuridics) and a core member of the Research Group Law Science Technology & Society (LSTS). He is an associated-professor at Tilburg University where he teaches "Privacy and Data Protection" at Master level at the Tilburg Institute of Law, Technology, and Society (TILT).

He is member of the editorial boards of several national and international scientific journals, including the *Inter-American and European Human Rights Journal* (Intersentia), *Criminal Law & Philosophy* (Springer) and *The Computer Law & Security Review* (Elsevier). He is co-editor in chief of the *Supranational Criminal Law Series* (Intersentia) and the *New Journal of European Criminal Law* (Intersentia). He is editor in chief of the Flemish human rights journal *Tijdschrift voor Mensenrechten*.

Katja de Vries is a Ph.D. student at the Centre for Law, Science, Technology, and Society (Vrije Universiteit Brussel). Her Ph.D. research is focused on the collisions and interactions between legal and technological modes of thinking. More in particular she studies probabilistic understandings of rationality and equality, and the differences and continuities between their functioning within advanced data technologies and within the legal semantics of privacy, data protection and anti-discrimination law.

De Vries publishes on a wide range of topics such as technology-mediated identity construction; Foucault's technologies of the self; Ambient Intelligence; the continuity between cybernetics and current algorithmic personalization technologies; the case law, practices and political debates with regard to online searches and data retention; the role of proportionality in privacy law; enlightened ways of relating to digital spaces like Second Life; legal semiotics and Latour's empirical-philosophical way of studying the mode of enunciation of law. In 2011 she co-organized a Philosophical Reading Panel on "Privacy and Due Process After the Computational Turn" (CPDP-conference, Brussels). She is a member of the European "Living in Surveillance Societies"-network.

De Vries studied at Sciences Po in Paris, obtained three masters degrees with distinction at Leiden University (Civil Law, Cognitive Psychology and Philosophy) and graduated in 2007 at Oxford University (Magister Juris). She has been a tutor in Roman Law and the history of Psychology.

Josep Domingo-Ferrer is a Full Professor of Computer Science and an ICREA-Acadèmia Researcher at Universitat Rovira i Virgili, Tarragona, Catalonia, where he holds the UNESCO Chair in Data Privacy. He received his M.Sc. and Ph.D. derees in Computer Science from the Autonomous University of Barcelona in 1988 and 1991 (Outstanding Graduation Award). He also holds an M.Sc. in Mathematics. His research interests are in data privacy, data security, statistical disclosure control and cryptographic protocols, with a focus on the conciliation of privacy, security and functionality.

He is a Fellow of IEEE. He won the 1st Edition of the ICREA Acadèmia Prize 2008 awarded by the Government of Catalonia, which distinguished him as one of the strongest 40 faculty members in public Catalan universities as far as research is concerned. Between 2007 and 2008, he was a co-recipient of four entrepreneurship prizes. In 2004 and 2003, he received two research prizes. He has authored 3 patents and over 280 publications. He has been the coordinator of EU FP5 project CO-ORTHOGONAL and of several Spanish funded and U.S. funded research projects. He currently co-ordinates the CONSOLIDER "ARES" team on security and privacy, one of Spain's 34 strongest research teams. He has chaired 12 international conferences and has served in the program committee of over 140 conferences on privacy and security. He is a co-Editor-in-Chief of "Transactions on Data Privacy", an Area Editor of "Computer Communications", and an Associate Editor of the "Journal of Official Statistics". He has held visiting appointments in Rome, Leuven, Princeton, Munich and Milwaukee.

Giusella Finocchiaro is full professor of Internet Law and Private Law at the University of Bologna. She was a member of the special National Commission on Copyright and New Technologies, which is part of the Permanent Consulting Committee on Copyright constituted by the Ministry of Cultural Heritage and Activities. She is the Italian correspondent for various EU Commission projects. She is a past member of the ENISA (European Network and Information Security Agency) "Permanent Stakeholders Group" (2008-2010) and of the ENISA "Working Group on Privacy and Technology". She is a former member of UNCITRAL (The United Nations Commission on International Trade Law) as an expert on legal issues regarding the digital electronic signature. In 2011, she has been designated as the Italian representative at UNCITRAL Working Group on electronic commerce. She is a current member of the international group of expert for the "Accountability-Based Privacy Governance" project of the Centre for Information Policy Leadership with the collaboration of the European Data Protection Commissioners

(2008-2012). Prof. Finocchiaro is the author of many publications in the field of Computer law and Internet law.

Raphaël Gellert is a Ph.D. student at the Centre for Law, Science, Technology, and Society (Vrije Universiteit Brussel). His PhD research focuses on privacy, data protection, and human rights theory in general. More particularly, he is undertaking an analysis of privacy impact assessments, focusing on their governance, or on the consequences of infusing risk thinking for the protection of privacy. Additionally, this research is fuelled by the mutual interactions between law and technology.

Raphaël Gellert has published on conceptualization of privacy and data protection, proportionality in the case-law of the European Court of Human Rights, precaution, equality and non-discrimination, or the principle of accountability in such reviews as the Revue belge de droit international, IEEE Technology & Society Magazine, or the International Review of Law, Computing and Technology. He will be a member of the jury of the Belgian 2012 "Big Brother Awards".

Raphaël Gellert studied Law at the Université Libre de Bruxelles (2008) with distinction. In 2009 he graduated from the European Master's Degree in Human Rights and Democratization with great distinction (Venice, Helsinki). He has worked for the European Parliament's sub-committee on human rights.

Serge Gutwirth is a professor of human rights, legal theory, comparative law and legal research at the Faculty of Law and Criminology of the Vrije Universiteit Brussel (VUB), where he studied law, criminology and also obtained a post-graduate degree in technology and science studies. Until 2010 he also held a part-time position of lecturer at the Faculty of law of the Erasmus University Rotterdam where he taught philosophy of law. Since October 2003 Gutwirth is holder of a 10 year research fellowship in the framework of the VUB-Research contingent for his project 'Sciences and the democratic constitutional state: a mutual transformation process'.

Gutwirth founded and still chairs the VUB-Research group Law Science Technology & Society. He publishes widely in Dutch French and English and participates (inter)national and interdisciplinary research projects (FP6, FP7). Gutwirth has been (co-)promoter of 8 publicly defended PhD's and he has often been invited to be an external member of PhD-juries in other universities in Belgium and abroad. He is a member of several editorial boards of scientific journals.

Currently, Serge Gutwirth is particularly interested both in technical legal issues raised by technology (particularly in the field of data protection and privacy) and in more generic issues related to the articulation of law, sciences, technologies and societies.

Sara Hajian is a PhD student of Computer Science at Universitat Rovira i Virgili, Tarragona, Catalonia, where she is a member of CRISES Research Group and is also affiliated with the UNESCO Chair in Data Privacy. She received her M.Sc. degree in Computer Science from Iran University of Science and Technology (IUST) in 2008. She also had been a member of APA-IUTcert, an academic research and development center in the area of Network Security Vulnerabilities and Incident Handling (2008-2010). Her research interests are data privacy, privacy preserving data mining, discrimination discovery and prevention in data mining, privacy preserving social network analysis. She has been a visiting student at the Knowledge Discovery and Data Mining Laboratory (KDD- Lab), a joint research group of the Information Science and Technology Institute of the Italian National Research Council (CNR) in Pisa and the Computer Science Department of the University of Pisa (2011).

Faisal Kamiran got his MSCS (Master in Science and Computer Science) degree from University of the Central Punjab (UCP), Lahore in 2006. He got the top position in UCP during his MSCS. He received his PhD degree from the Eindhoven University of Technology The Netherlands in October 2011. He has done his doctoral research in the Databases and Hypermedia (DH) group under the supervision of professor Toon Calders and professor Paul De Bra. His research interests includes constraints based classification, privacy preserving and graph mining.

Stan Matwin is a Distinguished University Professor of Electrical Engineering and Computer Science at the University of Ottawa in Canada, where he directs the Text Analysis and Machine Learning (TAMALE) lab. His research is in machine learning, data mining, text mining, data privacy and computer ethics. He is also affiliated with the Institute for Computer Science, Polish Academy of Sciences, Warsaw, Poland.

Author and co-author of some 230 research papers, Dr. Matwin has worked at universities in Canada, the U.S., Europe, and Latin America. Former president of the Canadian Artificial Intelligence Society and of the IFIP Working Group 12.2 (Machine Learning). Founding Director of the Graduate Certificate in Electronic Commerce at University of Ottawa. Fellow of the European Coordinating Committee for Artificial Intelligence. Fellow and recipient of the Distinguished Service Award of the Canadian Artificial Intelligence Society.

Dr. Matwin is one of the founders of Distil Interactive Inc. and Devera Logic Inc., and has significant experience and interest in innovation and technology transfer. He has an M.Sc. in Mathematics and, a Ph.D. in Computer Science from Warsaw University, and a D.Sc. from the Institute for Computer Science, Polish Academy of Sciences.

Dr. **Mykola Pechenizkiy** is Assistant Professor at the Department of Computer Science, Eindhoven University of Technology, the Netherlands. He received his PhD from the Computer Science and Information Systems department at the University of Jyvaskyla, Finland in 2005. He has broad expertise and research interests in data mining and data-driven intelligence, and their application to various (adaptive) information systems serving industry, commerce, medicine and education. Dr. Pechenizkiy has co-authored over 60 publications in these areas and has been organizing several workshops (HaCDAIS@ICDM2011, HaCDAIS@ECML/PKDD2010, LEMEDS@AIME2011) and conferences (IEEE CBMS 2008 and 2012, BNAIC 2009, EDM 2011). He was one of the tutorials at IEEE CBMS 2010, ECML/PKDD 2010 and PAKDD 2011. Recently, he co-edited the first Handbook of Educational Data Mining and served as a guest editor of two special issues in Elsevier DKE and AIIM journals. Currently, Dr. Pechenizkiy plays a leading role in nationally and internationally funded project including NWO HaCDAIS, STW CAPA, EIT ICT

Labs Stress@Work and NL Agency CoDAK, information on which can be found from his webpage at http://www.win.tue.nl/~mpechen/.

Dino Pedreschi is a full professor at the Department of Computer Science of the University of Pisa. He has been a visiting scientist and professor at the University of Texas at Austin (1989/90), at CWI Amsterdam (1993), at UCLA (1995), and at the Northeastern University at Boston (2009/10). His current research interests are in data mining and logic in databases, and particularly in data analysis, in the integration of data mining and databases, in spatio-temporal data mining, and data privacy techniques. He participated in the scientific committee of various conferences in the area of Logic Programming and Databases. He is one of the principal investigators of the research lab KDDLAB (http://www-kdd.isti.cnr.it). He is a member of the program committee of the main international conferences on data mining and knowledge discovery; he is an associate editor of the journal Knowledge and Information Systems. He was a co-chair of ECML/PKDD 2004, the European conference on Machine Learning and Knowledge Discovery in Databases. He served as the coordinator of the undergraduate studies in Computer Science at the University of Pisa, and as a vice-rector of the same university, with responsibility in teaching affairs.

Annarita Ricci is research assistant of Private Law at the University of Bologna. She holds a PhD in Civil Law and has been awarded a post doctoral research fellowship at the University of Bologna. Since 2005 she has been teaching Private Law and ICT Law in Master's programs and courses. She has published several articles and collaborated in various collective volumes on Private Law and ICT Law. She has published the monograph *Il criterio di ragionevolezza nel diritto privato*, Cedam, 2007.

Dr. ir. **Rutger Rienks** works for the Dutch National Policing Services Agency (KLPD). He is team leader of an analyst group researching criminal phenomena. He is chair of the KLPD sensing steering committee. He is innovation advisor to the senior management, member of the steering committee on ANPR, and member of the imaging and sensor board supporting the CIO. Rutger has a background in computer science, machine learning and human computing and is specialized in automated observation of human behavior. Rutger holds a M.Sc. and a PhD. degree from the computer science department of the University of Twente. His dissertation focused on the implications of progressing technology for business meetings and bridged a gap between social psychology and computer science. His publications appeared, amongst others, in the International Journal of Policing, the national Policing Magazine, ACM Transactions on Applied Perception, AI and Society, Perception, and The Visual Computer.

Andrea Romei received a laurea degree in Computer Science (University of Pisa, February 2004) and a Ph.D. degree in Computer Science (University of Pisa, December 2009). Currently, he is a researcher associate at the Department of Computer Science (University of Pisa), and a member of the research lab KDD-LAB ((http://www-kdd.isti.cnr.it). His research interests include XML, databases, business intelligence, data mining, knowledge discovery and representation. His main contribution is on inductive databases, and in particular on the design and the implementation of an inductive database system out of XML data. Among others, his main studies appeared in international journals such as "Data and Knowledge Engineering", "Concurrency and Computation: Practice and Experience", "Software: Practice and Experience" and "Knowledge and Information Systems". He has been one of the principal investigator on the design of the KDDML system (http://kdd.di.unipi.it/kddml). His resume also includes a two-years experience as teaching assistant at the Department of Computer Science, University of Pisa, and a six-months working experience as software designer and analyst at Hyperborea S.r.l.

Salvatore Ruggieri holds a laurea degree in Computer Science (1994) and a Ph.D. in Computer Science (1999). In 1995 he has been an ERCIM fellow visiting RAL in Oxford (UK). He is currently associate professor at the Department of Computer Science of the University of Pisa, Italy, participating to the KDDLAB research lab group ((http://www-kdd.isti.cnr.it). Current research interests are focused in the data mining and knowledge discovery area, including: discrimination discovery and prevention, languages and systems for modeling the process of knowledge discovery; sequential and parallel classification algorithms; applications. He has also investigated areas of (constraint) logic programming, software verification, type systems, and meta-programming. He has participated in more than fifteen national and EU funded projects. He is currently coordinator of "Enforce", a national FIRB (Italian Fund for Basic Research) young researchers project on "Computer science and legal methods for enforcing the personal rights of non-discrimination and privacy in ICT systems" (2010-2013).

Reinier Ruissen is a senior police officer with over 25 years of experience at the Dutch Police. Reinier is Business Process Manager Sensing and Profiling and is one of the founders of sensing and profiling within the Dutch National Policing Services Agency (KLPD). Originally trained as ship officer he joined the Maritime Police Force of the KLPD (DWP) as criminal investigator and intelligence officer, where he introduced Intelligence-Led Policing (ILP). Reinier's interest is on optimizing and innovating policing processes, including the incorporation of technical solutions in day-to-day policing practice. He advises the management on the application of new technology and tactics and their implications for police officers and policing practice. Reinier is head of the 'intelligence-funnel' project (iTrechter)', a complex event processing (CEP) system developed at the KLPD, which is used to evaluate live data-streams from a variety of networked sensors. Profiles used by the iTrechter are typically based on interviews that Reinier conducts with fellow police officers and criminal investigators.

Jan-Kees Schakel M.Sc. works for the Dutch National Policing Services Agency (KLPD). He has a dual role of researcher and senior advisor information and organization. As such Jan-Kees advices the senior management on organizational development issues and innovation. His work is focused on maximizing effectiveness in operations by introducing and researching innovative work practices, tactics, and technologies. His work included the introduction and design of sensing technologies and profiling methodologies, the design of an operational information coordination center, the design of flexible operational teams, and the design of an 'art of inspection' training program. Jan-Kees holds a M.Sc. degree (cum laude) in Information Science. Under the work-title 'organizing distributed knowledge for action' Jan-Kees is currently completing his PhD-research at the Amsterdam Business School, University of Amsterdam.

Mr. dr. **Bart Willem Schermer** is an assistant professor in Internet law at Leiden University (eLaw@Leiden) and a fellow at the E.M. Meijers Institute for Legal Studies. Apart from his work at the University Bart is partner and co-founder of research and consultancy firm Considerati. Bart is co-founder and editor of the Dutch Journal of Internet Law (Tijdschrift voor Internetrecht) and a member of the Cybercrime expert group for the Court of Appeal in The Hague.

Franco Turini was born in 1949 in Italy. He graduated in Computer Science (laurea in Scienze dell' Informazione) summa cum laude in 1973, at the University of Pisa. He is currently a full professor in the Department of Computer Science of the University of Pisa. In 78/80 he has been a visiting scientist of the Carnegie-Mellon University (Pittsburgh) and of the IBM Research Center S. Jose, afterwards. In 92/93 he has been visiting professor at the University of Utah. His research interests include programming languages design, implementation, and formal semantics, knowledge representation and knowledge discovery. He is one of the principal investigators of the research lab KDDLAB (http://www-kdd.isti.cnr.it) formed by researchers of the Computer Science Department of the University of Pisa and researchers of institute ISTI of the National Research Council in Pisa. Franco Turini is a member of both ACM and IEEE.

Dr. **Susan van den Braak** is a researcher at the Statistical Data and Policy Analysis Division of the Research and Documentation Centre (WODC) of the Ministry of Security and Justice in the Netherlands. She holds a PhD in Computer Science from Utrecht University and an MSc in Artificial Intelligence from Radboud University Nijmegen.

Dr. Van den Braak's doctoral dissertation focused on software support for crime analysts. Her research interests include e-government, law enforcement, argumentation and visualization, knowledge modeling, and human-computer interaction. She has published several papers in the field of Artificial Intelligence and Law.

For the Ministry of Security and Justice, Dr. Van den Braak carried out a project on the detection of fraud in companies and corporations. Currently, she is leading a project to monitor the execution of sentences in the Dutch criminal law chain through key performance indicators. Both projects involve combining and analyzing various (judicial) databases.

Bart van der Sloot is a researcher at the Institute for Information Law, University of Amsterdam, Netherlands (http://www.ivir.nl/medewerkers/vandersloot.html). He specializes in privacy, but is also interested in the liability of internet intermediaries, internet regulation and copyright issues on the web. He is the coordinator of the Amsterdam Platform for Privacy Research (APPR), which incorporates about 50 researchers from the University of Amsterdam, who in their daily research and teaching focus on privacy related issues. They do so from different perspectives, such as law, philosophy, economics, informatics, medicine, communication studies, political science, etc. In October 2012 APPR organized a four-day, interdisciplinary, international privacy conference in Amsterdam. See: www.apc2012.org and http://www.appr.uva.nl/appr-en/main.cfm

Dr. ir. **Sicco Verwer** is postdoctoral researcher in computer science at the Radboud University Nijmegen. He received his PhD. and MSc. In Computer Science from Delft University of Technology. Afterwards, he has worked as an assistant lecturer at Delft University of Technology, and as postdoctoral researcher at Eindhoven University of Technology and the Catholic University of Leuven. His research focuses on the theory and practice of machine learning, and grammatical inference in particular.

Dr. Verwer's interests within this focus area are diverse. He has published many papers on the theory and practice of learning timed state machines, discrimination-aware data mining, and applications of exact algorithms in machine learning. An example of his work is the development of a method that removes discrimination from probabilistic classifiers. Currently, he investigates how to learn state machine models for real-world software systems, applied to black-box software testing and malware analysis.

Dr. Verwer is also a researcher at the Statistical Data and Policy Analysis Division of the Research and Documentation Centre (WODC) of the Ministry of Security and Justice in the Netherlands. In this position, he works on intelligent data analysis with applications in fraud detection and cyber security.

Dr. **Tal Zarsky** is a Senior Lecturer at the University of Haifa - Faculty of Law. His research focuses on Information Privacy, Internet Policy, Telecommunications Law and Online Commerce, as well as Contract Theory. He also teaches Property Law and Secured Transactions.

Dr. Zarsky's publications appeared among others, in the Miami Law Review, Law and Contemporary Problems, Michigan Telecommunications and Technology Law Review, the Yale Journal of Law and Technology, the Penn State Law Review and NYU Academic Press. In addition to his scholarship, Dr. Zarsky has advised various regulators and legislators on technology-related issues.

Dr. Zarsky holds LL.M and J.S.D degrees from Columbia Law School. His doctorate dissertation focused on privacy and data mining in the internet society. He received an LL.B/B.A degree (summa cum laude) in Law and Psychology from the Hebrew University of Jerusalem. He was a Hauser Global Fellow at NYU Law School (2010-2011) and a fellow with the Information Society Project at Yale Law School (2003-2004).

Indre Zliobaite is a Lecturer in Computational Intelligence at at Smart Technology Research Centre, Bournemouth University UK. Prior to that she was a postdoctoral Researcher at Eindhoven University of Technology, the Netherlands. She received her PhD from Vilnius University, Lithuania in 2010. I. Zliobaite has six years of experience in credit analysis in banking industry.

Her research interests concentrate around online data mining, including learning from evolving streaming data, change detection, adaptive and context-aware learning, predictive analytics applications. Recently she has co-chaired workshops at ECMLPKDD 2010 and ICDM 2011, co-organized tutorials at CBMS 2010 and PAKDD 2011 on adaptive learning. She is a Research Task Leader within the INFER.eu project that is developing evolving and robust predictive systems. For further information see: http://zliobaite.googlepages.com.

Part I
Opportunities of Data Mining and Profiling

Chapter 1
Data Dilemmas in the Information Society: Introduction and Overview

Bart Custers

Abstract. This chapter provides and introduction to this book and an overview of all chapters. First, it is pointed out what this book is about: discrimination and privacy issues of data mining and profiling and solutions (both technological and non-technological) for these issues. A large part of this book is based on research results of a project on how and to what extent legal and ethical rules can be integrated in data mining algorithms to prevent discrimination. Since this is an introductory chapter, it is explained what data mining and profiling are and why we need these tools in an information society. Despite this unmistakable need, however, data mining and profiling may also have undesirable effects, particularly discriminatory effects and privacy infringements. This creates dilemmas on how to deal with data mining and profiling. Regulation may take place using laws, norms, market forces and code (i.e., constraints in the architecture of technologies). This chapter concludes with an overview of the structure of this book, containing chapters on the opportunities of data mining and profiling, possible discrimination and privacy issues, practical applications and solutions in code, law, norms and the market.

1.1 The Information Society

Vast amounts of data are nowadays collected, stored and processed. These data are used for making a variety of administrative and governmental decisions. This may considerably improve the speed, effectiveness and quality of decisions. However, at the same time, it is common knowledge that most databases contain errors. Data may not be collected properly, data may be corrupted or missing, and data may be biased or contain noise. In addition, the process of analyzing the data might include biases and flaws of its own. This may lead to discrimination. For instance,

---

Bart Custers
eLaw, Institute for Law in the Information Society, Leiden University, The Netherlands
e-mail: bartcusters@planet.nl

B. Custers et al. (Eds.): Discrimination & Privacy in the Information Society, SAPERE 3, pp. 3–26.
springerlink.com

when police surveillance takes place only in minority neighborhoods, their databases would be heavily tilted towards such minorities. Thus, when searching for criminals in the database, they will only find minority criminals.

As databases contain large amounts of data, they are increasingly analyzed in automated ways. Among others, data mining technology is applied to statistically determine patterns and trends in large sets of data. The patterns and trends, however, may easily be abused, as they often lead to unwanted or unjustified selection. This may result in the discrimination of particular groups.

Furthermore, processing huge amounts of data, often personal data, may cause situations in which data controllers know many of the characteristics, behavior and whereabouts of people. Sometimes to the extent of knowing (often based on statistics) more about individuals than these individuals know about themselves. Examples of such factors are life expectancies, credit default risks and probabilities of involvement in car accidents. Ascribing characteristics to individuals or groups of people based on statistics may create a digital world in which every person has several digital identities.[1] Whether these digital identities derived from data processing are a correct and sufficiently complete representation of natural persons or not, they definitely shed different light on our views of privacy. This book addresses the issues arising as a result of these practices.

In this chapter I will provide an introduction to this book and an overview of the chapters that will follow. In this first section I will briefly introduce the premise of this book and what triggered us to write it. Next, in Section 1.2, I will explain briefly what data mining and profiling are and why we need these tools in an information society. This is not a technical section: a more detailed overview of data mining techniques can be found in Chapter 2. In Section 1.3, I will explain why this book focuses on discrimination and privacy issues. In this section, I will also point out that this book is not only about identifying and describing possible problems that data mining and profiling tools may yield, but also about providing both technical and non-technical solutions. This will become clear in Section 1.4, where I sketch the structure of this book.

1.1.1 What This Book Is About

This book will deal with the ways in which new technologies, particularly data mining, profiling and other technologies that collect and process data, may prevent or result in discriminatory effects and privacy infringements. Focus of the book will also be on the question how and to what extent legal and ethical rules can be integrated into technologies, such as data mining algorithms, to prevent such abuse. Developing (legally and ethically) compliant technologies is increasingly important because principles such as "need to know" and "select before you collect" seem difficult to implement and enforce. Such principles focusing on access controls are increasingly inadequate in a world of automated and interlinked databases and information networks, in which individuals are rapidly losing grip on who is using

[1] Solove, D. (2004).

their information and for what purposes, particularly due to the ease of copying and disseminating information. A more integrated approach, not merely focusing on the collection of data, but also on the use of data (for instance using concepts like transparency and accountability) may be preferable.

Because of the speed with which many of the technological developments take place, particularly in the field of data mining and profiling, it may sometimes be difficult for people without a technological background to understand how these technologies work and what impact the may have. This book tries to explain the latest technological developments with regard to data mining and profiling in a manner which is accessible to a broad realm of researchers. Therefore, this book may be of interest to scientists in non-technical disciplines, such as law, ethics, sociology, politics and public administration. In addition, this book may be of interest to many other professionals who may be confronted with large amounts of information as part of their work.

1.1.2 Responsible Innovation

In 2009 the Netherlands Organization for Scientific Research (NWO) commenced a new research program on responsible innovation.[2] This program (that is still running) focuses on issues concerning technological developments that will have a dramatic impact (either positive or negative) on people and/or society. The program contributes to responsible innovation by increasing the scope and depth of research into societal and ethical aspects of science and technology.

A key element of the program is the interaction between research of technological sciences (such as computer science, mathematics, physics and chemistry) and non-technological sciences (such as law, ethics and sociology), to generate cooperation between these disciplines from the early stages of developing new technologies. When it comes to legal, ethical and social effects of new technologies, parties involved are sometimes tempted to shun specific responsibilities.[3] It is often the case that engineers and technicians assert that they only build a particular technology that others can use for better or for worse. The end users, however, often state from their perspective that they only use technologies for the purposes for which they were intended or designed. A value-sensitive design approach may contribute to incorporating legal, ethical and social aspects in the early stages of developing new technologies.[4]

Another key element of the program is the use of valorization panels. Valorization is the concept of disseminating and exploiting the results of scientific (particularly academic) research results to society (particularly industries and governments) to ensure the value of this knowledge is used in practice. For this purpose, research results of the projects are discussed with valorization panels, consisting of representatives of industries and governments.

As part of the NWO program, a project team which consisted of the editors of this book was granted funding for research with regard to responsible innovation of data

[2] http://www.nwo.nl/nwohome.nsf/pages/NWOA_73HBPY_Eng
[3] Vedder, A.H., and Custers, B.H.M. (2009).
[4] Friedman, B., Kahn, P.H., Jr., and Borning, A. (2006).

mining and profiling tools.[5] The aim of this project was to investigate how and to what extent legal and ethical rules can be integrated into data mining algorithms to prevent discrimination. For the practical testing of theories this project developed, data sets in the domain of public security made available by police and justice departments, were used for testing. The project's focus was on preventing an outcome according to which selection rules turn out to discriminate particular groups of people in unethical or illegal ways. Key questions were how existing legal and ethical rules and principles can be translated into formats understandable to computers and in which way these rules can be used to guide the data mining process. Furthermore, the technological possibilities were used as feedback to formulate concrete guidelines and recommendations for formalizing legislation. These concrete tasks also related to broader and abstract themes, such as clarifying how existing ethical and legal principles are to be applied to new technologies and what the limits of privacy are. Contrary to previous scholarly attempts to examine privacy in data mining, this project did not focus on (a priori) access limiting measures regarding input data. The project's focus rather was on (a posteriori) responsibility and transparency. Instead of limiting the access to data, which is increasingly hard to enforce, questions as to how data can and may be used were stressed.

The research project was scheduled to run from October 2009 to October 2010 and conclude at that time. In reality, it never did. The research results encouraged us to engage in further research, particularly when we discovered that simply deleting discrimination sensitive characteristics (such as gender, ethnic background, nationality) from databases still resulted in (possibly) discriminating patterns. In other words, things were far more complicated than everyone initially thought. New algorithms were developed to prevent discrimination and violations of privacy. Thus far, the research results were presented in several internationally acclaimed scientific journals, at international conferences in seven countries and in technical reports, book chapters and popular journals. A complete overview of the research results can be found at the wiki of the project.[6]

During one of the meetings with the valorization panel, the panel members suggested that the research results, particularly the more technical results, are very interesting for people with a non-technical background. Thus, the valorization panel asked us whether it would be possible to combine the research results in a book that explains the latest technological developments with regard to data mining and profiling in a manner which is comprehensible to a crowd which lacks a technological background. This book tries to achieve this. This book presents the research results of our project together with contributions of leading authors in this field, all written in a non-technical language. Complicated equations were avoided as much as possible or moved to the footnotes. Technological terminology is avoided in some places and carefully explained in other places. Similarly, the jargon of the legal and other non-technical chapters is avoided or carefully explained. All this should help non-technical readers to understand what is technologically already possible (or impossible) and how exactly it works. At the same time it should help technical readers to understand how end users really view, use and judge these technological tools and why they are sometimes

[5] http://www.nwo.nl/nwohome.nsf/pages/NWOP_8K6G4N_Eng
[6] http://wwwis.win.tue.nl/~tcalders/dadm/doku.php

criticized. A more thorough understanding of all these disciplines may help responsible innovation and technology use.

1.2 Data Mining and Profiling

This book addresses the effects of data mining and profiling, two technologies that are no longer new but still subject to constant technological developments. Data mining and profiling are often mentioned in the same breath, but they may be considered separate technologies, even though they are often used together. Profiling may be carried out without the use of data mining and vice versa. In some cases, profiling may not even involve (much) technology, for instance, when psychologically profiling a serial killer. There are many definitions of data mining and profiling. The focus of this book is not on definitions, but nevertheless, a description of what we mean by these terms may be useful.

Before starting, it is important to note that data mining refers to actions that go beyond a mere statistical analysis. Although data mining results in statistical patterns, it should be mentioned that data mining is different from traditional statistical methods, such as taking test samples.[7] Data mining deals with large databases that may contain millions of records. Statisticians, however, are used to a lack of data rather than to abundance. The large amounts of data and the way the data is stored make straightforward statistical methods inapplicable. Most statistical methods also require clean data, but, in large databases, it is unavoidable that some of the data is invalid. For some data types, some statistical operations are not allowed and some of the data may not even be numerical, such as image data, audio data, text data, and geographical data. Furthermore, traditional statistical analysis usually begins with an hypothesis that is tested against the available data. Data mining tools usually generate hypotheses themselves and test these hypotheses against the available data.

1.2.1 Data Mining: A Step in the KDD-Process

Data mining is an automated analysis of data, using mathematical algorithms, in order to find new patterns and relations in data. Data mining is often considered to be only one step, the crucial step though, in a process called Knowledge Discovery in Databases (KDD). Fayyad et al. define Knowledge Discovery in Databases as the nontrivial process of identifying valid, novel, potentially useful, and ultimately understandable patterns in data.[8] This process consists of five successive steps, as is shown in Figure 1.1. In this section, it is briefly explained how the KDD process takes place.[9] A more detailed account on data mining techniques is provided in Chapter 2.

---

[7] Hand, D.J. (1998).

[8] Fayyad, U.M., Piatetsky-Shapiro, G. and Smyth, P. (1996b), p. 6.

[9] Distinguishing different steps in the complex KDD process may also be helpful in developing ethical and legal solutions for the problems of group profiling using data mining.

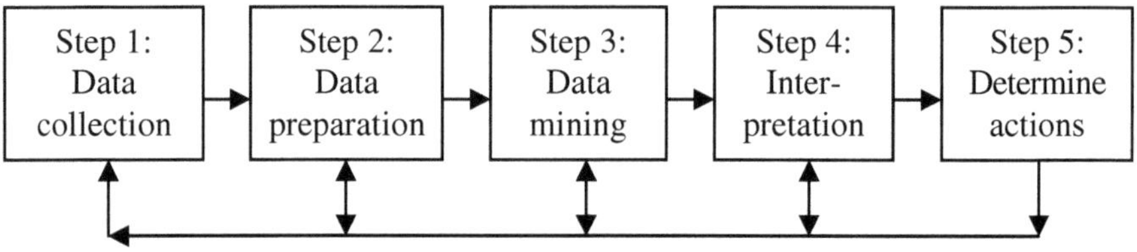

Fig. 1.1 Steps in the KDD process

Step 1: Data Collection

The first step in the KDD process is the collection of data. In the case of information about individuals, this may be done explicitly, for instance, by asking people for their personal data, or non-explicitly, for instance, by using databases that already exist, albeit sometimes for other purposes. The information requested usually consists of name, address and e-mail address. Depending on the purpose for which the information will be used, additional information may be required, such as credit card number, occupation, hobbies, date of birth, fields of interests, medical data, etc.

It is very common to use inquiries to obtain information, which are often mandatory in order to obtain a product, service, or price reduction. In this way, a take-it-or-leave-it situation is created, in which there is often no choice for a consumer but to fill in his personal data.[10] In most cases, the user is notified of the fact that privacy regulations are applied to the data. However, research shows that data collectors do not always keep this promise, especially in relation to information obtained on the Internet.[11] The same research also shows that customers are often not informed about the use that is made of the information, and in general much more information is asked for than is needed, mainly because it is thought that such data may be useful in the future.

Step 2: Data Preparation

In the second step of the KDD process, the data is prepared by rearranging and ordering it. Sometimes, it is desirable that the data be aggregated. For instance, zip codes may be aggregated into regions or provinces, ages may be aggregated into five-year categories, or different forms of cancer may be aggregated into one disease group. In this stage, a selection is often made of the data that may be useful to answer the questions set forth. But in some cases, it may be more efficient to make such a selection even earlier, in the data collection phase. The type of data and the structure and dimension of the database determine the range of data-mining tools that may be applied. This may be taken into account in selecting which of the available data will be used for data mining.

[10] These take-it-or-leave-it options are sometimes referred to as *conditional offers*.
[11] Artz, M.J.T. and Eijk, M.M.M. van (2000).

Step 3: Data Mining

The third step is the actual data-mining stage, in which the data are analyzed in order to find patterns or relations. This is done using mathematical algorithms. Data mining is different from traditional database techniques or statistical methods because what is being looked for does not necessarily have to be known. Thus, data mining may be used to discover new patterns or to confirm suspected relationships. The former is called a 'bottom-up' or 'data-driven' approach, because it starts with the data and then theories based on the discovered patterns are built. The latter is called a 'top-down' or 'theory-driven' approach, because it starts with a hypothesis and then the data is checked to determine whether it is consistent with the hypothesis.[12]

There are many different data-mining techniques. The most common types of discovery algorithms with regard to group profiling are clustering, classification, and, to some extent, regression. Clustering is used to describe data by forming groups with similar properties; classification is used to map data into several predefined classes; and regression is used to describe data with a mathematical function. Chapter 2 will elaborate on the data mining techniques.

In data mining, a *pattern* is a statement that describes relationships in a (sub)set of data such that the statement is simpler than the enumeration of all the facts in the (sub)set of data. When a pattern in data is interesting and certain enough for a user, according to the user's criteria, it is referred to as *knowledge*.[13] Patterns are interesting when they are novel (which depends on the user's knowledge), useful (which depends on the user's goal), and nontrivial to compute (which depends on the user's means of discovering patterns, such as the available data and the available people and/or technologies to process the data). For a pattern to be considered knowledge, a particular certainty is also required. A pattern is not likely to be true across *all* the data. This makes it necessary to express the certainty of the pattern. Certainty may involve several factors, such as the integrity of the data and the size of the sample.

Step 4: Interpretation

Step 4 in the KDD process is the interpretation of the results of the data-mining step. The results, mostly statistical, must be transformed into understandable information, such as graphs, tables, or causal relations. The resulting information may not be considered knowledge by the user: many relations and patterns that are found may not be useful in a specific context. A selection may be made of useful information. What information is selected, depends on the questions set forth by those performing the KDD process.

An important phenomenon that may be mentioned in this context is *masking*. When particular characteristics are found to be correlated, it may be possible to use trivial characteristics as indicators of sensitive characteristics. An example or this is indirect discrimination using redlining. Originally redlining is the practice

---

[12] SPSS Inc. (1999), p. 6.

[13] Adriaans, P. and Zantinge, D. (1996), p. 135.

of denying products and services in particular neighborhoods, marked with a red line on a map to delineate where not to invest. This resulted in discrimination against black inner city neighborhoods. For instance, when people living in a particular zip code area have a high health risk, insurance companies may use the zip code (trivial information) as an indication of a person's health (sensitive information), and may thus use the trivial information as a selection criterion. Note that refusing insurance on the basis of a zip code may be acceptable, as companies may choose (on the basis of market freedom) the geographic areas in which they operate. On the other hand, refusing insurance on the basis of sensitive data may be prohibited on the basis of anti-discrimination law. Masking may reduce transparency for a data subject, as he or she may not know the consequences of filling in trivial information, such as a zip code. In databases redlining may occur not necessarily by geographical profiling, but also by profiling other characteristics

Step 5: Acting upon Discovered Knowledge

Step 5 consists of determining corresponding actions. Such actions are, for instance, the selection of people with particular characteristics or the prediction of people's health risks. Several practical applications are discussed in Part III of this book. During the entire knowledge discovery process, it is possible –and sometimes necessary– to feedback information obtained in a particular step to earlier steps. Thus, the process can be discontinued and started over again when the information obtained does not answer the questions that need to be answered.

1.2.2 From Data to Knowledge

The KDD-process may be very helpful in finding pattern and relations in large databases that are not immediately visible to the human eye. Generally, deriving patterns and relations are considered creating added value out of databases, as the patterns and relations provide insight and overview and may be used for decision-making. The plain database may not (or at least not immediately) provide such insight. For that reason, usually a distinction is made between the terms data and knowledge. Data is a set of facts, the raw material in databases usable for data mining, whereas knowledge is a pattern that is interesting and certain enough for a user.[14] It may be obvious that knowledge is therefore a subjective term, as it depends on the user. For instance, a relation between vegetable consumption and health may be interesting to an insurance company, whereas it may not be interesting to an employment agency. Since a pattern in data must fulfill two conditions (*interestingness* and *certainty)* in order to become knowledge, we will discuss these conditions in more detail.

Interestingness

According to Frawley et al. (1991), interestingness requires three things: novelty, usefulness and non-triviality. Whether a pattern is *novel* depends on the *user's*

[14] Frawley, W.J., Piatetsky-Shapiro, G. and Matheus, C.J. (1993).

knowledge. A pattern that is not new may not be interesting. For instance, when a pattern is found according to which car accidents occur only in the group of people of over 18 years of age, this is not surprising, since the user may have already expected this.[15] Whether a pattern is already known to other people does not matter; what matters is that the pattern is new to the user.

A pattern is *useful* when it may help in achieving the *user's goals*. A pattern that does not contribute to achieving those goals may not be interesting. For instance, a pattern that indicates groups of people who buy many books is of no interest to a user who wants to sell CDs. Usefulness may be divided into an efficacy component and an efficiency component. *Efficacy* is an indication of the extent to which the knowledge contributes to achieving a goal or the extent to which the goal is achieved. *Efficiency* is an indication of the speed or easiness with which the goal is achieved.

Non-triviality depends on the *user's means*. The user's means have to be proportional to non-triviality: a pattern that is too trivial to compute, such as an average, may not be interesting. On the other hand, when the user's means are too limited to interpret the discovered pattern, it may also be difficult to speak of 'knowledge'. Looking at Figure 1.1 again, where the KDD process is illustrated, may clarify this, as a certain insight is required for Step 4, in which the results of data mining are interpreted.

Certainty

The second criterion for knowledge, certainty, depends on many factors. The most important among them are the integrity of the data, the size of the sample, and the significance of the calculated results. The *integrity* of the data concerns corrupted and missing data. When only corrupted data are dealt with, the terms *accuracy* or *correctness* are used.[16] When only missing data are dealt with, the term *completeness* is used. Integrity may refer to both accuracy and the completeness of data.[17]

Missing data may leave blank spaces in the database, but it may also be made up, especially in database systems that do not allow blank spaces. For instance, the birthdays of people in databases tend to be (more often than may be expected) on the 1st of January, because 1-1 is easiest to type.[18] Sometimes, a more serious effort is made to construct the values of missing data.[19]

The *sample size* is a second important factor influencing certainty. However, the number of samples that needs to be taken may be difficult to determine. In general, the larger the sample size, the more certain the results. Minimum sample sizes for acceptable reliabilities may be about 300 data items. These and larger samples, sometimes running up to many thousands of data items, used to be problematic for statistical research, but current databases are usually large enough to provide for enough samples.[20]

---

[15] In Europe, driving licenses may generally be obtained from the age of 18.
[16] Berti, L., and Graveleau, D. (1998).
[17] Stallings, W. (1999).
[18] Denning, D.E. (1983).
[19] Holsheimer, M., and Siebes, A. (1991).
[20] Hand, D.J. (1998).

A third important factor influencing certainty is *significance*. Significance indicates whether a discovered result is based on coincidence. For instance, when a coin is thrown a hundred times, it may be expected that heads and tails will each occur fifty times. If a 49-51 ratio were to be found, this may be considered a coincidence, but if a 30-70 ratio were found, it may be difficult to assume this is coincidental. The latter result is significantly different from what is expected. With the help of confidence intervals (see below), it is possible to determine the likelihood of whether a discovered result may be considered a coincidence or not.

Once the certainty of particular knowledge has been determined using a chosen mathematical method, it is up to the user to decide whether that certainty is sufficient for further use of that knowledge. The standard technique for calculating certainty in the case of regression techniques is the calculation of the *standard error*. The standard error indicates to what extent the data differs from the regression function determined. The larger the value of the standard error, the larger the spreading of the data. Using standard errors, it is possible to calculate *confidence intervals*. A confidence interval is a range of values with a given chance of containing the real value. In most cases, the user's confidence interval is chosen in such a way that confidence is fixed at 95 or 99 per cent.

Finally, it should be mentioned that for profiles, certainty is closely related to reliability. The reliability of a profile may be split into (a) the reliability of the profile itself, which comprises certainty, and (b) the reliability of the use of the profile. This distinction is made because a particular profile may be entirely correct from a technological perspective, but may still be applied incorrectly. For instance, when data mining reveals that 80 % of all motels are next to highways, this may be a result with a particular certainty. When all motels were counted, the certainty of this pattern is 100 %, but when a sample of 300 motels were taken in consideration, of which 240 turned out to lie next to highways, the certainty may be less because of the extrapolation. However, if a motel closes or a new motel opens, the reliability of the pattern decreases, because the pattern is based on data that are no longer up to date, yielding a pattern that represents reality with less reliability. The reliability of the use of a particular profile is yet another notion. Suppose a particular neighborhood has an unemployment rate of 80 %. When a local government addresses all people in this neighborhood with a letter regarding unemployment benefits, their use of the profile is not 100 % reliable, as they also address people who are employed.

1.2.3 Profiles of Individuals and of Groups

Profiling is the process of creating profiles. Although profiles can be made of many things, such as countries, companies or processes, in this book we focus on profiles of people or groups of people. Hence, we consider a profile a property or a collection of properties of an individual or a group of people. Several names exist for these profiles. Personal profiles are also referred to as *individual profiles* or *customer profiles*, while group profiles are also referred to as *aggregated profiles*. Others use the terms *abstract profiles* and *specific profiles* for group profiles and personal profiles, respectively.[21] Another common term is *risk profiles*, indicating the some kind of risk of an individual or group of people (such as the risk of getting a heart-attack, of not paying your mortgage or of being a terrorist).

A personal profile is a property or a collection of properties of a particular individual. A *property*, or a *characteristic*, is the same as an *attribute*, a term more used often in computer sciences. An example of a personal profile is the personal profile of Mr John Doe (44), who is married, has two children, earns 25,000 Euro a year, and has two credit cards and no criminal record. He was hospitalized only twice in his life, once for appendicitis and last year because of lung cancer.

A group profile is a property or a collection of properties of a particular group of people.[22] Group profiles may contain information that is already known; for instance, people who smoke live, on average, a few years less than people who do not. But group profiles may also show new facts; for instance, people living in zip code area 8391 may have a (significantly) larger than average chance of having asthma. Group profiles do not have to describe a causal relation. For instance, people driving red cars may have (significantly) more chances of getting colon cancer than people driving blue cars. Note that group profiles differ from individuals with regard to the fact that the properties in the profile may be valid for the group and for individuals as members of that group, though not for those individuals as such. If this is the case, this is referred to as *non-distributivity* or non-distributive properties.[23] On the other hand, when properties are valid for each individual member of a group as an individual, this is referred to as *distributivity* or distributive properties.

Several data mining methods are particularly suitable for profiling. For instance, classification and clustering may be used to identify groups.[24] Regression is more useful for making predictions about a known individual or group. More on these and other techniques can be found in Chapter 2.

1.2.4 Why We Need These Tools

The use of data mining and profiling is still on the increase, mainly because they are usually very efficient and effective tools to deal with the (also) ever increasing amounts of data that we collect and process in our information society. According to Moore's Law, the number of transistors on an integrated circuit (a 'chip' or 'microchip') for minimum component costs doubles every 24 months.[25] This more or less implies that storage capacity doubles every two years (or that data storage costs are reduced by fifty percent every two years). This empirical observation by Gordon Moore was made in 1965; by now, this doubling speed is approximately 18 months. From this perspective there is hardly any need to limit the amounts of

---

[21] See Bygrave, L.A. (2002), p. 303, and Clarke, R. (1997). See www.anu.edu.au/people/roger.clarke/dv/custproffin.html.

[22] Note that when the group size is 1, a group profile boils down to a personal profile.

[23] Vedder, A.H. (1999).

[24] SPSS Inc. (1999), p. 131.

[25] Schaller, R.R. (1997).

data we are collecting and processing. However, the amounts of data are enormous, so we do need tools to deal with these huge amounts of data. Data mining and profiling are exactly the type of technologies that may help us with analyzing and interpreting large amounts of data.

It is important to stress that due to Moore's Law we cannot get around the need for data mining and profiling tools. These tools, along with other tools for data structuring and analysis, are extremely important and it would be very difficult for an information society like ours if they would not be available. To stress this point we will provide here some major advantages of profiling. The advantages of profiling usually depend on the context in which they are used. Nevertheless, some advantages may hold for many or most contexts. At times group profiles may be advantageous compared to individual profiles. Sometimes profiling, whether it is individual profiling or group profiling, may be advantageous compared to no profiling at all. The main advantages of profiling, particularly of group profiling, concern *efficacy*, i.e., how much of the goal may be achieved, and *efficiency*, i.e., how easily the goal may be achieved. Data mining and profiling may process huge amounts of data in a short time; data that is often too complex or too great for human beings to process manually. When many examples are present in databases, (human) prejudices as a result of certain expectations may be avoided.

Profiling may be a useful method of finding or identifying target groups. In many cases, group profiling may be preferable to individual profiling because it is more cost efficient than considering each individual profile. This *cost efficiency* may concern lower costs in the gathering of information, since less information may be needed for group profiles than for individual profiles. Remember that if a group profile is based on less information, it is usually less reliable (see Section 1.2.2). But higher costs may also be expected in the time-consuming task of approaching individuals. While individuals may be approached by letter or by phone, groups may be approached by an advertisement or a news item. Take as an example baby food that is found to be poisoned with chemicals. Tracing every person who bought the baby food may be a costly process, it may take too much time, and some people may not be traced at all. A news item and some advertisements, for instance, in magazines for parents with babies, may be more successful.

Another advantage of group profiling over individual profiling is that group profiles may offer more possibilities for selecting targets. An individual may not appear to be a target on the basis of a personal profile, but may still be one. Group profiles may help in tracking down potential targets in such cases. For instance, a person who never travels may not seem an interesting target to sell a travel guide to. Still, this person may live in a neighborhood where people travel frequently. She may be interested in travel guides, not so much for using them for her own trips, but rather to be able to participate in discussions with her neighbors. A group profile for this neighborhood predicts this individual's potential interest in travel guides, whereas an individual profile may not do so. Such selection may also turn out to be an advantage for the targets themselves. For instance, members of a high-risk group for lung cancer may be identified earlier and treated, or people not interested in cars will no longer receive direct mail about them.

Profiling, regardless of whether individuals or groups are profiled, may be more useful than no profiling at all. Without any profiling or selection, the efficiency or 'hit ratio' is usually poor. For instance, advertising using inadequately defined target groups, such as on television, is less efficient than advertising only to interested and potentially interested customers.

1.3 Discrimination, Privacy and Other Issues

Despite all the opportunities described in the previous section, there are also concerns about the use of data mining and profiling. This book deals with the effects of data mining and profiling. By effects, we refer to a neutral term of what the use of these tools may result in. These effects can be positive (or at least positive to some people), as illustrated in the previous section and will be illustrated in Part III of this book. However, these effects can also be negative (or at least negative to some people). This book will deal with two major potentially negative effects of data mining and profiling, namely discrimination and privacy invasions. That is not to say that these are the only possible negative effects. Other negative effects, such as de-individualization,[26] possible loss of autonomy, one-sided supply of information, stigmatization and confrontation with unwanted information may be other examples of possible negative effects.[27] However, this book will focus on discrimination and privacy issues regarding data mining and profiling, since most progress has been made in the development of discrimination-aware and privacy preserving data mining techniques. Furthermore, even though discrimination and privacy may sometimes be difficult notions in law and ethics, they are still easier to grasp than notions like de-individualization and stigmatization, for which there hardly any legal concepts. For instance, most countries have laws regarding equal treatment (non-discrimination) and privacy, but laws against de-individualization or stigmatization are unknown to us.

1.3.1 Any News?

A New Book

Over the last years, many books and papers have been written on the possible effects of data mining and profiling.[28] What does this book to add to all this knowledge already available? First of all, most of these books focus on privacy issues, whereas this book explicitly takes discrimination issues into account. Second, we tried to include more technological background in this book, in a way that should be understandable to readers with a non-technical background. Third,

[26] Vedder, A.H. (1999).
[27] Custers, B.H.M. (2004), p. 77.
[28] Hildebrandt, M. and Gutwirth, S. (2008); Harcourt, B.E. (2007); Schauer, F. (2003); Zarsky, T. (2003); Custers, B.H.M. (2004).

this book provides technological solutions, particularly discrimination aware and privacy preserving data mining techniques. Fourth, this book explains state of the art technologies, an advantage over books published before, even though we realize that technological developments are very fast, outdating this book also within a few years.

A New Technology

Profiles were used and applied in the past without data mining, for instance, by (human) observation or by empirical statistical research. Attempts were often made to distinguish particular individuals or groups and investigate their characteristics. Thus, it may be asked what is new about profiling by means of data mining? Is it not true that we have always drawn distinctions between people?

Profiling by means of data mining may raise problems that are different from the problems that may be raised by other forms of statistical profiling such as taking test samples, mainly because data mining generates hypotheses itself. Empirical statistical research with self-chosen hypotheses may be referred to as *primary data analysis*, whereas the automated generating and testing of hypotheses, as with data mining, may be referred to as *secondary data analysis*. In the automated generating of hypotheses, the known problems of profiling may be more severe and new types of problems may arise that are related to profiling using data mining.[29] There are four reasons why profiling using data mining may be different from traditional profiling.

The first reason why profiling using data mining may cause more serious problems is a scale argument. Testing twice as much hypotheses with empirical research implies doubling the amount of researchers. Data mining is an automated analysis and does not require doubling the amount of researchers. In fact, data mining enables testing large numbers (hundred or thousands) of hypotheses (even though only a very small percentage of the results may be useful). There may be an overload of profiles.[30] Although this scale argument indicates that the known problems of group profiling are more severe, it does not necessarily imply new problems.

A second difference is that, in data mining, depending on the techniques that is used, every possible relation can be investigated, while, in empirical statistical research, usually only causal relationships are considered. The relations found using data mining are not necessarily causal. Or they may be causal without being understood. In this way, the scope of profiles that are discovered may be much broader (only a small minority of all statistical relations is directly causal) with unexpected profiles in unexpected areas. Data mining is not dependent on coincidence. Data-mining tools automatically generate hypotheses, independent of whether a relationship is (expected to be) causal or not.

---

[29] A distinction may be made between technology-specific and technology-enhanced concerns, because technology-specific concerns usually require new solutions, while conventional solutions may suffice for the technology-enhanced concerns. See also Tavani, H. (1999).

[30] See also Mitchell, T.M. (1999) and Bygrave, L.A. (2002) , p. 301.

Profiles based on statistical (but not necessarily causal) relationships may result in problems that are different from the problems of profiles based on causal relations, such as the aforementioned masking. Statistical results of data mining are often used as a starting point to find underlying causality, but it is important to note that merely statistical relations may already be sufficient to act upon, for instance, in the case of screening for diseases. The automated generation of hypotheses contributes to the scale argument as well: the number of profiles increases largely because non-causal relations can be found as well.

A third difference between data mining and empirical statistical research is that with the help of data mining trivial information may be linked (sometimes unintentionally) to sensitive information. Suppose data mining shows a relation between driving a red car and developing colon cancer. Thus, a trivial piece of information, the color of a person's car, becomes indicative of his or her health, which is sensitive information. In such cases the lack of transparency regarding data mining may start playing an important role: people who provide only trivial information may be unaware of the fact that they may also be providing sensitive information about themselves when they belong to a group of people about whom sensitive information is known. People may not even know to what groups they belong.

A fourth difference lies in a characteristic of information and communication technology that is usually referred to as the 'lack of forgetfulness of information technology'.[31,32] Once a piece of information has been disclosed, it is practically impossible to withdraw it. Computer systems do not forget things, unless information is explicitly deleted, but even then information can often be retrieved.[33] Since it is often difficult to keep information contained, it may spread through computer systems by copying and distribution. Thus, it may be difficult to trace every copy and delete it. This technological characteristic requires a different approach to finding solutions for the problems of profiling and data mining.

1.3.2 Problems and Solutions

This is a book about discrimination and privacy. That makes it a book on problems. However, instead of only discussing problems, we also provide solutions or directions for solutions to these problems. If data mining and profiling have undesirable effects, it may be regulated in several ways. Lessig distinguishes four different elements that regulate.[34] For most people, the first thing that comes to mind is to use legal constraints. Laws may regulate where and when and by whom data mining and profiling are allowed and under which conditions and circumstances. They operate as a kind of constraint on everyone who wants to use data mining and profiling.

---

[31] Blanchette, J.F., and Johnson, D.G. (1998).

[32] For this argument it should be noted that data mining is regarded as an information technology, contrary to empirical statistical research.

[33] It may be argued that paper files do not 'forget' either, but paper files are, in general, less accessible and thus there is generally less spreading of the information they contain.

[34] Lessig, L. (2006).

But laws are not the only, and often not the most significant constraint, to regulate something. Sometimes, things may be legal, but nevertheless considered unethical or impolite. Lessig mentions the example of smoking, something that is not illegal in many places, but may be considered impolite, at least without asking permission of others present in the same room. Examples of ethical issues that are strictly speaking not illegal that we will come across in this book are stigmatization of people, polarization of groups in society and de-individualization. Such norms have a certain constraint on behavior.

Apart from laws and norms, a third force is the market. Price and quality of products are important factors here. When the market supplies a wide variety of data mining and profiling tools (some of these tools may be less discriminating or more privacy friendly than others), there is more to choose from, reducing constraints. However, when there are only one or two options available, the market constrains the options. High prices (for instance, for data mining tools that do not discriminate or are privacy friendly) that may limit what you can buy.

The fourth and last constraint is created by technology. How a technology is built (its architecture) determines how it can be used. Walls may constrain where you are can go. A knife can be used for good purposes, like cutting bread, or for bad purposes, like hurting a person. Sometimes these constraints are not intended, but sometimes they are explicitly included in the design of a particular technology. Examples are copy machines that refuse to copy banknotes and cars that refuse to start without keys and, in some cases, without alcohol tests. In our case of data mining and profiling technologies, there are many constraints that can be built into the technologies. That is the reason why we separated these ‘solutions in code’ (Part IV of this book) from the other solutions (Part V of this book). Although this book has a strong focus on technological solutions, this does not mean, however, that this is the only (type of) solution. In some cases, what is needed are different attitudes, and in some cases new or stricter laws and regulations.

1.4 Structure of This Book

1.4.1 Part I: Opportunities of Data Mining and Profiling

Part I of this book explains the basics of data mining and profiling and discusses why these tools are extremely useful in the information society.

In Chapter 2, Calders and Custers explain what data mining is and how it works. The field op data mining is explored and compared with related research areas, such as statistics, machine learning, data warehousing and online analytical processing. Common terminology regarding data mining that will be used throughout this book is discussed. Calders and Custers explain the most common data mining techniques, i.e., classification, clustering and pattern mining, as well as some supporting techniques, such as pre-processing techniques and database coupling.

In Chapter 3, Calders and Žliobaitė explain why and how the use of data mining tools can lead to discriminative decision procedures, even if all discrimination sensitive data in the databases is removed or suppressed before the data mining is commenced. It is shown how data mining may exhibit discriminatory behavior towards particular groups based, for instance, upon gender or ethnicity. It is often suggested that removing all discrimination sensitive attributes such as gender and ethnicity from databases may prevent the discovery of such discriminatory relationships.[35] Without sensitive data it is impossible to find sensitive patterns or relations, it is argued. Calders and Žliobaitė show that this is not necessarily true. They carefully outline three realistic scenarios to illustrate this and explain the reasons for this phenomenon.

1.4.2 Part II: Possible Discrimination and Privacy Issues

Part II of this book explains the basics of discrimination and privacy and discusses how data mining and profiling may cause discrimination and privacy issues.

In Chapter 4, Gellert, De Vries, De Hert and Gutwirth compare and distinguish between European anti-discrimination law and data protection law. They show that both rights have the same structure and increasingly turn to the same mode of operation in the information society, even though their content is far from identical. Gellert, De Vries, De Hert and Gutwirth show that this is because both rights are grounded in the notion of negative freedom as evidenced by I. Berlin[36], and thus aim at safeguarding the autonomy of the citizen in the information society. Finally, they analyze two cases where both rights apply, and draw conclusions on how to best articulate the two tools.

In Chapter 5, Pedreschi, Ruggieri and Turini address the problem of discovering discrimination in large databases. Proving discrimination may be difficult. For instance, was a job applicant turned down because she was pregnant or because she was not suited for the job? In a single case, this may be difficult to prove, but it may be easier if there are many cases. For instance, if a company with over one thousand employees has no employees from ethnic minorities, this may be due to discrimination. Similarly, when all top management boards in a country consist of 90% of males, this may indicate possible discrimination. In Chapter 5, the focus is on finding discriminatory situations and practices hidden in large amounts of historical decision records. Such patterns and relations may be useful for anti-discrimination authorities. Pedreschi, Ruggieri and Turini discuss the challenges in discovering discrimination and present an approach for finding discrimination on the basis of legally-grounded interesting measures.

In Chapter 6, Romei and Ruggieri present an annotated bibliography on discrimination analysis. Literature on discrimination discovery and prevention is mapped in the areas of law, sociology, economics and computer sciences. Relevant legal and sociological concepts such as prejudices, racism, affirmative action (positive discrimination) and direct versus indirect discrimination are

---

[35] For instance, article 8 of the European Data Protection Directive (95/46/EC) explicitly limits the processing of special categories of data that is considered especially sensitive to data subjects, such as personal data revealing racial or ethnic origin, political opinions, religious or philosophical beliefs, trade-union membership, health and sex life.

[36] Berlin, I. (1969).

introduced guided by ample references. Furthermore, literature on economic models of labor discrimination, approaches for collecting and analyzing data, discrimination in profiling and scoring and recent work on discrimination discovery and prevention is discussed. This inventory is intended to provide a common basis to anyone working in this field.

In Chapter 7, Schermer maps out risks related to profiling and data mining that go beyond discrimination issues. Risks such as de-individualization and stereotyping are described. To mitigate these and other risks, traditionally the right to (informational) privacy is invoked. However, due to the rapid technological developments, privacy and data protection law have several limitations and drawbacks. Schermer discusses why it is questionable whether privacy and data protection legislation provide adequate levels of protection and whether these legal instruments are effective in balancing different interests when it comes to profiling and data mining.

1.4.3 Part III: Practical Applications

Part III of this book sets forth several examples of practical applications of data mining and profiling. These chapters intend to illustrate the added value of applying data mining and profiling tools. They also show several practical issues that practitioners may be confronted with.

In Chapter 8, Kamiran and Žliobaitė illustrate how self-fulfilling prophecies in data mining and profiling may occur. Using several examples they show how models learnt over discriminatory data may result in discriminatory decisions. They explain how discrimination can be measured and show how redlining may occur. Redlining originally is the practice of denying products and services in particular neighborhoods, marked with a red line on a map to delineate where not to provide credit. This resulted in discrimination against black inner city neighborhoods. In databases this effect may also occur, not necessarily by geographical profiling, but also by profiling other characteristics. Kamiran and Žliobaitė present several techniques to preprocess the data in order to remove discrimination, not by removing all discriminatory data or all differences between sensitive groups, but by addressing differences unacceptable for decision-making. With experiments they demonstrate the effectiveness of these techniques.

In Chapter 9, Schakel, Rienks and Ruissen focus on knowledge discovery and profiling in the specific context of policing. They observe that the positivist epistemology underlying the doctrine of information-led policing is incongruent with the interpretive-constructivist basis of everyday policing, and conclude that this is the cause of its failure to deliver value at the edge of action. After shifting focus from positivist information-led policing to interpretive-constructivist knowledge-based policing, they illustrate how profiling technologies can be used to design augmented realities to intercept criminals red-handedly. Subsequently, Schakel, Rienks and Ruissen discuss how the processing of data streams (rather than databases) can meet legal requirements regarding subsidiarity, proportionality, discrimination and privacy.

In Chapter 10, Van den Braak, Choenni and Verwer discuss the challenges concerning combining and analyzing judicial databases. Several organizations in the criminal justice system collect and process data on crime and law enforcement. Combining and analyzing data from different organizations may be very useful, for instance, for security policies. Two approaches are discussed, a data warehouse (particularly useful on an individual level) and a dataspace approach (particularly useful on an aggregated level). Though in principle all applications exploiting judicial data may violate data protection legislation, Van den Braak, Choenni and Verwer show that a dataspace approach is preferable with regard to taking precautions against such data protection legislation violations.

1.4.4 Part IV: Solutions in Code

Part IV of this book provides technological solutions to the discrimination and privacy issues discussed in Part II.

In Chapter 11, Matwin provides a survey of privacy preserving data mining techniques and discusses the forthcoming challenges and the questions awaiting solutions. Starting with protection of the data, methods for identity disclosure and attribute disclosure are discussed. However, adequate protection of the data in databases may not be sufficient: privacy infringements may also occur based on the inferred data mining results. Therefore, also model based identity disclosure methods are discussed. Furthermore, methods for sharing data for data mining purposes while protecting the privacy of people who contributed the data are discussed. Specifically, the chapter presents scenarios in which data is shared between a number of parties, either in a horizontal a or vertical partition. Then the privacy of individuals who contributed the data is protected by special-purpose cryptographic techniques that allow parties performing meaningful computation on the encrypted data. Finally, Matwin discusses new challenges like data from mobile devices, data from social networks and cloud computing.

In Chapter 12, Kamiran, Calders and Pechenizkiy survey different techniques for discrimination-free predictive models. Three types of techniques are discussed. First, removing discrimination from the dataset before applying data mining tools. Second, changing the learning procedures by restricting the search space to models that are not discriminating. Third, adjusting the models learned by the data mining tools after the data mining process. These techniques may significantly reduce discrimination at the cost of accuracy. The authors' experiments show that still very accurate models can be learned. Hence, the techniques presented by Kamiran, Calders and Pechenizkiy provide additional opportunities for policymakers to balance discrimination against accuracy.

In Chapter 13, Hajian and Domingo-Ferrer address the prevention of discrimination that may result from data mining and profiling. Discrimination prevention consists of inducing patterns that do not lead to discriminatory decision, even if the original data in the database is inherently biased. A taxonomy is presented for classifying and examining discrimination prevention methods. Next, preprocessing discrimination prevention methods are introduced and it is discussed how these methods deal with direct and indirect discrimination respectively. Furthermore, Hajian and Domingo-Ferrer present metrics that can be used to evaluate the performance of these approaches and show that discrimination removal can be done at a minimal loss of information.

In Chapter 14, Verwer and Calders show how positive discrimination (also known as affirmative action) can be introduced in predictive models. Three solutions based upon so-called Bayesian classifiers are introduced. The first technique is based on setting different thresholds for different groups. For instance, if there are income differences between men and women in a database, men can be given a high income label above $90,000, whereas women can be given a high income label above $75,000. Instead of income figures, the labels high and low income could be applied. This instantly reduces the discriminating pattern. The second techniques focuses on learning two separate models, one for each group. Predictions from these models are independent of the sensitive attribute. The third and most sophisticated model is focused on discovering the labels a dataset should have contained if it would have been discrimination-free. These latent (or hidden) variables can be seen as attributes of which no value is recorded in the dataset. Verwer and Calders show how decisions can be reverse engineered by explicitly modeling discrimination.

1.4.5 Part V: Solutions in Law, Norms and the Market

Part V of this book provides non-technological solutions to the discrimination and privacy issues discussed in Part II. These solutions may be found in legislation, norms and the market. Many of such solutions are discussed in other books and papers, such as (to name only a few) the regulation of profiling,[37] criteria for balancing privacy concerns and the common good,[38] self-regulation of privacy,[39] organizational change and a more academic approach,[40] and valuating privacy in a consumer market.[41] We do not discuss these suggested solutions in this book, but we do add a few other suggested solutions to this body of work.

In Chapter 15, Van der Sloot proposes to use minimum datasets to avoid discrimination and privacy violations in data mining and profiling. Discrimination and privacy are often addressed by implementing data minimization principles, restricting collecting and processing of data. Although data minimization may help to minimize the impact of security breaches, it has also several disadvantages. First, the dataset may lose value when reduced to a bare minimum and, second, the context and meaning of the data may get lost. This loss of context may cause or aggravate privacy and discrimination issues. Therefore, Van der Sloot suggests an opposite approach, in which minimum datasets are mandatory. This better ensures adequate data quality and may prevent loss of context.

In Chapter 16, Finocchiaro and Ricci focus on the opposite of being profiled, which is building one's own digital reputation. Although people have some

---

[37] See, for instance, Bygrave, L.A. (2002).
[38] Etzioni, A. (1999), p. 12/13.
[39] Regan, P.M. (2002).
[40] See, for instance, Posner, R.A. (2006), p. 210.
[41] See, for instance, Böhme (2009) and Böhme and Koble (2007).

choices in what information they provide about themselves to others (so-called informational self-determination),[42] this choice is limited to the data in databases and usually does not pertain to any results of data mining and profiling. Furthermore, due to the so-called lack of forgetfulness of information technology,[43] people have even less influence on their digital reputation. In order to reinforce the informational self-determination of people, Finocchiaro and Ricci propose the inverse of the right not to know,[44] which is the right to oblivion,[45] providing for the deletion of information which is no longer corresponds to an individual's identity.

In Chapter 17, Zarsky addresses the commonly heard complaint that there is a lack of transparency regarding the data that is collected by organizations and the ways in which these data are being used. Particularly in the context of data mining and profiling, transparency and transparency enhancing tools have been mentioned as important policy tools to enhance autonomy.[46] Transparency may also forward democracy, enhance efficiency and facilitate crowdsourcing, but it may also undermine policies and authority and generate stereotypes. While acknowledging that transparency alone cannot solve all privacy and discrimination issues regarding data mining and profiling, Zarsky provides a policy blueprint for analyzing the proper role and balance for transparency in data mining and profiling.

In Chapter 18, Zarsky considers whether the use of data mining can be conceptualized as a search (possibly an illegal search) and how this perspective can be used for policy responses. Illegal search is a common concept in criminal law, but applying this concept in the setting of data mining is novel. Three normative theories are introduced on illegal searches: these may be viewed as unacceptable psychological intrusions, as limits to the force of government or as limits to ´fishing expeditions´, i.e., looking through data of people who raise no suspicion. Zarsky shows how these theories can be used to understand data mining as illegal searches and how regulators and policymakers can establish which data mining practices are to be allowed and which must be prohibited.

1.4.6 Part VI: Concise Conclusions

Part VI of this book provides some concise conclusions. In Chapter 19, some general conclusions are drawn and the way forward is discussed. Throughout the book it becomes clear that a powerful paradigm shift is transpiring. The growing use of data mining practices by both government and commercial entities leads to both great promises and challenges. They hold the promise of facilitating an

[42] Westin, A. (1967).

[43] Blanchette, J.F., and Johnson, D.G. (1998).

[44] Chadwick, R., Levitt, M., and Shickle, D. (1997).

[45] The right to oblivion is sometimes referred to as the right to be forgotten. This right was also included in the EU proposal for revision of the EU data protection legislation that leaked end of 2011. See: https://www.privacyinternational.org/article/quick-review-draft-eu-data-protection-regulation

[46] Hildebandt, M. (2009).

information environment which is fair, accurate and efficient. At the same time, it might lead to practices which are both invasive and discriminatory, yet in ways the law has yet to grasp.

Chapter 19 starts with demonstrating this point by showing how the common measures for mitigating privacy concerns, such as a priori limiting measures (particularly access controls, anonymity and purpose specification) are mechanisms that are increasingly failing solutions against privacy and discrimination issues in this novel context.

Instead, we argue that a focus on (a posteriori) accountability and transparency may be more useful. This requires improved detection of discrimination and privacy violations as well as designing and implementing techniques that are discrimination-free and privacy-preserving. This requires further (technological) research.

But even with further technological research, there may be new situations and new mechanisms through which privacy violations or discrimination may take place. This is why Chapter 19 concludes with a discussion on the future of discrimination and a discussion on the future of privacy. With regard to discrimination, it is worth mentioning that a shift to automated predictive modeling as means of decision making and resource allocation might prove to be an important step towards a discrimination-free society. Discriminatory practices carried out by officials and employees could be detected and limited effectively. Nevertheless, two very different forms of discrimination-based problems might arise in the future. First, novel predictive models can prove to be no more than sophisticated tools to mask the "classic" forms of discrimination of the past, by hiding discrimination behind new proxies for the current discriminating factors. Second, discrimination might be transferred to new forms of population segments, dispersed throughout society and only connected by one or more attributes they have in common. Such groups will lack political force to defend their interests. They might not even know what is happening.

With regard to privacy, the adequacy of the current legal framework is discussed with regard to the technological developments of data mining and profiling discussed in this book. The European Union is currently revising the data protection legislation. The question whether these new proposals will adequately address the issues raised in this book is dealt with.

References

Adriaans, P., Zantinge, D.: Data mining. Addison Wesley Longman, England (1996)

Artz, M.J.T., van Eijk, M.M.M.: Klant in het web. In: Privacywaarborgen voor Internettoegang. Achtergrondstudies en verkenningen, vol. 17. Registratiekamer, Den Haag (2000)

Berlin, I.: Four Essays on Liberty. Oxford University Press, Oxford (1969)

Berti, L., Graveleau, D.: Designing and Filtering On-line Information Quality: New Perspectives for Information Service Providers. In: Ethicomp 1998, 4th International Conference on Ethical Issues of Information Technologies, Rotterdam (1998)

Blanchette, J.F., Johnson, D.G.: Data retention and the panopticon society: the social benefits of forgetfulness. In: Introna, L. (ed.) Computer Ethics: Philosophical Enquiry

(CEPE 1998). Proceedings of the Conference held at London School of Economics, December 13-14, pp. 113–120. London ACM SIG/London School of Economics (1998)
Böhme: Valuating Privacy with Option Pricing Theory. In: Berthold, S. (ed.) Workshop on the Economics of Information Security (WEIS 2009), June 24-25. University College London, London (2009)
Böhme, Koble: On the Viability of Privacy-Enhancing Technologies in a Self-regulated Business-to-consumer Market: Will Privacy Remain a Luxury Good? In: Proceedings of Workshop on the Economics of Information Security (WEIS), June 7-8. Carnegie Mellon University, Pittsburgh (2007)
Bygrave, L.A.: Data protection law; approaching its rationale, logic and limits. Information law series, vol. 10. Kluwer Law International, The Hague (2002)
Chadwick, R., Levitt, M., Shickle, D.: The right to know and the right not to know. Avebury Ashgate Publishing Ltd., Aldershot (1997)
Clarke, R.: Customer profiling and privacy implications for the finance industry (1997)
Custers, B.H.M.: The Power of Knowledge; Ethical, Legal, and Technological Aspects of Data Mining and Group Profiling in Epidemiology, p. 300. Wolf Legal Publishers, Tilburg (2004)
Denning, D.E.: Cryptography and Data Security. Addison-Wesley, Amsterdam (1983)
Etzioni, A.: The Limits of Privacy. Basic Books, New York (1999)
Fayyad, U.M., Piatetsky-Shapiro, G., Smyth, P.: From Data Mining to Knowledge Discovery: An Overview. In: Fayyad, U.M., Piatetsky-Shapiro, G., Smyth, P., Uthurusamy, R. (eds.) Advances in Knowledge Discovery and Data Mining. AAAI Press/The MIT Press, Menlo Park, California (1996b)
Frawley, W.J., Piatetsky-Shapiro, G., Matheus, C.J.: Knowledge Discovery in Databases; an overview. In: Piatetsky-Shapiro, G., Frawley, W.J. (eds.) Knowledge Discovery in Databases. AAAI Press/The MIT Press, Menlo Park, California (1993)
Friedman, B., Kahn Jr., P.H., Borning, A.: Value Sensitive Design and information systems. In: Zhang, P., Galletta, D. (eds.) Human-Computer Interaction in Management Information Systems: Foundations, pp. 348–372. M.E. Sharpe, Armonk (2006)
Hand, D.J.: Data mining: statistics and more? The American Statistician 52(2), 112–118 (1998)
Harcourt, B.E.: Against Prediction: Profiling, Policing and Punishing in an Actuarial Age. University of Chicago Press, Chicago (2007)
Hildebandt, M.: Behavioral Biometric Profiling and Transparency Enhancing Tools. Scientific Report of EU-program Future of Identity in the Information Society (FIDIS), WP7-D7.12 (2009), http://www.fidis.net
Hildebrandt, M., Gutwirth, S.: Profiling the European Citizen. Springer, Heidelberg (2008)
Holsheimer, M., Siebes, A.: Data Mining: the Search for Knowledge in Databases. Report CS-R9406 Centrum voor Wiskunde en Informatica, Computer Science/Department of Algorithmics and Architecture (1991)
Lessig, L.: Code Version 2.0. Basic Books, New York (2006)
Mitchell, T.M.: Machine Learning and Data Mining. Communications of the ACM 42(11) (1999)
Posner, R.A.: Uncertain Shield. Rowman & Littlefield Publishers, Inc., New York (2006)
Regan, P.M.: Privacy and commercial use of personal data: policy developments in the United States. Paper Presented at the Rathenau Institute Conference on Privacy, Amsterdam (January 17, 2002)
Schaller, R.R.: Moore's Law: Past, Present and Future. IEEE Spectrum 34, 52–59 (1997)
Schauer, F.: Profiles, Probabilities and Stereotypes. Harvard University Press, Cambridge (2003)

Solove, D.: The Digital Person; Technology and Privacy in the Information Age. University Press, New York (2004)
SPSS Inc. Data Mining with Confidence. SPSS Inc., Chicago (1999)
Stallings, W.: Cryptography and Network Security; principles and practice. Prentice Hall, Upper Saddle River (1999)
Tavani, H.: Internet Privacy: some distinctions between Internet-specific and Internet-enhanced privacy concerns. In: Proceedings of the 4th Ethicomp International Conference on the Social and Ethical Impacts of Information and Communication Technologies, Ethicomp 1999 (1999)
Vedder, A.H.: KDD: The challenge to individualism. Ethics and Information Technology 1, 275–281 (1999)
Vedder, A.H., Custers, B.H.M.: Whose responsibility is it anyway? Dealing with the consequences of new technologies. In: Sollie, P., Düwell, M. (eds.) Evaluating New Technologies: Methodological Problems for the Ethical Assessment of Technology Developments, pp. 21–34. Springer, New York (2009) (The international library of ethics, law and technology, 3)
Westin, A.: Privacy and Freedom. Bodley Head, London (1967)
Zarsky, T.: Mine Your Own Business! Making the Case for the Implications of the Data Mining of Personal Information in the Forum of Public Opinion. Yale Journal of Law and Technology 5, 57 (2002-2003)

Chapter 2
What Is Data Mining and How Does It Work?

Toon Calders and Bart Custers

Abstract. Due to recent technological developments it became possible to generate and store increasingly larger datasets. Not the amount of data, however, but the ability to interpret and analyze the data, and to base future policies and decisions on the outcome of the analysis determines the value of data. The amounts of data collected nowadays not only offer unprecedented opportunities to improve decision procedures for companies and governments, but also hold great challenges. Many pre-existing data analysis tools did not scale up to the current data sizes. From this need, the research filed of data mining emerged. In this chapter we position data mining with respect to other data analysis techniques and introduce the most important classes of techniques developed in the area: pattern mining, classification, and clustering and outlier detection. Also related, supporting techniques such as pre-processing and database coupling are discussed.

2.1 Introduction

In this chapter, we explain what data mining is and how it works. In Section 2.2 we start with exploring data mining as a research area and comparing it with related research areas, such as statistics, machine learning, data warehousing and online analytical processing. In Section 2.3 we discuss some common terminology regarding data mining that will be used throughout this book. In Section 2.4 we explain some basic discovery algorithms: classification, clustering and pattern mining. In Section 2.5 some supporting techniques are explained. These include pre-processing techniques (such as discretization, missing value imputation, dimensionality reduction and feature extraction and construction) and database

Toon Calders
Eindhoven University of Technology, The Netherlands
e-mail: t.calders@tue.nl

Bart Custers
eLaw, Institute for Law in the Information Society, Leiden University, The Netherlands
e-mail: bartcusters@planet.nl

B. Custers et al. (Eds.): Discrimination & Privacy in the Information Society, SAPERE 3, pp. 27–42.
springerlink.com

coupling. In Section 2.5 the two main questions of this chapter (what is data mining and how does it work?) are answered.

2.2 Data Mining and Related Research Areas

Data Mining emerged as a field only recently, with as a notorious milestone, the first ACM Conference on Knowledge Discovery in Databases held in August 1995 in Montreal, Canada[1]. The data mining research community grew out of many related areas, including machine learning, artificial intelligence, visualization, statistics, and analytics.

Data mining is often defined as *the automated or convenient extraction of patterns representing knowledge implicitly stored or catchable in large databases, data warehouses, the Web, other massive information repositories, or data streams*[2]. Unlike in statistics, where the data is collected specially with the purpose of testing a particular hypothesis, or estimating the parameters of a model, in data mining one usually starts with historical data that was not necessarily collected with the purpose of analysis, but rather as a by-product of an operational system. In this context, data mining is often referred to as *secondary data-analysis*[3]. Another major difference with traditional statistical methods is that data mining aims at *data-driven discovery*; instead of the user stating which hypothesis needs to be checked against the data, the data itself is used to generate the hypotheses. As such, hypotheses generated by data mining do not have the same status as those in statistics. The following example illustrates this difference using the concept of a *p-value* from statistics.

Example 1. *Suppose one throws a coin 10 times, and 9 times the coin falls head up. Under the hypothesis that the coin is fair (equal probability of heads and tails), the probability of seeing an outcome being so skewed; i.e., the chance of having nine or more of heads or nine or more tails, is approximately 2%. This value is called the p-value of the observation; it expresses how likely it is to see an outcome as extreme as observed, under the assumption that the hypothesis holds. If the p-value falls below a threshold, the level of significance, we deem the observation to be so extreme, that we reject the hypothesis. To continue the example, a data mining equivalent of this hypothesis test would be that we analyze a dataset consisting of the outcomes of 1,000 coins that have been tossed, each 10 times. Even if all coins are fair, the data mining algorithm would mark approximately 20 coins as being "suspicious", because their tosses show a disproportionally high number of tails or heads. Indeed, looking at the statistics, it is likely that among the 1,000 coin toss experiments, some will have an exceptional outcome. For those 20 suspicious coins, if we would run a statistical test on our dataset, the hypothesis that they are fair coins would be rejected. The problem with this setup is, however, that in order for a statistical test to be valid,*

---

[1] Fayyad, U.M., Uthurusamy, R. (1995).
[2] Han, J. and Kamber, M. (2006).
[3] Hand, D., Mannila, H., Smyth, P. (2001).

the data used in the test should be independent from the data that was used to generate the hypothesis.

From a methodological point of view, another difference with statistics is that in the data mining research field there is a much stronger focus on scalable techniques that work for very large datasets; for instance, techniques that scale linear in the dataset size in the sense that their running time is proportional to data size. Many statistical techniques do not scale well as they were developed initially to work on small datasets.

Most closely related to data mining is without doubt *machine learning*. There is a big overlap between the two communities, and over time the difference became less relevant and boundaries are beginning to blur. Traditionally, machine learning is about learning to perform a task, whereas data mining is more about "finding knowledge from the data". Both are tightly connected; on the one hand, in general, useful knowledge extracted from given examples of a task will allow for performing the task better, whereas on the other hand, during the learning process of a task, knowledge about the task will have to be accumulated in one form or another, from the examples, and be stored in the system. Given its task-oriented nature, historically one can see the ML community having a strong focus on supervised tasks, whereas data mining is more concerned with unsupervised tasks. One important challenge the data mining community is faced with in this perspective, is that often it is difficult to quantify the quality of a result. In a supervised context with a well-described task the quality of a solution is much easier to assess, but in an unsupervised context questions like "When does a discovered pattern represent useful knowledge?" are less obvious to answer. Another notorious field having similar problems is that of data visualization; also there it is hard to unambiguously determine if a particular visualization is informative.

Another area closely related to data mining is that of *data warehousing* and *online analytical processing (OLAP)*. In the field of online analytical processing, a myriad of highly performant data analysis techniques have been developed. A main concept here is that of a *data cube*[4], a conceptual model of the data as a multidimensional cube that can be seen as an extension of a cross-table. OLAP, however, is user-driven; it merely provides the user with the tools to quickly generate the aggregates in the data he or she selects to be displayed and presents them in a convenient display. Unlike data mining, in OLAP there is no notion of exploratory search performed by the computer algorithm; the exploration is completely determined by the user.

2.3 Database Terminology

In this section we will provide an overview of some common terminology used throughout the book. Unless stated differently, throughout the book it will be assumed that data to be analyzed is available in a structured format, such as a

---

[4] Gray, J. et al. (1997).

relational database. In a relational database, data is organized in tables that are linked together. See, for instance, the following example database consisting of three tables, Student, Course, and Grade.

Student

SID	Fname	Sname	Dob
0001	John	Williams	10/05/1985
0002	Peter	Peterson	08/09/1984
0003	Ann	Van Hee	07/05/1986

Course

Code	Name	Lecturer
2II15	Datamining	T. Calders
2ID45	Databases	G. Fletcher

Grade

SID	Code	Grade
0001	2II15	7
0002	2II15	6

Fig. 2.1 Example of a relational database consisting of three relations

Every row in a table will be called a *tuple* or a *record*, and every column corresponds to a specific characteristic, or *attribute* of the tuple.

Example 5 (Relational Database). *In the table Student in Figure 2.1, every tuple corresponds to one particular student. For every student, the name (attributes fname and sname), the student identity (attribute SID), and date of birth (attribute dob) are recorded.*

Every table has one or more attributes, that together uniquely define the identity of the objects stored in the table. Such a specially designated combination of attributes is called the *primary key* of the table. For example, in students SID is the primary key, in courses Code is the primary key, and in the table Grade, the combination of the attributes SID and Code form a primary key. The primary keys are used to establish the links between the tables. For instance, in table Grade, the primary key of Student, i.e., SID, is used to link to a particular student for which the grade is being recorded. A good database design reduces the *redundancy* and *inconsistency* in the data. Redundancy refers to the unnecessary repetition of information. Suppose, for instance that next to the SID, the table grade would also include the other attributes of students, then we would repeat a student's name and date of birth for every course the student has a recorded grade. Not only would this be wasteful, it would also lead to inconsistencies in the data; there could be different tuples with the same identity, but a different date of birth. With a good database design many of such problems can be avoided automatically.

One major problem when coupling different databases is that not all databases would use the same primary keys to identify objects. For example, suppose that we want to combine the student database of example above with a database of the financial department registering which students paid their tuition fees. It could be the case that in the dataset from the financial department the Social Security Number of the students is used to identify them, and the student number is not recorded. In such a situation, when we want to link both databases, we could only rely on common attributes in both datasets, such as first name and second name, and maybe the data of birth. Add now some misspellings or different conventions on how to treat composite names such as "Van Hee" versus "Hee, Van" to the mix, and linking the two databases may become a far from trivial problem. Resolving such linking problems is often called *entity resolution* and it often requires *disambiguation*.[5]

Therefore, often a first step in data analysis, the combination of different datasets, is far from trivial and may require itself the application of data mining or learning techniques.

2.4 Basic Techniques

In this section, several basic discovery algorithms are explained and the kinds of group profiles that may result from them are discussed. We do not present a detailed description, nor do we give an exhaustive enumeration of all methods. Only the data-mining techniques that may be relevant to group profiling, namely, classification, clustering and pattern mining are discussed.[6,7] Figure 2.2 illustrates these types of discovery algorithms.

The purpose of pattern mining is to find patterns, for instance *regression* patterns that describe data using a function. In Figure 2.2A, the data is represented by a linear function. A typical example of a linear relation is the relation between shoe size and tallness: taller persons have, in general, larger feet. And the taller the person, the larger his or her feet will be. *Clustering* is used to describe data by forming groups with similar properties. In Figure 2.2B, three different groups are identified, marked by stars (*), open dots (o) and crosses (x). After identification, descriptions of these groups may be found, indicated by the ellipses drawn. Note that the groups may overlap. *Classification* is used to map data into several predefined classes. In Figure 2.2C, a predefined class boundary is drawn (a non-linear curve), creating two classes (one to the left of the curve and one to the right of the curve). After the class boundary is defined, each data subject is classified into one of the two classes. Once it is clear to which class each data subject belongs, it is possible to attach labels, which is done by attaching crosses (x) and open dots (o).[8]

---

[5] For more on this problem, see also Subsection 2.5.2 and Chapter 10.
[6] Fayyad, U.M., Piatetsky-Shapiro, G. and Smyth, P. (1996a).
[7] Fayyad, U.M., Piatetsky-Shapiro, G. and Smyth, P. (1996b).
[8] Note that overlap is not possible in the case of classification.

In the literature, many other types of algorithms are mentioned, but most of them may be accounted for by the three types mentioned here.[9] These three types of data mining may be relevant to group profiling.

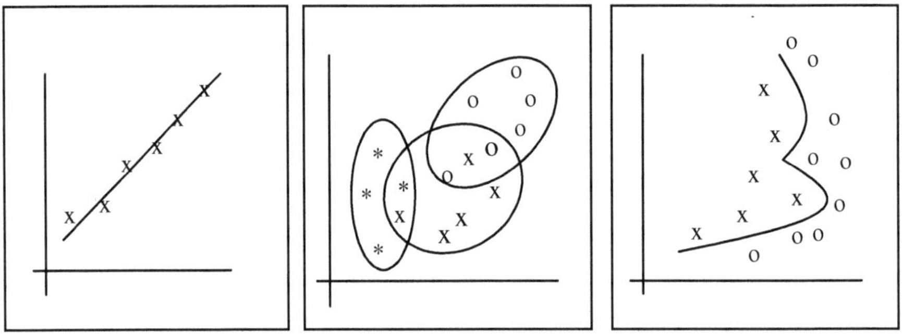

(A) (B) (C)

Fig. 2.2 Examples of different types of discovery algorithms: Pattern mining with a linear regression function (A), clustering (B), and classification (C)

2.4.1 Classification

Classification in its simplest form is the ordering of data into groups or classes on the basis of their similarity.[10] Similarity is usually determined using distance scores (see the previous subsection). The difference between clustering and classification is that classification uses predefined classes, while clustering is used to establish such classes or groups. Two basic requirements of classification are that the classes must be both *exhaustive* and *mutually exclusive*.[11] This means that all data can be assigned to one class and one class only. Of course, there may be some classes to which no data are assigned, but there is no data that cannot be assigned to any class. *Classifier induction* stands for the task of learning a classification model based on training data. In order to learn to classify based on training data, the correct labels need to be given in the database table containing the training data. This information is provided by adding a special dedicated *class attribute* that records the class a record belongs to. The value of the class attribute for a record is often referred to as the *class, class label* or *label* of the record.

Example 2 (Classification) *Based upon historical client records, an insurance company wants to learn a model for predicting the risk category of a new customer applying for car insurance, based upon his or her gender, type of car to*

[9] E.g., *decision trees* may be further divided into classification trees and regression trees; see Berry, M.J.A., and Linoff, G.S. (2000).

[10] Bailey, K.D. (1994).

[11] Note that these requirements need not be fulfilled in the case of clustering.

be insured, residential area, age, etc. As an outcome, a classification algorithm could learn for example the decision tree given in Figure 2.3.

Important in the example is the observation that we do not postulate a model on beforehand to be validated against the data, but rather use the data to guide our search for a good decision model.

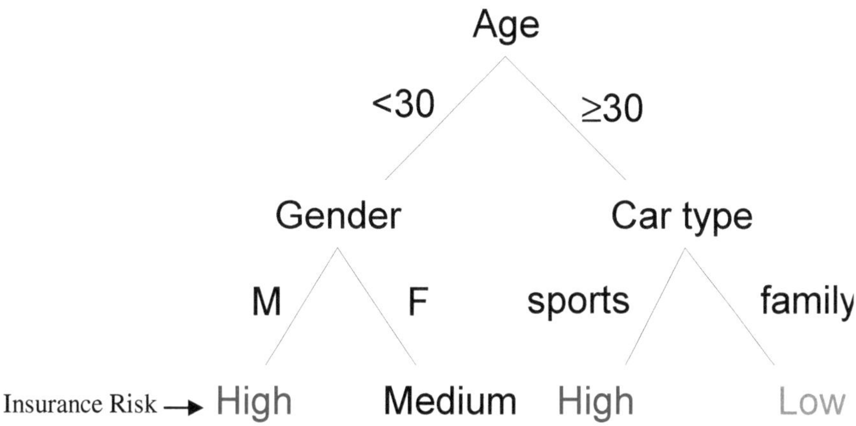

Fig. 2.3 Example of a decision tree learned by an insurance company on historical data of its customers regarding car insurance risks

Two important factors in classification methods are the number of classes and the class boundaries. The number of classes must not be confused with the dimension of the classification method, i.e., the number of attributes used in the classification. For instance, a classification may consist of four classes: male with children, male without children, female with children, and female without children, while the used classification has only two dimensions, gender and children.

The class boundaries are described using equations. The simplest form of classification is, therefore, *linear classification*, in which the data set is divided by linear equations. A single line divides the data into two classes. Such opposite classes are usually referred to as *polar types*.[12] Note that the use of linear equations does not determine the dimension of the classification. A linear equation may involve several attributes (see Figure 2.4 A for the case of linear classification for two attributes). A *threshold* (see Figure 2.4 B) is a special case of linear classification where only one attribute is considered.

The advantage of linear models is that they are relatively easy to comprehend for a user, but they may not always present a good classification of the data. Thus, in some cases, a more complex (i.e., non-linear) function (illustrated by Figure 2.4 C) may be needed to describe the data better.

[12] Bailey, K.D. (1994).

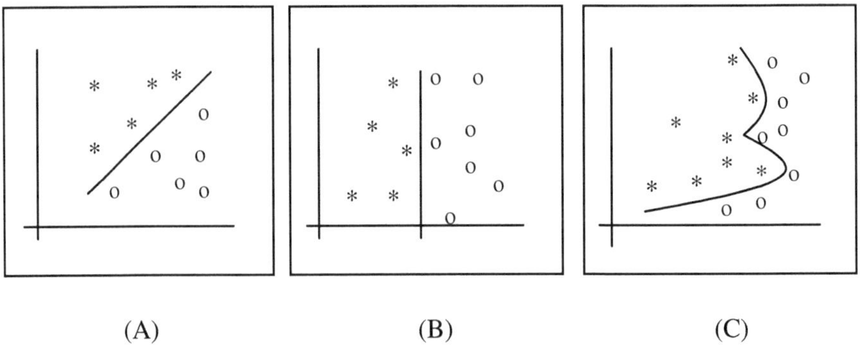

Fig. 2.4 Several methods of classification. A: Linear classification; B: A threshold, a particular type of linear classification; C: Non-linear classification.

An important point is the way in which the class boundaries are set in the first place. This may be done on the basis of an existing model or with the help of an example-based method. Existing models are dependent on the context of the data. Often, classes are chosen in such a way that they are similar in size or that they contain equal numbers of persons or an equal amount of data. An often-used example of equally sized classes is the five-year classes used in the classification of ages. For classification based on equal numbers of persons or equal amounts of data per class, the usual method is to determine the average and standard deviation of the distribution and then determine the class boundaries in these terms. The *standard deviation* is an indication of the extent to which the persons or the data differ from the average.[13]

Example-based methods determine class boundaries on the basis of a *sample* of the data. This sample should be *representative* of the data, which means that the composition of the sample should be comparable to the composition of the data. Usually, when the sample is large enough and taken at random, this is the case. Class boundaries may be determined on the basis of a sample using the clustering techniques described in the previous subsection, or on the basis of an ad-hoc hypothesis.

2.4.2 Clustering

The second large class of techniques is that of *clustering*. In clustering the goal is to divide a given dataset into homogeneous subsets. As the application of clustering does not require a set of pre-classified examples, it is often called an unsupervised technique. Whereas classification requires a "teacher" supervising

[13] The standard deviation of an attribute x is expressed as: $s = \sqrt{\frac{\sum_{i=1} (x_i - \bar{x})^2}{n-1}}$ where $\bar{x}$ is the average of all x's and n is the total number of x's.

the process by giving examples of the different classes hoping that the classifier will learn to generalize them, in clustering such supervision is not required.

Example 3 (Clustering) *Consider the set of all web-pages returned by a keyword search "bush" on the Web. The resulting set of documents will contain documents about the former president Bush Sr., of former president George W. Bush, of a grunge band named Bush, the brand of beer with the same name, and maybe also documents about vegetation. A clustering algorithm would divide, without interaction of the user or a pre-defined taxonomy, group similar documents together. Partitional clustering methods would do so by dividing the data into disjoint groups, whereas hierarchical clustering algorithms give a complete taxonomy.*

Closely related to clustering is *outlier detection*. In outlier detection, one tries to identify those objects that are unlike many other objects. Such outliers could indicate for instance, errors in the data (e.g., outside temperatures of over 60 degrees Celsius), or potentially interesting exceptional cases. Conceptually, outliers could be considered points not belonging to a large cluster, or forming a cluster by themselves.

An important factor in clustering is the order in which data points are compared with each other. Some important methods for determining this are hierarchical clustering, k-means clustering, and neural network clustering.[14] *Hierarchical clustering* starts by combining cases and clusters that are similar to each other, one pair at a time. In each step, a pair of closest cases/clusters is merged. This is repeated until the closeness of the clusters is larger than the determined threshold. In *k-means clustering*, it is assumed that the data falls into a known number (k) of clusters. First, a random profile is defined for each cluster. These profiles are called cluster centres. Next, each data point is assigned to the cluster centre to which it is most similar. *Neural network clustering* starts from so-called nodes that work similarly to the neurons in the human brain. Each node computes the weighted sum of its inputs (e.g., the distance of other nodes) and after a certain threshold is subtracted, the result is passed to a non-linear function, e.g., a sigmoid function.[15] The result of this function determines the importance of the node as a clustering centre. Neural networks are constructed by connecting the output of a node to the input of one or more other nodes.[16] It is important to select appropriate weights and thresholds. The network can also 'learn', i.e., weights and thresholds may be adjusted after several examples are compared with the desired output. In this way, strong connections are kept and weak connections are disposed of.

---

[14] SPSS Inc. (1999).

[15] Hence, each node computes a function $y = f\left(\sum_{i=1} w_i x_i - \theta\right)$ where

$f(x) = \dfrac{1}{1+e^{-x}}$ is the sigmoid function and w_i are the weights.

[16] Holsheimer, M., and Siebes, A. (1991).

Most clustering methods use *distance scores* for the calculation of *similarity*. It is important to realise that distance scores, which express relative distances between data objects, may be calculated in different ways. First of all, it depends on the data type whether distances can be calculated at all. If this is possible, the distances first need to be normalised (i.e., expressed in terms of a particular standard distance) and may then be calculated using the multidimensional equation of Pythagoras, usually called the *Euclidian distance*.[17] A weighing of the distances is also possible, if particular attributes are considered more important.

It should be mentioned that the number of dimensions n included in the clustering method might need to be limited for several reasons. For instance, the complexity of the clustering method should not be too high, in order to retain reasonable calculation times.[18] But high-dimensional spaces also make it difficult to interpret the results, since it may be hard to apply intuition. And, finally, the distance scores between any two data points in high-dimensional spaces will not really be different from the scores in lower-dimensional spaces if the extra dimensions are not relevant.[19]

The calculation of distance scores usually requires several assumptions. For instance, when the data concerns persons, it is assumed that persons of the same type are close together in the data space. Another assumption may be that persons of the same type show the same behaviour.

As a by-product of clustering, often isolation points, so-called *outliers* can be identified in the dataset. Although there also exist techniques that directly find outliers, most techniques are based upon first finding a strong clustering, and then reporting those points that do not conform to any of the found clusters.

2.4.3 Pattern Mining

The third and last class is that of the *pattern mining* techniques. Pattern mining is also unsupervised as no labels are required. Whereas clustering and classification techniques try to build global models of the data, pattern mining aims at the identification of locally valid, surprising patterns. Although technically speaking, a large collection of many small patterns could be considered a global model of the data, the quality of the patterns is not measured in terms of how well together

---

[17] The multidimensional (n dimensions) equation of Pythagoras states that the distance (d) between x and y is $\sqrt{\sum_{i=1}^{n}(x_i - y_i)^2}$ or, with weights w_i and normalization on x_i, $\sqrt{\sum_{i=1}^{n} w_i\left(\frac{x_i - y_i}{x_i}\right)^2}$.

[18] In general, the complexity should be no higher than $n\log n$, where n is the number of records; see Adriaans, P. and Zantinge, D. (1996).

[19] Irrelevant dimensions may be added, but for these dimensions $x_i \approx y_i$, which means that the resulting distance calculated by the multidimensional equation of Pythagoras is hardly influenced by these extra dimensions.

they represent the complete dataset or the relation of a response variable to the other variables. Rather the interestingness of a pattern is determined by how surprising it is;[20] in that respect it is even more likely that a pattern diverging from the global structure of the dataset is surprising. The standard example of pattern mining is association rule mining. In association rule mining a single table of 0-1 data is given and the goal is to detect relations between the columns that hold in many rows.

Example 4 (Pattern Mining) *Consider sales data of a supermarket; the rows in the dataset correspond to clients, and the columns to products. 1 in the column of a product A for a customer C indicates that the customer C bought the product A. An association rule could be that customers that buy diapers, also buy beer in 60% of the cases. Such a rule would be surprising if this 60% deviates a lot from the overall frequency of beer among all customers.*

Another example of pattern mining is finding relationships in the database that can be used to describe the data and/or predict attributes of data subjects. This is usually done with *regression*, i.e., finding a function to describe the data. The simplest regression is *linear regression*, which is used to find the line that best fits the data. Linear regression is done using the *least squares method.*[21] Non-linear regression is also possible, but is mostly done when it may reasonably be expected that the data are better described using non-linear functions. Examples of non-linear regression are exponential functions (for instance, for increasing growth), cyclical functions (for instance, for seasonal influences), and Gaussian functions (for normal distributions). Combinations of these functions are also possible, such as a combination of linear growth and seasonal influences.

One of the main concerns when using regression is whether the chosen function is a good description of the data. The quality of such a fit is often expressed by the so-called *correlation coefficient.*[22] The value r of the correlation coefficient is always between –1 and +1.[23] When r = 1, the line is a perfect fit for the data, i.e., all data points are on the line. This is called perfect positive correlation. In the

---

[20] See Subsection 1.1.2.

[21] The least squares method involves a minimization procedure of distances of the data values from the regression function. In the case of a linear fit $y=\alpha x+\beta$, the values of α and β are calculated by minimizing $\sum_{i=1} (y_i - \alpha x_i - \beta)^2$.

[22] The correlation coefficient is independent of the type of regression. In its simplest form, the correlation coefficient between two parameters x and y is

$$r = \frac{\sum_{i=1} x_i y_i - \frac{1}{n}\sum_{i=1} x_i \sum_{i=1} y_i}{\sqrt{\sum_{i=1} x_i^2 - \frac{1}{n}\left(\sum_{i=1} x_i\right)^2}\sqrt{\sum_{i=1} y_i - \frac{1}{n}\left(\sum_{i=1} y_i\right)^2}}.$$

[23] The value is often expressed as a percentage, but since this is usually done in positive values, it is impossible to distinguish between positive and negative correlation.

case of r = –1, there is perfect negative correlation. Negative correlation exists when one parameter increases while the other decreases, and vice versa. In the case of positive correlation, parameters decrease or increase simultaneously. In the case of r = 0, there is no correlation at all. No line can be found that gives a good description of the data; any line is as good or bad as any other. In practice, a correlation is seldom perfect, i.e., r = 1. Depending on the context, correlations of roughly 0.75 to 0.95 are considered high.[24] When the correlation coefficient lies between roughly –0.5 and +0.5, it is assumed that there is no correlation.

2.5 Supporting Techniques

The previous section discussed data mining techniques that aim directly at discovering patterns and relations. This section discusses some additional techniques that are not directly aimed at discovering patterns and relations, but that may nevertheless significantly enhance the results of the data mining techniques discussed in the previous section. We will distinguish pre-processing techniques and database coupling techniques.

2.5.1 Pre-processing Techniques

An important first step when analyzing data is to make sure that the input data is suitable for mining. Here we will briefly explain some common pre-processing techniques:

- **Discretization**: Some data mining methods are developed to work with *nominal attributes* only; i.e., attributes that are non-numerical and do not have any natural order. An example of such an attribute could be the brand of a car. If the dataset does contain numerical attributes, we cannot directly apply the data mining method as the data mining method will assume that the attributes are nominal and contain only a limited number of distinct values. Discretization is the process of dividing up the values of a numerical attribute into a limited number of non-overlapping ranges. For example, an attribute age could be divided into the ranges 0-10, 11-20, 21-30, and so on. The exact numerical values are then replaced by the range it falls in, effectively reducing the number of distinct values and making it useable for the nominal data mining method. In this process, necessarily some of the accuracy of the data is lost, but at the same time the dataset becomes suitable for more methods, and if the ranges are carefully chosen, the final results may even be more interpretable for a human user.

- **Missing value imputation**: In many datasets there are sporadic values missing in the records. For example, for some people in a dataset we might not know their age, and record "*null*"-unknown in database speech-instead of a value. In the discretization example we already saw

[24] This goes for negative correlation as well: between –0.95 and –0.75 the correlation is (depending on the context) considered high.

that not all data mining methods can deal with all types of data, and missing values are a notorious example of a reality with which many algorithms have difficulties to deal with. Missing value imputation techniques circumvent this problem by completing the missing field and filling in an appropriate substitute value. Ideally, the imputed values should be such that they do not disturb the overall distribution of the data in a significant way, such that the final outcome of the data mining process is not affected by the imputed values.

- **Dimensionality reduction**: Often the attributes in a dataset are highly inter-correlated and redundant. Consider for example a dataset to learn to distinguish spam email from regular mail. Suppose that the dataset contains for every mail, and for every single word that appeared in any of the mails, whether or not it appears in that mail, and if so, how many times. Such a dataset would have a tremendous *dimensionality* leading to very high running times and very complex models which will be difficult to interpret for a specialist. Dimensionality reduction techniques deal with this problem by applying transformations of the data into a lower dimensional space. Objects close to each other in the lower dimension are also close in the high dimensional space, and vice versa. In the spam emails, one dimension in the reduced space could be if the mail contains a lot of "medicine-related" words, such as "Viagra", "aspirin", "pain", etc.

- **Feature extraction and construction**: A last type of preprocessing technique is feature extraction; the process of making new features or attributes from combinations of other attributes already present in the dataset. An example would be to transform an attribute date-of-birth to an attribute age, which could be much more informative for the learning algorithm, or to combine two attributes height and weight to create a new one, the body-mass index.

2.5.2 Database Coupling

Database coupling may enhance the possibilities of data mining. When the underlying database is larger, more relations may be found than in separate databases. Figure 2.5 illustrates this, showing two very small databases. For large databases, the coupling of two databases may result in twice as many (dual) relationships as when the databases are not coupled.[25] This form of database

---

[25] For the mathematicians, two separate databases of size n can make up $2\binom{n}{2}$ dual relations, whereas the coupled database of size 2n can make up $\binom{2n}{2}$ dual relations. For $n \to \infty$, the quotient in the number of dual relations can be calculated using basic mathematics and results in a factor of 2.

coupling is referred to as *integration*. The integration of databases leads to a new, larger database.[26]

There are two basic forms of database coupling. First, several records may be integrated. For instance, suppose that a particular type of cancer is rare, and there are many hospitals with only a few patients suffering from this disease. Since every hospital database contains little data, epidemiological research is difficult. Combining the records of these patients may allow or enhance such research.

Second, attributes may be integrated. In this case, not the number of patients, but the number of attributes per patient increases. Extending the previous example, this may become possible when the medical insurance database is coupled with the medical database of a hospital. Attributes concerning insurance and medical attributes are now integrated. This can be done, of course, only when the databases contain information on the same individuals.

The coupling of databases requires that the databases have the same identifier system, something that is not always the case.[27] However, combinations of integrating records and integrating attributes are possible as well. In this way, integration may be full, i.e., all the data are in the new database, or integration may be partial, i.e., only parts of the data are in the new database. Partial integration may be used to find missing data.[28]

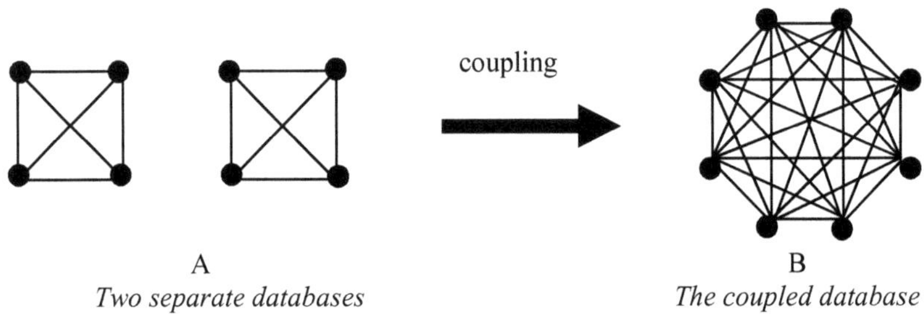

Fig. 2.5 Separate databases result in far fewer (dual) relationships (represented by lines) than do coupled databases: two separate databases of four data items (represented by dots) result in twelve (two times six) relationships (A), whereas the coupled database with a total of eight data items results in 28 relationships (B).

The different methods of database coupling are illustrated in Figure 2.6. The blocks in this figure are to be interpreted as relational database matrix structures, where the rows represent the records and the columns represent the attributes. The coloured sections are filled with data; the blank sections are empty.

It should be mentioned that in coupling, a distinction is usually drawn between computer matching, verification,[29] and computer profiling.[30] These forms of

[26] Note that the integration may be temporary, since it is possible to retain copies of the separate databases and to destroy the integrated database after use.
[27] National Research Council (1997), p. 118.
[28] Running a check of the data against another database is called verification.
[29] Verification is sometimes referred to as *computer-assisted front-end verification.*
[30] OTA Report (1986).

coupling, however, refer to the coupling of data, not to the coupling of databases.[31] Matching and verification are not closely related to enhancing data mining and are, therefore, beyond the scope of this chapter. For more on combining database and identity resolution issues, see Chapter 10.

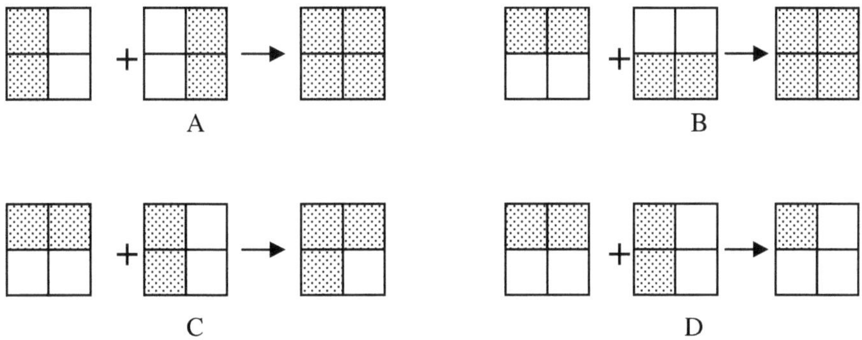

Fig. 2.6 Different forms of database coupling. The dotted parts are filled with data and the blank parts are empty. A: The coupling of records; B: The coupling of attributes per record; C: The combination of coupling records and attributes; full integration; D: The combination of coupling records and attributes; partial integration. The horizontal length represents the number of records and the vertical length represents the number of attributes.

2.6 Conclusion

In this chapter we introduced data mining as a technique to build models on huge amounts of data. The need for data mining is motivated by the challenges posed by the huge amounts of data available nowadays. Data mining offers many different tools for the automatic analysis of data. In the chapter we discussed two unsupervised techniques: pattern mining to find local patterns, each describing a particular trend or regularity in the data, and clustering which aims at building a global model of the data by dividing the dataset into clusters of homogeneous data records. The third technique, classification, was a supervised technique as it required the availability of records extended with a class attribute that holds the label of the group to which the record belongs.

In the data mining community many algorithms were developed for these three main tasks, providing governments and companies with new tools to build better profiles and make more accurate predictions in the future, extrapolating from information extracted from the past.

As will be discussed in the next chapters of this book, however, these new data mining techniques also harbour some dangers. When collecting data for data mining, often data from many different databases needs to be coupled and combined, often leading to privacy problems for the individuals whose personal

[31] Although verification is not really a method of database coupling, it may enhance the results of data mining, as was mentioned above.

data may be present in the database. Surprisingly, the blind application of data mining may also lead to discrimination, when data mining methods start over-generalizing negative properties. The different threats by data mining to privacy and anti-discrimination will be discussed in depth in the next chapters, as well as techniques to detect and avoid undesirable side-effects of applying data mining.

References

Adriaans, P., Zantinge, D.: Data mining. Addison Wesley Longman, Harlow (1996)

Bailey, K.D.: Typologies and Taxonomies; an introduction to classification techniques. In: Quantitative Applications in the Social Sciences, vol. (102). SAGE Publications, Thousand Oaks (1994)

Berry, M.J.A., Linoff, G.S.: Mastering Data Mining; the Art and Science of Customer Relationship Management. Wiley Computer Publishing, John Wiley & Sons, Inc., New York (2000)

Fayyad, U.M., Uthurusamy, R.: Proceedings of the First International Conference on Knowledge Discovery and Data Mining (KDD 1995), Montreal, Canada, August 20-21. AAAI Press (1995)

Fayyad, U.M., Piatetsky-Shapiro, G., Smyth, P.: The KDD Process for Extracting Useful Knowledge from Volumes of Data. Communications of the ACM 39(11) (1996a)

Fayyad, U.M., Piatetsky-Shapiro, G., Smyth, P.: From Data Mining to Knowledge Discovery: An Overview. In: Fayyad, U.M., Piatetsky-Shapiro, G., Smyth, P., Uthurusamy, R. (eds.) Advances in Knowledge Discovery and Data Mining. AAAI Press/The MIT Press, Menlo Park, California (1996b)

Gray, J., Chaudhuri, S., Bosworth, A., Layman, A., Reichart, D., Venkatrao, M., Pellow, F., Pirahesh, H.: Data Cube: A Relational Aggregation Operator Generalizing Group-By, Cross-Tab, and Sub-Totals. Journal Data Mining and Knowledge Discovery 1(1) (1997)

Han, J., Kamber, M.: Data Mining: Concepts and Techniques. In: Gray, J. (Series ed.) The Morgan Kaufmann Series in Data Management Systems, 2nd edn. Morgan Kaufmann Publishers (March 2006)

Hand, D., Mannila, H., Smyth, P.: Principles of Data Mining. MIT press (2001)

Holsheimer, M., Siebes, A.: Data Mining: the Search for Knowledge in Databases. Report CS-R9406 Centrum voor Wiskunde en Informatica, Computer Science/Department of Algorithmics and Architecture (1991)

National Research Council. For the Record; protecting electronic health information, Computer Science and Telecommunications Board, National Research Council. National Academic Press, Washington, DC (1997)

OTA Report. Computer Profiling. In: Electronic Record Systems and Individual Privacy. OTA Report, Congress of the United States (1986)

SPSS Inc. Data Mining with Confidence. SPSS Inc., Chicago (1999)

Chapter 3
Why Unbiased Computational Processes Can Lead to Discriminative Decision Procedures

Toon Calders and Indrė Žliobaitė

Abstract. Nowadays, more and more decision procedures are supported or even guided by automated processes. An important technique in this automation is data mining. In this chapter we study how such automatically generated decision support models may exhibit discriminatory behavior towards certain groups based upon, e.g., gender or ethnicity. Surprisingly, such behavior may even be observed when sensitive information is removed or suppressed and the whole procedure is guided by neutral arguments such as predictive accuracy only. The reason for this phenomenon is that most data mining methods are based upon assumptions that are not always satisfied in reality, namely, that the data is correct and represents the population well. In this chapter we discuss the implicit modeling assumptions made by most data mining algorithms and show situations in which they are not satisfied. Then we outline three realistic scenarios in which an unbiased process can lead to discriminatory models. The effects of the implicit assumptions not being fulfilled are illustrated by examples. The chapter concludes with an outline of the main challenges and problems to be solved.

3.1 Introduction

Data mining is becoming an increasingly important component in the construction of decision procedures (See Chapter 2 of this book). More and more historical data is becoming available, from which automatically decision procedures can be derived. For example, based on historical data, an insurance company could apply

Toon Calders
Eindhoven University of Technology, The Netherlands
e-mail: t.calders@tue.nl

Indrė Žliobaitė
Bournemouth University, UK
e-mail: izliobaite@Bournemouth.ac.uk

B. Custers et al. (Eds.): Discrimination & Privacy in the Information Society, SAPERE 3, pp. 43–57.
springerlink.com

data mining techniques to model the risk category of customers based on their age, profession, type of car, and history of accidents. This model can then be used to advise the agent on pricing when a new client applies for car insurance.

In this chapter we will assume that a data table is given for learning a model, for example, data about past clients of an insurance company and their claims. Every rows of the table represent an individual case, called an *instance*. In the insurance company example, every row could correspond to one historical client. The instances are described by their characteristics, called *attributes* or *variables*. The attributes of a client could for example be his or her gender, age, years of driving experience, a type of car, a type of insurance policy. For every client the exact same set of attributes is specified. Usually there is also one special *target attribute*, called the *class* attribute that the company is interested to predict. For the insurance example, this could, e.g., be whether or not the client has a high accident risk. The value of this attribute can be determined by the insurance claims of the client. Clients with a lot of accidents will be in the high risk category, the others in the low risk category. When a new client arrives, the company wants to predict his or her risk as accurately as possible, based upon the values of the other attributes. This process is called *classification*. For classification we need model the dependency of the class attribute on the other attributes. For that purpose many classification algorithms have been developed in machine learning, data mining and pattern recognition fields, e.g. a decision tree, a support vector machine, logistic regression. For a given classification task *a model* that relates the value of the class attribute to the other attributes needs to be learned on *the training data*; i.e., instances of which the class attribute is known. A *learned* model for a given task could be for example a set of rules such as:

IF Gender=male and car type=sport THEN risk=high.

Once a model is learned, it can be deployed for classifying new instances of which the class attribute is unknown. The process of learning a classifier from training data is often referred to as *Classifier induction*. For a more detailed overview of classifiers and how they can be derived from historical data, see Chapter 2.

In this chapter we will show that data mining and classifier induction can lead to similar problems as for human decision makers, including basing their decisions upon discriminatory generalizations. This can be particularly harmful since data mining methods are often seen as solidly based upon statistics and hence purely rational and without prejudice. *Discrimination* is the prejudiced treatment of an individual based on their membership in a certain group or category. In most European and Northern-American countries, it is forbidden by law to discriminate against certain protected-by-law groups (See Chapter 4 of this book for an overview). Although we do not explicitly refer to the anti-discrimination legislation of a particular country, most of our examples will directly relate to EU directives and legislation. The European Union has one of the strongest anti-discrimination legislations (See, e.g., Directive 2000/43/EC, Directive 2000/78/EC/ Directive 2002/73/EC, Article 21 of the Charter of Fundamental Rights and Protocol 12/Article 14 of the European Convention on Human Rights), describing discrimination on the basis of race, ethnicity, religion, nationality, gender, sexuality, disability, marital status, genetic features, language and age. It does so in a number of settings, such as employment and training, access to housing, public services, education and health care; credit and insurance; and adoption. European efforts on the non-discrimination front make clear the fundamental importance for Europe's citizens of the effective implementation and enforcement of non-discrimination norms. As a recent European Court of Justice case-law on age discrimination suggests, non-discrimination norms constitute fundamental principles of the European legal order. (See, e.g., Case 144/04 [2005] ECR I-9981 (ECJ), Judgment of the Court of 22 November 2005, Werner Mangold v Rüdiger Helm; Case C-555/07 [2010], Judgment of the Court (Grand Chamber) of 19 January 2010, Seda Kücükdeveci v Swedex GmbH & Co. KG.) Therefore it is in the interest of banks, insurance companies, employment agencies, the police and other institutions that employ computational models for decision making upon individuals to ensure that these computational models are free from discrimination. In this chapter, discrimination is considered to be present if for two individuals that have the same characteristic relevant to the decision making and differ only in the sensitive attribute (e.g., gender or race) a model results in different decisions.

The main reason that data mining can lead to discrimination is that the computational model construction methods are often based upon assumptions that turn out not to be true in practice. For example, in general it is assumed that the data on which the model is learned follows the same distribution as the data on which the classifier will have to work; i.e., the situation will not change. In section 4.2 we elaborate on the implicit assumptions made during classifier construction and illustrate with fictitious examples how they may be violated in real situations. In Section 4.3 we move on to show how this mismatch between reality and the assumptions could lead to discriminatory decision processes. We show three types of problems that may occur: sampling bias, incomplete data, or incorrect labeling. We show detailed scenarios in which the problems are illustrated. In Section 4.4 we discuss some simple solutions to the discrimination problem, and show why these straightforward approaches do not always solve the problem. Section 4.5 then concludes the chapter by giving an overview of the research problems and challenges in discrimination-aware data mining and connects them to the other chapters in this book.

We would like to stress that all examples in this chapter are purely fictitious; they do not represent our experiences with discrimination in real life, or our belief of where these processes are actually happening. Instead this chapter is a purely mechanical study of *how* we believe such processes occur.

3.2 Characterization of the Computational Modeling Process

Computational models are mathematical models that predict an outcome from characteristics of an object. For example, banks use computational models (classifiers) for credit scoring. Given characteristics of an individual, such as age, income, credit history, the goal is to predict whether a given client will repay the loan. Based on that prediction a decision whether to grant a credit is made. Banks build their models using their historical databases of customer performance. The objective is to achieve as good accuracy as possible on unseen new data. Accuracy is the share of correct predictions in the total number of predictions.

Computational models are built and trained by data mining experts using historical data. The performance and properties of a model depend, among other factors, on the historical data that has been used to train it. This section provides an overview of the computational modeling process and discusses the expected properties of the historical data. The next section will discuss how these properties translate into models that may result in biased decision making.

3.2.1 Modeling Assumptions

Computational models typically rely on the assumptions, that (1) the characteristics of the population will stay the same in the future when the model is applied, and (2) the training data represents the population well. These assumptions are known as the *i.i.d.* setting, which stands for *independently identically distributed* random variables (see e.g. Duda, Hart and Stork, 2001).

The first assumption is that the characteristics of the population from which the training sample is collected are the same as the characteristics of the population on which the model will be applied. If this assumption is violated, models may fail to perform accurately (Kelly, Hand and Adams, 1999). For instance, the repayment patterns of people working in the car manufacturing industry may be different at times of economic boom as compared to times of economic crisis. A model trained at times of boom may not be that accurate at times of crises. Or, a model trained on data collected in Brazil may not be correct to predict the performance of customers in Germany.

The second assumption is satisfied if our historical dataset closely resembles the population of the applicants in the market. That means, for instance, that our training set needs to have the same share of good and bad clients as the market, the same distribution of ages as in the market, the proportion of males and females, and the same proportion high-skilled and low-skilled labor. In short, the second assumption implies that our historical database is a small copy of a large population out there in the market. If the assumption is violated, then our training data is incomplete and a model trained on such data may perform sub-optimally (Zadrozny, 2004).

The representation of the population in our database may be inaccurate in two ways. Either the selection of people to be included may be biased or the selection of attributes by which people are described in our database may be incomplete. Suppose that a bank collects a dataset consisting only of people that live in a major city. A model is trained on this data and then it is applied to all incoming customers, including the ones that live in remote rural areas, and have different employment opportunities and spending habits. The model may not perform well on the rural customers, since the training was forced to focus on the city customers. Or suppose that a bank collects a representative sample of clients, but does not ask about the stability of income of people, which is considered to be one of the main factors in credit performance. Without this information the model will treat all the individuals as if they earn the same and thus lose the opportunity to improve upon accuracy for people with very high and very low income stability.

If the two assumptions are satisfied, it is reasonable to expect that models will transfer the knowledge from the historical data to the future decision making. On the other hand, however, if the historical data is prejudiced, the models trained on this data can be expected to yield prejudiced decisions. As we will see in the following subsection the assumptions may not hold in reality due to the origins of data. If the *i.i.d.* assumptions are not satisfied, the computational models built in such settings might still be valid; however, possible effects of these breaches need to be taken into account when interpreting the results.

3.2.2 Origins of Training Data

In order to identify the sources of possible discrimination in trained models we need to analyze the origins and the characteristics of the training data.

Data Collection

First of all, *the data collection process* may be intentionally or unintentionally biased. For instance, Turner & Skidmore (1999) discuss different stages of the mortgage lending process that potentially may lead to racial discrimination. Advertising and promotions can be sent to selected neighborhoods. Pre-application consultancy may be offered on a biased basis. These actions may lead to a situation when the historical database of applicants does not represent the potential clients. Other examples of biased data collection include racial profiling of crime suspects or selecting people for further security checks at airports. If people of particular ethnic backgrounds are stopped for searches more often, even if they were never convicted for carrying forbidden items, the historical database will contain a skewed representation of a population.

Relations between Attributes in Data

Second, the attributes that characterize our subjects may *not be independent* from each other. For example, a postal code of a person may be highly correlated with ethnicity, since people may tend to choose to live close to relatives, acquaintances or a community (see Rice, 1996 for more examples in lending). A marital status may be correlated with gender, for instance, the statuses as "wife" or "husband" directly encode gender, while "divorced" does not relate to gender.

If the attributes are independent, every attribute contributes its separate share to the decision making in the model. If variables are related to each other, it is not straightforward to identify and control which variable contributes to what extent to the final prediction. Moreover, it is often impossible to collect all the attributes of a subject or take all the environmental factors into account with a model. Therefore our data may be *incomplete*, i.e., missing some information and some hidden information may be transferred indirectly via correlated attributes.

Data Labeling
Third, the historical data to be used for training a model contains the true *labels*, which in certain cases may be incorrect and contain prejudices. Labels are the targets that an organization wants to predict for new incoming instances. The true labels in the historical data may be *objective or subjective*. The labels are objective when assigning these labels, no human interpretation was involved; the labels are *hard* in the sense that there can be no disagreement about their correctness between different human observers. Examples of objective labels include the indicators weather an existing bank customer *repaid* a credit or not, whether a suspect *was wearing* a concealed weapon, or whether a driver *tested* positive or negative for alcohol intoxication. Examples of subjective labels include the assessment of a human resource manager if a job candidate is suitable for a particular job, if a client of a bank should get a loan or not, accepting or denying a student to a university, the decision whether or not to detain a suspect. For the subjective labels there is a gray area in which human judgment may have influenced the labeling resulting in a bias in the target attribute. In contrast to the objective labels, here there may be disagreement between different observers; different people may assess a job candidate or student application differently; the notion of what is the correct label is fuzzy.

The distinction between subjective and objective labels is important in assessing and preventing discrimination. Only the subjective labels can be incorrect due to biased decision making in the historical data. For instance, if females have been discriminated in university admission, some labels in our database saying whether persons should be admitted will be incorrect according to the present non-discriminatory regulations. Objective labels, on the other hand, will be correct even if our database is collected in a biased manner. For instance, we may choose to detain suspects selectively, but the resulting true label whether a given suspect actually carried a gun or not will be measurable and is thus objectively correct.

The computational modeling process requires an insightful analysis of the origins and properties of training data. Due to origins of data the computational models trained on this data may be based on incorrect assumptions, and as a result, as we will see in the next section, may lead to biased decision making.

3.3 Types of Problems

In this section we discuss three scenarios that show how the violation of the assumptions sketched in the previous section may affect the validity of models learned on data and lead to discriminatory decision procedures. In all three scenarios we explicitly assume that the only goal of data mining is to optimize accuracy of predictions, i.e. there is no incentive to discriminate based on taste. Before we go into the scenarios, we first recall the important notion of accuracy of predictions and we explain how we will assess discrimination of a classifier. Then we will deal with three scenarios illustrating the following situations:

- Labels are incorrect: due to historical discrimination the labels are biased. Even though the labels accurately represent decisions of the past, for the future task

they are no longer appropriate. Reasons could be, e.g., explicit discrimination, or a change in labeling in the future. This corresponds to assumption 1 of Section 4.2.1 being violated.
- The sampling procedure is biased: the labels are correct and unbiased, but particular groups are under- or overrepresented in the data, leading to incorrect inferences by the classifier induction. This corresponds to assumption 2 (first principled way) of Section 4.2.1 being violated.
- The data is incomplete; there are hidden attributes: often not all attributes that determine the label are being monitored. Often because of reasons of privacy or just because they are difficult to observe. In such a situation it may happen that sensitive attributes are used as a proxy and indirectly lead to discriminatory models. This corresponds to assumption 2 (second principled way) of Section 4.2.1 being violated.

3.3.1 Accuracy and Discrimination

Suppose that the task is to learn a classifier that divides new bank customers into two groups: *likely to repay* and *unlikely to repay*. Based on historical data of existing customers and whether or not they repaid their loans, we learn a classifier. A classifier is a mathematical model that allows us to extrapolate based on observable attributes such as gender, age, profession, education, income, address, and outstanding loans to make predictions. Recall that the *accuracy* of a classifier learned on such data is defined as the percentage of predictions of the classifier that are correct. To assess this key performance measure before actually deploying the model in practice, usually some labeled data (i.e., instances of which we already know the outcome) is used, that has been put aside for this purpose and not been used during the learning process.

Our analysis is based upon the following two assumptions about classification process.

Assumption 1: the classifier learning process is only aimed at obtaining an accuracy as high as possible. No other objective is strived for during the data mining phase.
Assumption 2: A classifier discriminates with respect to a sensitive attribute, e.g. gender, if for two persons which only differ by their gender (and maybe some characteristics irrelevant for the classification problem at hand) that classifier predicts different labels.

Note that the two persons in assumption 2 only need to agree on *relevant* characteristics. Otherwise one could easily circumvent the definition by claiming that a person was not discriminated based on gender, but instead because she was wearing a skirt. Although people "wearing a skirt" do not constitute a protected-by-law subpopulation, using such an attribute would be unacceptable given its high correlation with gender and that characteristics such as "wearing a skirt" are considered to be irrelevant for credit scoring. Often, however, it is far less obvious to separate relevant and irrelevant attributes. For instance, in a mortgage application an address may at the same time be important to assess the intrinsic value of a property, and reveal information about the ethnicity of a person. As we will see in Chapter 8 on explainable and non-explainable discrimination, however, it is not at all easy to measure and assess such possibilities for indirect discrimination in practical cases. The legal review in Chapter 4 shows that our definition of discrimination is in line with current legislation forbidding direct as well as indirect discrimination. Article 2 of Directive 2000/43/EC by the European commission explicitly deals with indirect discrimination: "*indirect discrimination shall be taken to occur where an apparently neutral provision, criterion or practice would put persons of a racial or ethnic origin at a particular disadvantage compared with other persons, unless that provision, criterion or practice is objectively justified by a legitimate aim and the means of achieving that aim are appropriate and necessary.*"

3.3.2 Scenario 1: Incorrect Labels

In this scenario the labels do not accurately represent the population that we are interested in. In many cases there is a difference in the labels in the training data and the labels that we want to predict on the basis of test data.

- *The labels in the historical data are the result of a biased and discriminative decision making process.* Sample selection bias exists when, instead of simply missing information on characteristics important to the process under study, the researcher is also systematically missing subjects whose characteristics vary from those of the individuals represented in the data (Blank et al, 2004). For example, an employment bureau wants to implement a module to suggest suitable jobs to unemployed people. For this purpose, a model is built based upon historical records of former applicants successfully acquiring a job by linking characteristics such as their education and interests to the job profile. Suppose, however, that historically women have been treated unfairly by denying higher board functions to them. A data mining model will pick up this relation between gender and higher board functions and use it for prediction.
- *Labeling changes in time.* Imagine a bank wanting to make special offers to its more wealthy customers. For many customers only partial information is available, because, e.g., they have accounts and stock portfolios with other banks as well. Therefore, a model is learned that, based solely upon demographic characteristics, decides if a person is likely to have a high income or not. Suppose that one of the rules found in the historical data states that, overall, men are likely to have a higher income than women. This fact can be exploited by the classifier to deny the special offer to women. Recently, however, gender equality programs and laws have resulted in closing the gender gap in income, such that this relation between gender and income that exists in the historical data is expected to vanish, or at least become less apparent than in the historical data. For instance, the distance Learning Center (2009) provides data indicating the earning gap between male and female employees. Back in 1979 women earned 59 cents for every dollar of income that men earned. In 2009 that figure has risen to 81 cents for every dollar of income that men earned. In this example, the

target attribute changes between the training data and the new data to which the learned model is applied, i.e. the dependence on the attribute gender decreases. Such background knowledge may encourage an analyst to apply discrimination-aware techniques that try to learn the part of the relation between the demographic features and the income that is independent of the gender of that person. In this way the analyst kills two birds with one stone: the classifier will be less discriminatory and at the same time more accurate.

3.3.3 Scenario 2: Sampling Bias

In this scenario training data may be biased, i.e. some groups of individuals may be over- or underrepresented, even though the labels themselves are correct. As we will show, such a *sample bias* may lead to biased decisions.

Let us consider the following example of over- and underrepresented groups in studies. To reduce the number of car accidents, the police increases the number of alcohol checks in a particular area. It is generally accepted that young drivers cause more accidents than older drivers; for example, a study by Jonah (1986) confirms that *young (16–25) drivers (a) are at greater risk of being involved in a casualty accident than older drivers and (b) this greater risk is primarily a function of their propensity to take risks while driving*). Because of that, the police often specifically targets this group of young drivers in their checks. People in the category "*over 40*" are checked only sporadically, when there is a strong incentive or suspicion of intoxication. After the campaign, it is decided to analyze the data in order to find specific groups in society that are particularly prone to alcohol abuse in traffic. A classification model is learned on the data to predict, given the age, ethnicity, social class, car type, gender, whether a person is more or less likely to drive while being intoxicated. Since only the labels are known for those people that were actually checked, only this data is used in the study. Due to data collection procedure there is a clear sample bias in the training data: only those people that were checked are in the dataset, while this is not a representative sample of all people that participate in the traffic. Analysis of this dataset could surprisingly conclude that particularly women of over 40 represent a danger of being intoxicated while driving. Such a finding is explainable by the fact that according to the examples presented to the classifier, middle aged women are more intoxicated than on average. A factor that was disregarded in this analysis, however, is that middle-aged women were only checked by the police when there was a more than serious suspicion of intoxication. Even though in this example it is obvious what went wrong in the analysis, sample bias is a very common and hard to solve problem. Think, e.g., of medical studies only involving people exhibiting certain symptoms, or enquiries by telephone that are only conducted for people whose phone number appeared on the list used by the marketing bureau. Depending on the source of the list that may have been purchased from other companies, particular groups may be over- or underrepresented.

3.3.4 Scenario 3: Incomplete Data

In this scenario training data contains only partial information of the factors that influence the class label. Often important characteristics are not present because of, e.g., privacy reasons, or because that data is hard to collect. In such situations a classifier will use the remaining attributes and get the best accuracy out of it, often overestimating the importance of the factors that are present in the dataset. Next we discuss an example of such a situation.

Consider an insurance company that wants to determine the risk category of new customers, based upon their age, gender, car type, years of driving experience etc. An important factor that the insurance company cannot take into account, however, is the driving style of the person. The reason for the absence of this information is obvious: gathering it; e.g., by questioning his or her relatives, following the person while he or she is driving, getting information on the number of fines the person had during the last few years, would not only be extremely time-consuming, but would also invade that person's privacy. Therefore, as a consequence, the data is often incomplete and the classifier will have to base its decisions on other available attributes. Based upon the historical data it is observed that in our example next to the horsepower of the car, age and gender of a person are highly correlated to the risk (the driving style is hidden for the company), see Table 1.

Table 1 Example (fictitious) dataset on risk assessment for car insurances based on demographic features. The attribute *Driving style* is hidden for the insurance company.

Customer no.	Gender	Age	Hp	Driving style	Risk
#1	Male	30 years	High	Aggressive	+
#2	Male	35 years	Low	Aggressive	-
#3	Female	24 years	Med.	Calm	-
#4	Female	18 years	Med.	Aggressive	+
#5	Male	65 years	High	Calm	-
#6	Male	54 years	Low	Aggressive	+
#7	Female	21 years	Low	Calm	-
#8	Female	29 years	Med.	Calm	-

From this dataset it is clear that the true decisive factor is the driving style of the driver, rather than gender or age; all high risk drivers have an aggressive driving style, and vice versa, only one aggressive driver does not have a high risk. There is an almost perfect correlation between being an aggressive driver and presenting a high accident risk in traffic. The driving style, however, is tightly connected to gender and age. Young male drivers will thus, according to the insurance company, present a higher danger and hence receive a higher premium. In such a situation we say that the gender of a person is a so-called *proxy* for the difficult to observe attribute driving style. In statistics, a proxy variable describes something that is probably not in itself of any great interest, but from which a variable of interest can be obtained.[1] An important side effect of this treatment, however, will be that a calm male driver will actually receive a higher insurance premium than an aggressive female driving the same car and being of the same age. The statistical discrimination theory (see Fang and Moro, 2010) states that inequality may exist between demographic groups even when economic agents (consumers, workers, employers) are rational and non-prejudiced, as stereotypes may be based on the discriminated group's average behavior.[2] Even if that is rational, according to anti-discrimination laws, this may constitute an act of discrimination, as the male person is discriminated on the basis of a characteristic that pertains to males as a group, but not to that person individually. Of course, a classifier will have to base its decisions upon some characteristics, and the incompleteness of the data will inevitably lead to similar phenomena; e.g., an exaggerated importance in the decision procedure on the color of the car, the horsepower, the city the person lives in, etc. The key issue here, however, is that some attributes are considered by law to be inappropriate to generalize upon, such as gender, age, religion, etc., but others, such as horsepower or a color of a car are not.

3.4 Potential Solutions for Discrimination Free Computation

We argued that unbiased computational processes may lead to discriminatory decisions due to historical data being incorrect or incomplete. In this section we discuss the main principles how to organize computational modeling in such a way that discrimination in decision making is prevented. In addition, we outline the main challenges and problems to be solved for such modeling.

3.4.1 Basic Techniques That Do Not Solve the Problem

We start with discussing the limitations of several basic solutions for training computational models.

Removing the Sensitive Attribute

Table 2 Example (fictitious) dataset on lending decisions

Customer no.	Ethnicity	Work exp.	Postal code	Loan decision
#1	European	12 years	1212	+
#2	Asian	2 years	1010	-
#3	European	5 years	1221	+
#4	Asian	10 years	1011	-
#5	European	10 years	1200	+
#6	Asian	5 years	1001	-
#7	European	12 years	1212	+
#8	Asian	2 years	1010	-

[1] Wikipedia: Proxy (statistics).
[2] Wikipedia: Statistical discrimination (economics).

The first possible solution is to remove the sensitive attribute from the training data. For example, if gender is the sensitive attribute in university admission decisions, one would first think of excluding the gender information from the training data. Unfortunately, as we saw in the previous section (Table 1), this solution does not help if some other attributes are correlated with the sensitive attribute.

Consider an extreme example on a fictitious lending decisions dataset in Table 2. If we remove the column "Ethnicity" and learn a model over the remaining dataset, the model may learn that if the postal code starts with 12 then the decision should be positive, otherwise the decision should be negative. We see that, for instance, customers #4 and #5 have identical characteristics except the ethnicity, and they will be offered different decisions. Such a situation is generally considered to be discriminatory.

The next step would be to remove the correlated attributes as well. This seems straightforward in our example dataset; however, it is problematic if the attribute to be removed also carries some objective information about the label. Suppose a postal code is related to ethnicity, but also carries information about real estate prices in the neighborhood. A bank would like to use the information about the neighborhood, but not information about the ethnicity in deciding for a loan. If the ethnicity is removed from the data, a computational model still can predict the ethnicity (internally) indirectly, based on the postal code. If we remove the postal code, we also remove the objective information about real estate prices that would be useful for decision making. Therefore, more advanced discrimination handling techniques are required.

Building Separate Models for the Sensitive Groups

The next solution that comes to mind is to train separate models for individual sensitive groups, for example, one for males, and one for females. It may seem that each model is objective, since individual models do not include gender information. Unfortunately, this does not solve the problem either if the historical decisions are discriminatory.

Table 3 Example (fictitious) dataset on university admissions

Applicant no.	Gender	Test score	Level	Acceptance
#1	Male	82	A	+
#2	Female	85	A	+
#3	Male	75	B	+
#4	Female	75	B	-
#5	Male	65	A	-
#6	Female	62	A	-
#7	Male	91	B	+
#8	Female	81	B	+

Consider a simplified example of a university admission case in Table 3. If we build a model for females using only data from females, the model will learn that every female that scores at least 80 in the test, should be accepted. Similarly, a model trained only on male data will learn that every male that scores over 70 in the test should be accepted. We see that, for instance, applicants #3 and #4 will have identical characteristics except the gender, yet they will be offered different decisions. This situation is generally considered to be discriminatory as well.

3.4.2 Computational Modeling for Discrimination Free Decision Making

Two main principles can be employed for making computational models discrimination free when historical data is biased. A data miner can either correct the training data or impose constraints on the model during training.

Correcting the Training Data
The goal of correcting the training data is to make the dataset discrimination free and/or unbiased. If the training data is discrimination free and unbiased, then we expect a learned computational model to be discrimination free.
Different techniques or combinations of those techniques can be employed for modifying data that include, but are not limited to:

1. modifying labels of the training data,
2. duplicating or deleting individual samples,
3. adding synthetic samples,
4. transforming data into new representation space.

Several existing approaches for discrimination free computational modeling use data correction techniques (Kamiran & Calders, 2010) (Kamiran & Calders, 2009). For more information see Chapter 12, where selected data correcting techniques are discussed in more detail.

Imposing constraints on the model training
Alternatively to correcting the training data, a model training process can be directed in such a way that anti-discrimination constraints are enforced. The techniques how to do that will depend on specific computational models employed. Several approaches for imposing such constraints while training exist (Calders & Verwer, 2010) (Kamiran, Calders, & Pechenizkiy, 2010). For more information see Chapter 14, where selected techniques for model training with constraints are discussed in more detail.

3.5 Conclusion and Open Problems

We discussed the mechanisms may produce computational models that may produce discriminatory decisions. A purely statistics-based, unbiased learning algorithm may produce biased computational models if our training data is biased, incomplete or incorrect due to discriminatory decisions in the past or due to properties of the data collection. We have outlined how different implicit assumptions in the computational techniques for inducing classifiers are often violated, and how this leads to discrimination problems. Because of the opportunities presented by growing amounts of data available for analysis automatic classification gains importance. Therefore, it is necessary to develop classification techniques that prevent this unwanted behavior.

Building discrimination free computational models from biased, incorrect or incomplete data is in its early stages, however, in spite of the fact that a number of case studies *searching* for discrimination evidence are available (see e.g. Turner & Skidmore, 1999). Removing discrimination from computational models is challenging. Due to incompleteness of data and underlying relations between different variables it is not sufficient to remove the sensitive attribute or apply separate treatment to the sensitive groups.

In the last few years several non discriminatory computational modeling techniques have been developed but there are still large challenges ahead: In our view two challenges require urgent research attention in order to bring non-discriminatory classification techniques to deployment in applications. The first challenge is how to measure discrimination in real, complex data with a lot of attributes. According to the definition, a model is discriminatory if it yields different predictions for candidates that differ only in the sensitive attribute and otherwise are identical. If real application data is complex, it is unlikely for every data point to find the "identical twin" that would differ only in the value of the sensitive attribute. To solve this problem, legally grounded and sensible from data mining perspective notions and approximations of similarity of individuals for non-discriminatory classification need to be established. The second major challenge is how to find out which part of information carried by a sensitive (or correlated) attribute is sensitive and which is objective, as in the example of a postal code carrying the ethnicity information and the real estate information. Likewise, the notions of partial explainability of decisions by individual or groups of attributes need to be established, and they need to be legally grounded and sensible from data mining perspective.

References

Blank, R., Dabady, M., Citro, C.: Measuring Racial Discrimination. Natl Academy Press (2004)

Jonah, B.A.: Accident risk and risk-taking behavior among young drivers. Accident Analysis & Prevention 18(4), 255–271 (1986)

Calders, T., Verwer, S.: Three Naive Bayes Approaches for Discrimination-Free Classification. Data Mining and Knowledge Discovery 21(2), 277–292 (2010)

Distance Learning Center. Internet Based Benefit and Compensation Administration: Discrimination in Pay, ch. 26 (2009), `http://www.eridlc.com/index.cfm?fuseaction=textbook.chpt26` (accessed: November 2011)

Duda, R.O., Hart, P.E., Stork, D.G.: Pattern Classification, 2nd edn. John Wiley & Sons (2001)

Fang, H., Moro, A.: Theories of Statistical Discrimination and Affirmative Action: A Survey. In: Benhabib, J., Bisin, A., Jackson, M. (eds.) Handbook of Social Economics, pp. 133–200 (2010)

Kamiran, F., Calders, T.: Classification with no discrimination by preferential sampling. In: Proceedings of the 19th Annual Machine Learning Conference of Belgium and the Netherlands (BENELEARN 2010), pp. 1–6 (2010)
Kamiran, F., Calders, T.: Classifying without Discrimination. In: IEEE International Conference on Computer, Control and Communication (IEEE-IC4), pp. 1–6 (2009)
Kamiran, F., Calders, T., Pechenizkiy, M.: Discrimination Aware Decision Tree Learning. In: Proceedings of IEEE ICDM International Conference on Data Mining (ICDM 2010), pp. 869–874 (2010)
Kelly, M.G., Hand, D.J., Adams, N.M.: The Impact of Changing Populations on Classifier Performance. In: Proceedings of the Fifth ACM SIGKDD International Conference on Knowledge Discovery and Data Mining (KDD 1999), pp. 367–371 (1999)
Rice, W.: Race, Gender, "Redlining", and the Discriminatory Access to Loans, Credit, and Insurance: An Historical and Empirical Analysis of Consumers Who Sued Lenders and Insurers in Federal and State Courts. San Diego Law Review 33, 637–646 (1996)
Turner, A., Skidmore, F.: Introduction, Summary, and Recommendations. In: Turner, A., Skidmore, F. (eds.) Mortgage Lending Discrimination: A Review of Existing Evidence (Urban Institute Monograph Series on Race and Discrimination), pp. 1–22. Urban Institute Press, Washington, DC (1999)
Widmer, G., Kubat, M.: Learning in the presence of concept drift and hidden contexts. Machine Learning 23(1), 69–101 (1996)
Zadrozny, B.: Learning and Evaluating Classifiers under Sample Selection Bias. In: Proceedings of the 21st International Conference on Machine Learning (ICML 2004), pp. 903–910 (2004)

Part II

Possible Discrimination and Privacy Issues

Chapter 4
A Comparative Analysis of Anti-Discrimination and Data Protection Legislations

Raphaël Gellert, Katja de Vries, Paul de Hert, and Serge Gutwirth*

Abstract. Departing from the ECJ's Huber case where Germany was condemned for discriminatory processing of personal data and which suggests that there is a strong kin between data protection and discrimination issues, this chapter is an attempt to further compare the two fundamental rights - non-discrimination, and data protection.

Beyond their place in the EU legal order, their respective object or scope, this chapter will contend that these two human rights increasingly turn to the same mode of operation, including, inter alia, reliance upon administrative structures and procedures, and the endowment of citizens with a bundle of individual rights. We will argue that this similarity can be understood in the light of their nature as regulatory human rights, that is, embodying the logic of negative freedom.

The final section will examine situations of overlap between the rights, building upon the Huber and Test-Achats cases. This will lead to final conclusions on how to best articulate these rights.

4.1 The *Huber* Case: How the German Register of Foreign Nationals (AZR) Raises Both Questions of Data Protection and Anti-Discrimination

In recent years automated data mining and profiling on large amounts of retained data has become an increasingly important tool in both the public and private sector. One salient example of the legal controversies arising from such practices is the contestation of the German *Ausländerzentralregister* (AZR) in the *Huber*

Raphaël Gellert · Katja de Vries · Paul de Hert · Serge Gutwirth
Vrije Universiteit Brussel, Belgium
e-mail: {rgellert,edevries,paul.de.hert,serge.gutwirth}@vub.ac.be

* The authors would like to thank Sebastian Abrahamsson, Antoinette Rouvroy, and Rosamunde Van Brakel for their salient comments and suggestions.

B. Custers et al. (Eds.): Discrimination & Privacy in the Information Society, SAPERE 3, pp. 61–89.
springerlink.com

case (ECJ, 2008). The contested register is a central, nation-wide, automated database in which all foreigners who live or have lived in Germany for more than three months are registered. At the moment (2011) the AZR contains data about more than 20 million individuals, both relating to asylum seekers and to foreigners holding a German residence permit. Approximately a quarter of these data relates to EU citizens. A wide range of officials can access the database: apart from the German Immigration authorities and the Secret Services approximately 6.500 other public bodies (e.g. courts, social services, police) can consult it.

The facts leading to the *Huber* case began in 1996, when Mr. Huber, an Austrian national, moved to Germany. As an EU national there was no impediment for him to work and live in another member state but, as prescribed by the AZR law, his personal data had to be processed in the AZR. In 2000 Mr. Huber contested the presence of his data in the database as discriminatory and requested their deletion: a register like the AZR does not exist for German nationals and the AZR data are also subject to secondary use for purposes of criminal investigation and population statistics. In the legal proceedings that followed, the national judge felt compelled to pose several preliminary questions to the European Court of Justice (ECJ). Before the Court, he questioned the compatibility of such a database with the prohibition of nationality-based discrimination among Union citizens, and its legitimacy and necessity from the point of view of data protection. Second, the question was put forth as to whether the secondary use fell within the scope of the Data Protection Directive. In its ruling, the ECJ stated that the use of a central register like the AZR can be legitimate in principle, but only in as far as it is necessary to support authorities in a more effective application of legislation on the right of residence, and personal data should not be stored for other purposes, such as criminal investigations and the creation of population statistics (§§58-59). For the latter purpose anonymized data should be used. The ECJ referred the case back to the national court (Higher Administrative Court for the State of North Rhine-Westphalia 24 June 2009), which decided that in the case of Mr. Huber the storage of data in the AZR was legitimate.[1]

Most interesting, for us, is the question concerning the legal concepts the ECJ used to address the issues at stake. Whereas contested storage of data in databases is normally addressed in terms of privacy and data protection, it appears that the issue of discrimination is at the core of this case, and that the Court established a very interesting link between data protection and non-discrimination. Indeed, the Court addressed the issue of the presence of a non-national in a database for secondary purposes of crime fighting, from the perspective of discrimination (and thus not solely data protection, §§ 78-79).[2]

---

1 According to the authorities responsible for the AZR, Huber's data are necessary for the application of the law concerning his right of residence on German territory and are only used for this purpose. Based on this statement the national judge (Higher Administrative Court for the State of North Rhine-Westphalia, 24 June 2009) rejected the request to remove Huber's data from the AZR, where they are probably still kept until present day.

2 Advocate General Poiares Maduro (*Opinion Huber*, C-524/06, 2008, §§ 5 and 21) reached the same conclusion by stating that although the purpose of crime fighting is *prima facie* legitimate, it does not justify such a difference in treatment with regard to the processing of personal data, which, ultimately, casts a "unpleasant shadow" over non-national EU citizens.

Therefore, beyond the crucial data protection issue of the secondary use of personal information available in specific databases, the issue at stake here is the discriminatory consequences of data processing operations.

Departing from the link made by the ECJ between discrimination and data processing, this article will further explore the relation between the rights to data protection and anti-discrimination, and will undertake a comparative analysis between them.

The first part of this chapter will be dedicated to a comparison of the legal architecture of the two rights.[3] Beyond the similar fashion in which they are integrated into the EU legal order, we will focus our attention on the object of the two legal frameworks. We will show that whereas the object of data protection legislation (i.e., the processing of personal data) is a fairly straightforward notion, the same cannot be said concerning discrimination. Closely linked to this first remark, is the scope of both legislations. Here too, contrarily to data protection's scope, which is evenly distributed, the scope of anti-discrimination is scattered, not least because of the different Directives that have been adopted and that each protects a specific ground. Finally, we will embark on a comparison of the legal regimes (LR) of the two rights. This exercise will evidence the presence in both legislations of an administrative body as well as a bundle of subjective (i.e., individual) rights granted to the concerned legal subjects. We will argue that the differences between the two legal regimes can be traced back to a fundamental difference, that is, whereas data protection concerns one particular action, anti-discrimination concerns one precise legal outcome no matter the action it stems from. However, we will also argue that these differences are not as fundamental as they might appear *prima facie*, and that future legislation might even severely mitigate them.

In the second part of this chapter, we will try to make sense of the comparison between the legal regimes by going back to the theoretical underpinnings of the two rights. As human rights, they are fundamentally bound to the democratic constitutional state, and hence to the notion of freedom. Building upon Berlin's dichotomy between positive and negative freedom, we will make the case that both data protection and anti-discrimination embody the logic of negative freedom, which (at least partly) accounts for their similar legal regimes, and justifies that we qualify them as "regulatory human rights".

The third and last part of the chapter will be dedicated to situations of overlap. We will show how one given legal situation can be simultaneously apprehended through the two lenses, by building upon the *Huber* case already mentioned in the introduction, and the *Test Achat* case.

We conclude by proposing how to best articulate these rights.

4.2 Place of the Two Rights in the EU Legal Order

The protection of both rights follows the same pattern from the perspective of the hierarchy of norms: both rights are enshrined in the *EU Charter of Fundamental Rights* (EUCFR), and can therefore be considered as autonomous fundamental

[3] A comparison of the theoretical underpinnings of the legal similarities and differences is beyond the scope of this chapter.

rights with a general scope and direct effect. Article 8 of the Charter guarantees the protection of personal data, and Title III is dedicated to equality and is composed of a general provision on anti-discrimination (art. 21 EUCFR), and of provisions regarding specific[4] groups of people (art. 22-26 EUCFR). Furthermore, both rights are also incorporated in specific provisions of the *Treaty on the Functioning of the European Union* (data protection in art. 16 TFEU; anti-discrimination in art. 18-25 TFEU).

In addition to this, these two rights are further developed and implemented by specific legislations in a similar design. As far as data protection law is concerned, the main piece of legislation is Directive 95/46/EC commonly known as the *Data Protection Directive*. Since the publication of this seminal piece of legislation data protection has evolved significantly, which has resulted in the adoption of several additional instruments such as the *Data Protection Regulation* (45/2001/EC), the *e-privacy Directive* (2002/58/EC), or the *Council Framework Decision on Third Pillar (police and judicial cooperation) Data Processing* (2008/977/JHA). Similarly, anti-discrimination legislation in the EU has undergone a long evolution of expansion giving content to the general principle of equality and non-discrimination.[5] The EU legislative framework is composed of the *Race Equality Directive* (2000/43/EC), the *Employment Equality Directive* (2000/78/EC); the *Gender Recast Directive* (2006/54/EC), prohibiting gender discrimination in employment and occupation and, also regarding gender, the *Gender Goods and Services Directive* (2004/113/EC). Finally, and to a lesser extent, one can mention the *Council Framework Decision 2008/913/JHA of 28 November 2008 on combating certain forms and expressions of racism and xenophobia by means of criminal law*, and the *Proposal for a Council Directive of 2 July 2008 (Proposal COM (2008) 426) on implementing the principle of equal treatment between persons irrespective of religion or belief, disability, age or sexual orientation.*

4.3 Discrimination, a Concept in Search of Unity; Data Protection, a Fairly Stabilised Notion

In the field of EU anti-discrimination law one has to distinguish between legislation relating to discrimination on *specific* protected grounds (e.g., race, gender, disability or age) and the *general* principle of equal treatment. This general principle can be understood as rooted in the classic Aristotelian idea that similar situations must not be treated differently and different situations must not be treated in the same way unless such treatment is objectively justified. However, the conformity to this general principle, which follows from the constitutional traditions of the member states, international human right treaties, in particular the

[4] Articles 22-26 EUCFR are respectively dedicated to cultural, religious and linguistic diversity; equality between women and men; the rights of the child, of the elderly, and of persons with disabilities.

[5] For a comprehensive description of this evolution, see Bribosia (2008). When mentioning anti-discrimination legislation, we will designate any of the aforementioned directives, since their structure and their provisions are identical as far as our argument is concerned.

European Convention on Human Rights (ECHR), and since 2009 the EU Charter of Fundamental Rights (EUCFR), is only assessed by a marginal test: as long as a difference in treatment has *some* rationality to it and is not completely arbitrary, the quality of the underlying reasoning is not further assessed ('equality as rationality', McCrudden and Prechal 2009). This approach was recently restated in *Arcelor* (ECJ, C-127/07, 16 December 2008).[6] Next to the general principle of equality, the European Union has also developed anti-discrimination law relating to *specific* grounds. In the following section (4.4) we will take a closer look at the scopes and particularities of the different Directives regarding specific forbidden grounds of discrimination, but first we must point out that here, in contrast to the general principle of equality, a more conceptually refined notion of discrimination is presented, broken down into different types of discrimination.

A first important conceptual distinction is the one between direct and indirect discrimination, both of which are protected in all of the recent Directives. Direct discrimination occurs when a person is treated in a less favourable way than another person and this difference is based directly on a forbidden ground. For instance, the Race Equality (*RE*) Directive states that "direct discrimination shall be taken to occur where one person is treated less favourably than another is, has been or would be treated in a comparable situation on grounds of racial or ethnic origin" (art. 2(2a)). Indirect discrimination makes a conceptual shift from consistency to substance (Fredman, 2002) by providing protection from apparently neutral provisions, criteria or practices which have the 'side effect' of discriminating against one of the specific forbidden grounds. Discrimination based on a neutral 'proxy' that disadvantages a protected group[7] is thus prevented, "unless that provision, criterion or practice is objectively justified by a legitimate aim and the means of achieving that aim are appropriate and necessary" (art. 2(2b) *RE*). Next to direct and indirect discrimination there is another form of discrimination, referred to as harassment: "when an unwanted conduct related to racial or ethnic origin takes place with the purpose or effect of violating the dignity of a person and of creating an intimidating, hostile, degrading, humiliating or offensive environment" (art. 2(3) *RE*). However, this definition is not uniform across the different anti-discrimination directives and is open to varying interpretations: "the concept of harassment may be defined in accordance with the national laws and practice of the Member States" (art. 2(3) *RE*).

As the previous analysis demonstrates, the legal concept of discrimination is multi-layered and sometimes contentious. Indeed, because discrimination is a

---

[6] The steel company *Arcelor* contested a European Directive by arguing that imposing certain measures to reduce CO_2 emissions on the iron and steel industry but not on the aluminium and plastic industry, amounted to an infringement on the general principle of equality. The ECJ concurred that a difference in treatment of comparable industrial sectors which is not based on an objective and reasonable criterion, amounts to arbitrariness and thus infringes on the general principle of equality. However, the legislator has a broad discretion to introduce complex legislation in a gradual way for different sectors. Therefore the ECJ held that the difference in treatment was not unjustified.

[7] For example, a disproportionate low salary for part-time work can be considered discriminatory against women if it's predominantly women who work part-time.

complex social phenomenon that is sometimes hard to grasp, the European legislator has tried to define it in the most precise possible manner in a series of legal instruments. However, this very precision may have jeopardized the unity (and consequent understanding) of the concept. As a result, that which is considered to be an instance of forbidden discrimination differs depending on which protected ground (e.g., race or age). We refer to this varying conceptualization and protection as the *asymmetrical scope* of EU anti-discrimination law.

In comparison, the object of data protection legislation (i.e., personal data) appears to be much clearer. In the EU legal order, its definition can be traced back to the *Data Protection Directive*. Here, personal data is "*any information relating to an identified or identifiable natural person*",[8] whereas the processing of personal data can be defined as "*any operation or set of operations which is performed upon personal data*".[9] Hence, the processing of personal data must respect the several principles enshrined in the Directive. However, like any legal concept, the notion of personal data is not void of controversies.[10]

As will be further explored (*infra*, section 4.5), one possible explanation for the conceptual controversies surrounding anti-discrimination is that this legal regime deals with the qualification of a difference of treatment, and *not* with the specificities of the practice leading up to the discriminatory or non-discriminatory 'end result'. Thus, anti-discrimination law is not tied to only one specific locus or field. A forbidden differential treatment can take many shapes and materialise itself in virtually any type of action, which is why anti-discrimination law is not limited to a certain kind of practice or behaviour: there are many roads that can lead to an instance of 'prohibited discrimination'. Moreover, anti-discrimination law is not one unified entity but a landscape filled with a variety of 'towns' and 'villages' of different size, shape and constitution. Data protection, on the contrary, is tied to one particular practice, namely the processing of personal data. Its focus is processual (it will prohibit the *process* of opaque handling of personal data without any legitimate aim, even if there are seemingly no direct adverse *effects*) and oriented towards one particular, clearly defined field.

One could object that in the case of reverse discrimination, i.e. when a differential treatment of a protected group follows from a so-called 'affirmative' or 'positive' action (art. 5 *RE*) the focus is *not* on the end result but on the preceding actions: though there is no unfair result the preceding action can be qualified as 'discriminatory'. Following this line of thought the case of reverse discrimination seems to be an exception to the rule that anti-discrimination law (*AD*) is more engaged with the result rather than the process. However, one could also argue the opposite: that reverse discrimination confirms the focus of *AD* on the *end result*, as it focuses on the enhancement of *substantial* equality ("equality of results" on a group level, which is opposed to formal equality, that is, the "consistent treatment of likes" on an individual level). (Fredman, 2002, p. 11) Yet it should be noted that, although the promotion of substantive equality (e.g. by

8 Article 2(a).
9 Article 2(b).
10 Cf. *infra*.

positive action, proactive measures and the prohibition of indirect discrimination) slowly gains in importance (see e.g. Fredman, 2009), formal equality is still the dominant approach in anti-discrimination law.

4.4 Differences in the Scope of EU Data Protection and Anti-Discrimination Legislation

The scope of data protection law is not as difficult to define as that of anti-discrimination law. In principle, the point of departure within the Data Protection (*DP*) Directive is that it applies to any "processing of personal data wholly or partly by automated means, and to the processing otherwise than by automated means of personal data which form part of a filing system or are intended to form part of a filing system" (art. 3(1)). There are two main exceptions to this general rule: firstly the scope of *DP* does not include "processing operations concerning public security, defence, State security [...] and the activities of the State in areas of criminal law" (art. 3(2))[11], and secondly there is the so-called 'household exception,' which exempts any processing "by a natural person in the course of a purely personal or household activity"[12] (art. 3(2)).

Why is data protection conceptually unified, while anti-discrimination law consists of a patchwork of legislative documents with asymmetrical protective scopes? As mentioned above (4.3), next to the general principle of equality, the EU has developed anti-discrimination laws relating to *specific* grounds. In the early days of the European Community such specialized anti-discrimination laws were not conceived as a fundamental rights in themselves, but as tools to facilitate mobility and the functioning of the internal market: combating discrimination among EU-citizens based on nationality (art. 18 *TFEU*) and gender in labour related matters were ways to enhance the efficiency of the common market and to prevent discrimination on grounds that are economically inefficient (More, 1999). However, in the last decade the scope of anti-discrimination law has been broadened beyond mere economic considerations and the list of grounds for unlawful discrimination has been extended with the entry into force in 1999 of article 13 *TEC*[13] (Meenan, 2007). This provision has given rise to several new directives. These directives have differing protective scopes, which we will now look at in more detail.

Firstly, with regard to race, there is Directive 2000/43/EC (*Race Equality Directive*) which provides a very wide protection against discrimination based on

---

[11] However, this is the very object of Council Framework Decision 2008/877/JHA of 27 November 2008. Furthermore, these processing operations are also encompassed by Council of Europe Convention 108 (1981), which is applicable in the legal order of every EU member state.

[12] A "purely personal or household activity" should be interpreted in a restrictive manner. See *Lindquist*, ECJ, C-101/01, 6 November 2003.

[13] The *Lisbon Treaty* (2009) has amended the *Treaty Establishing the European Community* (*TEC*, 1997) into the *Treaty on the Functioning of the Union* (*TFEU*, 2008), and consequently ex article 13 *TEC* has become article 19 of the *TFEU*.

race or ethnic origin: such discrimination is forbidden with regard to employment, occupation and vocational training, and the non-employment fields of social welfare (such as education, social security, health care) and access to goods and services, which includes housing. Council Framework Decision 2008/913/JHA of 28 November 2008 on combating certain forms and expressions of racism and xenophobia by means of criminal law, even extends this already wide scope. Secondly, with regard to gender, there are Directive 2006/54/EC[14], on equal treatment for men and women in the field of employment and occupation (*Gender Recast Directive*), and Directive 2004/113/EC prohibiting sex discrimination concerning access to and supply of goods and services (*Gender Goods and Services Directive*). It follows from these, and the earlier gender related Directives[15] from the first (1970s) and second wave (1990s), that the range of prohibited gender discrimination is narrower than that of racial discrimination, as it neither covers the areas of education, media and advertising (Directive 2004/113/EC, art. 3(3)), nor taxation and, in all likelihood, health care. Gender discrimination is not prohibited with regard to goods and services provided by public bodies that are not part of the common market (preamble of Directive 2004/113/EC, §11), and only covers social security – which is not as broad as the social welfare protected by racial anti-discrimination law (Fribergh and Kjaerum 2011). The difference in protective scope against racial and gender discrimination has been criticized (see e.g. Caracciolo di Torella 2005; Van Drooghenbroeck and Lemmens 2010). Finally, there is Directive 2000/78/EC (*Employment Equality Directive*), which prohibits discrimination on grounds of religion and belief, age, disability and sexual orientation, but only with regard to employment, occupation and vocational training.

It follows from the above that at present the scope of anti-discrimination legislation varies widely according to the protected grounds. It is to be pointed however that the Proposal for a Council Directive of 2 July 2008 is meant to overcome some of the asymmetries by extending the prohibition of discrimination based on grounds of religious, disability, age or sexual orientation beyond labour market issues (see for a critical discussion: Van Drooghenbroeck and Lemmens, 2010).

4.5 A Legal Regime Comprising Both an Administrative Structure and a Bundle of Subjective Rights

This section will give a closer look at the legal regimes of the two rights. A common feature of the two types of legal regimes is that they do not merely consist of legal principles, but also contain administrative bodies and a series of so-called ‘subjective’ rights: concrete, individual rights granted to the legal subjects they aim to protect, which can be mobilised at will (Dabin, 1952).

[14] This directive actualises Directive 2002/73/EC.

[15] Most of the earlier directives on gender, which were introduced in the 1970s and 1990s, have been superseded by the *Gender Recast Directive*, but for instance, Directive 79/7/EEC, the *Gender Social Security Directive*, and Directive 92/85/EEC, the *Pregnancy Directive*, are still in force and binding the member states.

Both the EU data protection and anti-discrimination frameworks rely upon the existence of supervisory bodies: Data Protection Authorities (DPAs) and Equality bodies.

Data Protection Authorities are independent supervisory authorities that have several, sometimes different, powers and responsibilities (depending on the national legislations implementing the Data Protection Directive). Thus, apart from keeping a processing register, they can offer advice, investigate issues, handle complaints, make a certain number of decisions concerning determinate data processing operations, provide authorisations, take a case to court, or even institute binding rules/regulations (Gutwirth 2002, p. 93). This does not mean however that judicial processes are totally absent from data protection law: member states are obliged to ensure the existence of judicial remedies that can grant compensation to data subjects (art. 23 *DP*).

Equality bodies have similar powers and responsibilities as they must provide independent assistance to victims of discrimination in pursuing their complaints before the Courts, conduct independent surveys concerning discrimination, publish independent reports and make recommendations on any issue relating to discrimination (art. 13(2) *RE*). Also, depending on the country, their powers will often include competences to provide advice, or handle complaints in the framework of alternative dispute settlement mechanisms (see e.g. De Hert and Ashiagbor 2011).[16]

It is interesting to note that the differences between the supervisory bodies are only marginal, which is not the case for the subjective rights featured by the two legal regimes.

As far as data protection is concerned, data subjects have the right to be informed that their data is being used in a processing operation (art. 10 and 11 *Data Protection (DP) Directive 95/46/EC*). They also have the right to access their data when these have been processed, e.g., they can investigate how the processing operation is carried out, whether databases exist, what their purpose is, and who is responsible for the processing (art. 12(a) *DP*; Gutwirth, 2002, p. 102). Furthermore, in case the data appear to be incomplete, inaccurate, or processed in a manner that is incompatible with the other data protection principles, the data subject has the right to ask for the rectification, or even the erasure of his or her data (art. 12(b) *DP*; Gutwirth, 2002, p. 102). Data subjects are also entitled to object to the processing of their personal data provided there are "compelling legitimate grounds" (art. 14(a) *DP*). Finally, data subjects have the right "not to be subject to a decision which produces legal effects concerning him or significantly affects him and which is based solely on automated processing of data" (art. 15 *DP*), which means that important decisions concerning them cannot be taken solely on the automated processing of data, and that they have a right to actively participate in those very decisions (Gutwirth, 2002, p. 104).

Anti-discrimination legislation also warrants individual rights to the subjects they aim to protect. Those rights are mostly intended to guarantee access to justice that is as efficient as possible (Fredman 2009). In this respect, some provisions aim at improving the ability of "discrimination subjects" to defend their rights, since they foresee that "Member States shall ensure that judicial and/or administrative

[16] Also, on the role of the Article 29 Working Party, see Poullet and Gutwirth (2008).

procedures (…) are available to all persons who consider themselves victims of discrimination". Also, associations that have a legitimate interest can help discrimination subjects to file a complaint, or even act on their behalf (art. 7(2) of *Race Equality Directive (RE) 2000/43/EC*). In order to enhance the chances of success of an action any contract, or provision of a contract, which is discriminatory can automatically be declared null and void by a judge (art. 14 *RE*); alleged victims benefit from a reversal of the burden of proof provided there are sufficient presumptions (i.e., according to art. 8 *RE*, it is up to the respondent to prove that there has been no discrimination), and they are entitled to an effective, proportionate and dissuasive remedy (art. 15 *RE*). Victims of discrimination are also guaranteed the right to be protected against retaliation in case of a successful procedure (art. 9 *RE*). For an example at the national level, Belgium has implemented this requirement by setting up a special procedure called *action en cessation* (action for injunction), which guarantees victims that their case will be swiftly examined (6 months), and that they will automatically receive a lump sum, if discrimination is proven (Closset-Marchal & Van Drooghenbroeck, 2008, p. 363).[17]

When taking a closer look at what *kind* of subjective rights each of these two regimes contain, some interesting contrasts come to the fore. The rights granted by data protection, such as the right to access one's own data, are very concrete actions that each data subject can undertake in an autonomous way (even though in practice only a limited amount of data subjects bother or manage to mobilise them). In comparison, what we have qualified as subjective rights in the field of anti-discrimination legislation does not refer to fully-fledged subjective rights, but rather to guarantees that aim at making action before court successful, thereby ensuring a real judicial efficiency to anti-discrimination principles. Thus it would not seem unfair to argue that data protection rights correspond closer to the notion of subjective rights: it could be argued that the data subject's rights are part of the very essence of data protection, i.e. that data protection is about granting prerogatives to the person whose personal information is being processed, whereas in the case of anti-discrimination the prerogatives merely represent an ancillary tool in order to ensure the efficiency of the legal framework.

In order to make sense of this distinction, one has to take into account the *object* of each of these legal regimes. Data protection is fundamentally different from anti-discrimination law, in that it regulates one[18] kind of action (the processing of personal data), independently of its actual consequences.[19] By

---

[17] See also, Belgian Equality Act, art. 20(1); Belgian Gender Discrimination Act, art. 25(1); Belgian Anti-Racism Act, art. 18(1).

[18] There are two exceptions (art. 3(2) *DP*): data processed in the context of the household or criminal law enforcement do not fall under the scope of the *DP* Directive. See *supra*, section 4.3.

[19] One should, however, distinguish between the *actual consequences* and the *aim* of the data processing *as inscribed in the process* of data handling. According to the *DP* Directive the latter is of great importance in assessing the overall legitimacy of the processing of data. Thus, data protection does not look into the actual outcomes of data processing, but it does assess whether the reasons and interests (art. 7(f) *DP*) for a particular instance of data processing were legitimate. Of course, in practice this conceptual distinction might turn out to be permeable. See *infra*, section 4.7.

contrast, anti-discrimination legislation concerns one[20] determinate legal consequence (a breach of equality between citizens), no matter what action it stems from. Data protection is, from this particular point of view, less contentious than anti-discrimination. Indeed, data protection is about one particular operation (the processing of personal data), the status of which is unproblematic.[21] Discrimination goes a step further because it does not regulate an action as such (e.g., data processing), but a legal consequence of any actions (thus, also including eventually data processing), which inherently entails operating a (legal) qualification of the facts. While the question of what qualifies as data processing might have some of its own legal intricacies, clearly, the question as to what counts as an unwarranted discriminatory action is a more contentious one (cf. *supra,* 4.3 and 4.5). Asking the latter question automatically entails operating a legal qualification of facts.

It could thus be argued that in data protection, data subjects are more empowered (and hence more autonomous) because of the inherently less contentious nature of the type of actions they are concerned with. In contrast, making an appeal to anti-discrimination law requires the intervention of a third party endowed with the legitimacy to undertake the legal hermeneutics to decide about the discriminatory nature of the consequences of the contested action. Hence, the level of contentiousness, which is higher in anti-discrimination than in data protection, could explain why subjective rights are ancillary in the former and substantial in the latter.

However, the difference between the two sets of subjective rights may not be as fundamental as it appears. A historical analysis of anti-discrimination legislation could lead us to mitigate an overly essentialist understanding of the divide and show the historical contingencies which gave rise to it.

To the extent that data protection can be traced back, be it in the OECD data protection guidelines (1980), the Council of Europe Convention 108 (1981), or the UN Guidelines concerning Computerized Personal Data Files (1990), it has always existed as a set of 'Fair Information Practices'.[22] This is hardly the case for anti-discrimination legislation, since it has not always featured such procedural characteristics (including both subjective rights and supervisory administrative structures). Indeed, much has been written on the changing approaches to the fight against discrimination (Bribosia, 2008; Fredman, 2005, 2006). The first approach, which can be qualified as an *ex post* (or post active) approach, consists in prohibiting discriminations whilst correlatively foreseeing a judicial sanction aimed at enforcing this ban. This is the classical human rights approach, which is

[20] This needs to be mitigated, however, given the varying scopes of the different directives. See *supra*, section 4.4.

[21] However, there are some controversies regarding the definition of personal data. See, Article 29 Data Protection Working Party, Opinion 4/2007 on the concept of personal data.

[22] That is, practices concerning fairness, transparency and legitimacy of the processing of personal data. Some authors disagree on this point. For instance, Mayer-Schönberger (2001) argues that the content of Data Protection legislation has undergone major evolutions.

still applicable to the other fundamental freedoms (except precisely for data protection, see also (Fredman 2006, p. 41). For a wide range of reasons, this approach has not been as successful in the case of discrimination as with other fundamental rights (Ringelheim 2010, p. 163; Fredman 2005, p. 372; 2009). Pursuant to these unsatisfactory results, the EU has decided to complement the first approach with an *ex ante* (or proactive) approach[23], leading to the adoption of new principles and mechanisms, i.e., the so-called mainstreaming approach,[24] and the different administrative procedures and mechanisms. Rather than fighting discrimination by repression, i.e. by imposing a judicial sanction upon infringements, the ultimate goal of the new preventive or proactive approach is to put an end to systemic factors of discrimination; therefore creating the necessary conditions whereby it is no longer possible for discriminating practices to exist (Fredman, 2009, p. 3). Hence, the need for policies that tackle the root factors of discrimination and for a binding decentralised administrative system that guarantees the equality between citizens in a quasi-automatic manner. The current EU anti-discrimination legal framework is therefore composed of policies that promote equality within society on the one hand (mainstreaming), and on the other hand, of a set of procedural mechanisms that strive for the immediate stop of discriminatory behaviour, *inter alia*, by empowering discrimination subjects and by relying upon a supervisory body (see Gellert & De Hert, 2012).[25]

Understanding the logic at work in the evolution of anti-discrimination legislation leads us to support the affirmation that the current differences between the two legislations are not irremediable, and they could be mitigated in the future.[26] Future developments of anti-discrimination legislation might thus feature new types of subjective rights that are fully-fledged, and not simply ancillary. Such a stance is supported by the fact that in both cases supervisory bodies have been granted similar powers.

Furthermore, the convergence between the two rights can also be observed from the reversed perspective. As far as data protection is concerned, it seems that the recent focus has been put upon the enforcement of the legislative framework. So whereas anti-discrimination appears to be going in the direction of more subjective rights, data protection appears to emphasise the need for enforcement

[23] This move has not only been undertaken at EU level; see also, e.g., the UN Convention on the rights of people with disabilities (2006).

[24] According to the EU, mainstreaming can be defined as "*a social justice-led approach to policy making in which equal opportunities principles, strategies and practices are integrated into the every day work of government and public bodies*", available on the following website, http://ec.europa.eu/social/main.jsp?catId=421&langId=fr.

[25] It seems to us that data protection and anti-discrimination share the awareness that part of the solutions lie in changing the very structures. In the case of anti-discrimination, it concerns social structures, and it is achieved through policies of mainstreaming, whereas in the case of data protection it concerns technical structures and is achieved through design (e.g., privacy by design).

[26] According to De Hert and Ashiagbor (2011), equality bodies devote an important part of their workload into activities of counselling to discrimination victims, and/or into dispute settlement, thereby importantly reducing the role of traditional courts and tribunals in matter of discrimination.

procedures.[27] This stance seems to be confirmed by the new draft Regulation on Data Protection (25 January 2012) which includes provisions for the accountability of the data controllers, provisions strengthening the powers of the supervisory bodies. Its chapter dedicated to remedies, liability and sanctions, contains an article on judicial assistance that is similar to what is provided by anti-discrimination legislation (art. 73).[28]

In conclusion, data protection and anti-discrimination legislation increasingly turn to the same mode of operation. However, the comparison is not symmetrical, due to reasons stemming from the different characteristics of the two rights at stake. In the next section, we will therefore argue that this similarity in the legal regimes of the two rights can be explained by their common nature, which we will qualify as being regulatory.

4.6 Data Protection and Anti-Discrimination: Two Regulatory Human Rights

In order to better understand the proposition according to which data protection and anti-discrimination are human rights of a regulatory nature, it is necessary to turn to the broader framework within which (all) human rights operate: the democratic constitutional State. Contrary to political systems characterised by an authoritarian ruler, the very aim of democratic regimes is to guarantee personal freedom and self-determination while at the same time preserving order. This regime is thus in constant tension, as it has to preserve simultaneously two antagonistic values - individual liberty *and* order (Gutwirth 1998; De Hert and Gutwirth 2008).

In order to realize this objective, democratic constitutional states have created a political structure wherein power is limited and non absolute, and which resorts to a double constitutional architecture. On the one hand, fundamental freedoms empower citizens with a set of individual rights that limit and counterbalance the power of the state. It is crucial to understand that human rights protect individuals from the State insofar as they create a sphere of autonomy or self-determination. On the other hand, the power of the State is subject to a set of constitutional rules holding the government to its own rules and to a system of mutual checks and balances. Furthermore, governments will be legitimate if and only if they can be considered as an expression of the "will of the people" (i.e., representation through elections) (De Hert and Gutwirth 2006).

Such architecture is thus not only based upon the assumption that citizens are "indigenous" (they were already "there" before the state) and autonomous

[27] See Article 29 Working Party Joint contribution to the Consultation of the European Commission on the legal framework for the fundamental right to protection of personal data, WP 169 adopted on 01 December 2009, or Opinion 3/2010 on the principle of accountability, WP 173 adopted on 13 July 2010.

[28] Also, on a national perspective again, the Belgian Act for the protection of personal data contains a provision setting up a specific judicial procedure similar to the one concerning anti-discrimination. However, no use has ever been made of it. See, Belgian Act on the protection of privacy regarding the processing of personal data, article 14.

political actors, but it also constitutionally enforces it. By shielding individuals from abuses of power through human rights, and by controlling this power with checks and balances, transparency and accountability, this architecture has contributed to the constitutional creation of the political *private* sphere. By comparison, the political *public* sphere is the political space where government and State intervention are legitimate (Gutwirth 1998; De Hert and Gutwirth 2006). In other words, the political private sphere is the political space wherein individuals can exercise their liberty/self-determination. Moreover, it can be argued that each different human right is the legal materialisation (or translation) of a given aspect of the political private sphere.

So far, we have purported that the project of the democratic constitutional state is built upon the idea of individual liberty, and it is to this end that it has instituted a so-called "political private sphere", which is the *locus* of political liberty, and which is shielded by human rights. The entire spectrum of human rights is mobilised for the protection of the political private sphere, including such different rights as the prohibition of torture, freedom of assembly, data protection (assuming it is a fundamental right, which we do), and anti-discrimination.

Since liberty seems to be at the core of the *raison d'être* of human rights, it seems to us that exploring different meanings and conceptions of liberty might give us some indications as to the mode of operation of the two legal regimes that we have evidenced *supra*, in section 4.5.

In this respect, the seminal work of Berlin appears as crucial. In his essay on the two concepts of liberty (1969), Berlin makes the distinction between "positive" and "negative" freedom. Negative freedom answers to the question "What is the area within which the subject is or should be left to do or be, without interference by other persons?" (p. 121-122). Negative freedom is thus the freedom not to be interfered with by others (p. 123), that is, ultimately, "freedom from" (p. 131). Positive freedom, on the contrary, "derives from the wish on the part of the individual to be his own master" (p. 131), or "freedom as self-mastery" (p. 134). Ultimately, this is a *freedom to* (to lead one's preferred way of life) (p. 131).

Accordingly, negative freedom is about the determination of the boundaries of individual freedom, whereas positive freedom is about the substantiation of this very freedom. This entails that negative freedom needs to take into account the behaviour of others, since the subject must be free from them in his area of freedom that has been deemed as legitimate. However, freedom in the positive sense is not concerned about the actions of others, but solely with that of the individual, as it is concerned with the "empowerment" of the latter.

Keeping in mind that human rights are the legal translation of the political private sphere of individual liberty, the foregoing dichotomy between negative and positive freedom can be of use as far as human rights are concerned. Indeed, it seems to us that a distinction can be made between human rights that literally empower the subject by granting him/her a prerogative (such as freedom of assembly, freedom of opinion, which Berlin refers to as a "catalogue of individual liberties", p. 126), and human rights that aim precisely at guaranteeing this "catalogue of individual liberties" against the deeds of others, be it other subjects, or the state.

That is the reason why we consider it relevant to introduce the distinction between *substantial* and *regulatory* human rights. From this perspective, substantial human rights empower the subject by granting him/her one of the individual liberties that constitute freedom in its positive sense, that is, centred around the (possibilities of) actions of the individual. Hence, they are about the substance, the content of one's freedom. Regulatory human rights on the other hand, embody the logic of negative freedom and hence aim at regulating, channelling the actions of others, so as to make sure they do not infringe upon, and consequently respect, the freedom of the subject.[29] We are of the opinion that such is the case for the rights to data protection and anti-discrimination. In both cases their very aim consists of assuring that the actions of others remain within boundaries that prevent them from infringing upon the freedom of their fellow subjects (one by regulating all data processing operations, the other by making sure that all actions respect the core principle of equality among citizens).

Consequently, the legal regimes of these rights should reflect their nature as regulatory human rights. Is this the case? As announced at the end of the preceding section, we believe that similar traits in both regimes we have evidenced (cf. the bundle of subjective rights and the supervisory bodies) are characteristic of this regulatory nature. By granting a bundle of subjective rights and relying upon (administrative) supervisory bodies, they strive towards a proactive judicial approach that aims less at sanctioning the violation of a right than at preventing this violation from taking place. In doing so, they thus channel and regulate the actions of others (precisely through the two means we have evidenced: subjective rights and supervisory bodies).

[29] Of course it might be argued that other rights do also have *ex ante* measures. For instance, in the case of freedom of speech and expression, there exists some regulations that ensure that the channels of expression are open, that guarantee the plurality of political ideas on the media, or that protect the sources of journalists. However, we believe that the two issues do not proceed from the same logic and thus need to be distinguished. In the first case we are confronted with human rights that correspond to the logic embodied by negative freedom, and thus the very aim of the latter is to guarantee the freedom of the subject regarding the actions of others. Their primary aim is to make the individual *free from*. In the second case we are facing measures that are encompassed by what is known in human rights theory as positive obligations. Positive obligations theory aims at guaranteeing that third parties do not violate a given substantial human right (freedom of expression in our example), and thus aim at guaranteeing the enjoyment of the right by its legitimate holder. Enjoying one's right indeed entails to some extent to be free from these actions that will violate the right in question, and in that sense positive obligations can be related to the logic underpinning negative freedom, since, ultimately, one needs to be "free from" in order to be "free to". However this does not affect the validity of our analysis, according to which it is clearly possible to differentiate two types of human rights. This distinction is not merely theoretical. For practical implications, see *supra*, 4.5 and *infra*, 4.7 on how to simultaneously protect negative freedom from several perspectives.

Although data protection and anti-discrimination are both about the channelling of the actions of others, they do so departing from two different perspectives: whereas data protection focuses on one particular action, anti-discrimination solely envisages a specific legal outcome (cf. *supra*, 4.3 and 4.5). One might then ask whether the possibility exists that these two perspectives coincide. In other words, whether there are potential overlaps between the two rights, that is, whether the protection offered by the two rights might apply to one very same action.

This will be the topic of the next section where we will examine the potentialities of overlap.

4.7 Overlaps: At the Crossroad between Data Protection and Anti-Discrimination

In this section we will examine the possibilities of overlap between data protection and anti-discrimination through the lens of two cases. Both deal with the inclusion of citizens in databases and the ensuing violation of rights.

What does this mean in practice? When we have a database in which personal data are processed there are two ways in which this database can give rise to a differential treatment: either the difference is made between those who are included and those who are excluded (*inter*), or the differentiation is made within the database (*intra*). For instance, an insurance company can decide that all the persons in a certain database have to pay 50% higher fees compared to those who are not (*inter*), or differentiate within (*intra*) a database by deciding that all persons with attribute X pay 50% more than those with attribute Y. To further explore these two situations we look at two recent decisions made by the ECJ: *Huber v. Germany* (2008), regarding a disadvantageous inclusion in a database, and *Test-Achats v. Council* (2011), regarding gender differentiation of insurance fees based on statistical profiling.

a. Huber v. Germany (2008): disadvantageous inclusion in a database

There are many instances when one's presence in a database is disadvantageous[30] compared to those who are not included (see e.g. González Fuster et al. 2010). This was the case in *Huber* (ECJ, 2008).[31] Clearly, the inclusion in the German Register of Foreign Nationals (AZR) is disadvantageous as it increases the likelihood of being suspected, falsely or correctly, of criminal activities. Hence, the lawfulness of such inclusion in a database is dubious. Of interest to us, it can be approached from both a data protection and an anti-discrimination perspective.

[30] Not every inclusion in a database is necessarily disadvantageous – it might also clear a person in some cases. See further on this issue *infra*, our discussion in section 4.8.

[31] Cf, *supra*, 4.1. There are other examples such as the *Marper* case (ECtHR, 4 December 2008), where it was contested that the DNA sample of any arrested individual in the UK was stored for an indefinite period of time in the National DNA Database, even if the individual was acquitted or never charged.

From the point of view of data protection the pivotal question is whether the processing of one's data in a particular database is legitimate. Is there a reason that legitimizes one's presence in the database? Article 7 of the *DP* Directive gives several reasons that could make data processing legitimate, the most important[32] one being Article 7(f): when it is "necessary for the purposes of the *legitimate interests* pursued by the controller".[33] Thus, the data protection perspective looks at the legitimacy of one's presence in the database *in itself.*

Contrary to data protection, anti-discrimination would take a *comparative* point of view: it looks at the difference in treatment between those who are included in the database and those who are not. Data protection asks: is the goal for which the data are being processed legitimate? Anti-discrimination asks: is the difference in treatment legitimized by a related and proportionate difference in the respective situations?

The interesting move in *Huber* (2008) is that the ECJ interconnects these two legal regimes. The 'magical' words that link them together are *necessity* and *proportionality*. With regard to data protection, *necessity* is embodied in the *purpose specification principle*, which requires that data must be "collected for specified, explicit and legitimate purposes and not further processed in a way incompatible with those purposes" (art. 6(b)*DP*), and *proportionality* is embodied in the *data quality principle*, which requires, *inter alia*, that the data must be "adequate, relevant and not excessive in relation to the purposes for which they are collected and/or further processed"(art. 6(c)*DP*). This entails that even when the aims of an instance of data processing are legitimate according to art. 7, this particular instance will only effectively be fully legitimate if the data collected and the way they are processed are in line with the requirements of art. 6 *DP*. In other words, a processing of data will be lawful if it is legitimate (according to article 7). However, this same processing would lose its legitimacy if it were not, *additionally*, necessary and proportional to the aim pursed (Gutwirth, 2002).

Thus, in *Huber*, it is not disputed that the processing of data of foreign residents serves objectives of public interest – applying the laws of residence and producing population statistics, but it is questioned whether these acts of processing are proportionate to the pursued objectives. In a move that is not uncontested, the Court engages the *necessity/proportionality* discussion only on the basis of art. 7, without any additional reference to art. 6. As a result, the Court links art. 7(e) of the Data Protection Directive to the prohibition of anti-discrimination based on

[32] Contrary to what is often argued, we do not believe that the consent criterion of Article 7(a) *DP* is the most important. Since art. 7(e) and (f) do already justify any processing of personal data tending to the realisation of a legitimate aim of the data controller, the legitimacy by consent criterion foreseen by art. 7(a) will often, if not always, be superfluous. If the consent criterion could supersede the other "legitimate aims" criteria this would perversely imply that consent could legitimize processing for "illegitimate aims", which would be an unacceptable outcome.

[33] We underline that in *Huber*, the article at stake was Article 7(e), which can be considered a sub category of article 7(f), as it states that data may be processed if the "processing is necessary for the performance of a task carried out in the public interest or in the exercise of official authority vested in the controller or in a third party to whom the data are disclosed".

nationality (the former art. 12(1) TEC), by interpreting the former *in the light* of the latter (section 66 of the *Huber* judgment). Keeping a register like the AZR purely for the purpose of population statistics would be disproportionate, because anonymous data would serve that purpose equally well (section 65 and 68), and processing non-anonymised data for the purpose of population statistics is thus both unnecessary, in the meaning of *DP*, and discriminatory, in the sense of art. 12(1) TEC.

Surprisingly, when anti-discrimination considerations (section 75) are applied independently of data protection considerations[34], a comparable proportionality test seems to be implied. However, in Huber the question of proportionality is not explored, because the fight against crime "in the general sense" (section 78) is, unlike the application of the right of residence for foreigners, not something that is only related to foreigners. In other words, the discrimination was so blatant that the Court did not consider it relevant to engage in a proportionality test of the measure at stake. Nevertheless one could cautiously argue that the prohibition of arbitrariness derived from anti-discrimination law can encompass a proportionality test: in the aforementioned section 75, the Court considers that disproportionate differences in treatment, based on a protected ground like nationality, qualify as arbitrary discriminations. Only if there is a legitimate, proportionate objective for distinguishing among German citizens and citizens of other member states, discrimination on grounds of nationality can be allowed. Advocate General Maduro seems to go along those lines when he states that although there is of course a difference between German citizens and non-German Union citizens, this does not allow for any discrimination whatsoever, because "the difference in treatment must relate and be proportionate to the difference in [...] situations".

Concluding, we see in the *Huber* case that both data protection and anti-discrimination have the possibility to address a difference of treatment following from the disadvantageous inclusion in a database. In *DP*, differential treatment is approached through the question of legitimacy, which entails proportionality, which in turn prohibits disproportionate differences in treatment. However, in the case of *DP* the question of disproportional differential treatment is only *one* of the criteria that will help determining whether a given instance of data processing is proportional, and hence legitimate (and lawful) or not. Therefore the '*bite*' of *DP* with regard to infringements will be comparatively small in relation to the more direct approach of *AD*, to which the difference of treatment is the core concern.

Yet a drawback of the *AD* approach is that it only concerns a limited set of protected grounds. In the *Huber* case the disproportionate differentiation was viewed through the lens of an *AD* provision, prohibiting discrimination on one particular forbidden ground (i.e., nationality). However, one could also imagine that the Court, were it to be confronted with a similar case wherein the differential treatment *did not* concern one of the grounds protected by *AD*, could link the

[34] In assessing the legitimacy of the secondary use of AZR data for *purposes of criminal investigation*, the Court cannot ground its decision on the *DP* directive because any data related to the enforcement of criminal law are excluded from its scope (art. 3(2) *DP*). See *supra*, section 4.4.

provisions from art. 6 and 7 *DP* to the *general* principle of equal treatment.[35] Even though such proportionality assessment would be marginal and lenient –especially in comparison with the protection granted by the *AD* on protected grounds!–(cf. previous paragraph), what we refer to as the "art. 6 *DP* + general equality"-route could be a useful tool to supplement any too strict limitations in the scope of *AD*.[36]

b. Test-Achats v. Council (2011), discrimination based on statistical profiling

A second situation of possible overlap between *DP* and *AD* in the field of data mining and profiling can occur when a differentiation is made within a database, resulting from an analysis of the data. Often such analysis will involve statistical profiling. At present many cases in this vein take place in the field of insurance. In such instances (De Hert et al. 2007) data protection may give the data subject certain subjective rights (cf., *supra* 4.4). However, it is contested whether the use of data that are not derived from the data subject but that are applied to him or her, can be considered as personal data as defined within the Data Protection Directive. In *Opinion 4/2007*, the Article 29 Working Party[37] (*WP* 29) has answered this question affirmatively:

> "Also a purpose element can be responsible for the fact that information 'relates' to a certain person. That purpose element can be considered to exist when the data are used or are likely to be used, taking into account all the circumstances surrounding the precise case, with the purpose to evaluate, treat in a certain way or influence the status or behaviour of an individual".

When a statistical profile functions as the basis for unequal treatment of similar cases, considerations of anti-discrimination can also play a role (Gandy 2008, 2009). The fact that a differentiation in treatment is not arbitrary but based on reliable statistics does not necessarily exclude it from the category of prohibited discriminations (Rüegger 2007). In *Lindorfer* (ECJ, C-227/04, 11 September 2007) Advocate-General Sharpston stated:

> "[i]n order to see such discrimination [based on sex] in perspective, it might be helpful to imagine a situation in which (as is perfectly plausible) statistics might show that a member of one ethnic group lived on average longer than those of another. To take those differences into account when determining the correlation between contributions and

[35] For the general principle of equal treatment see *supra*, cf. 4.3. For the link between this principle and art. 6 and 7 of *DP*, see *supra*, previous paragraph.

[36] For instance, certain companies take some credit decisions based upon whether a person has a contract for a mobile phone or not, or whether a person is surfing the Internet at three a.m. or not, see http://www.economist.com/node/18396166. In these cases, the differences of treatment are based upon grounds that are not protected by anti-discrimination legislation.

[37] The Opinions of WP 29 are not binding. If the issue ever became the matter of dispute in a real case, the court could interpret the notion of personal data differently.

> entitlements under the Community pension scheme would be wholly unacceptable, and I cannot see that the use of the criterion of sex rather than ethnic origin can be more acceptable."

Recently, the ECJ addressed this kind of discrimination in *Test-Achats* (ECJ, C-236/09, 1 March 2011), wherein the Belgian consumer organisation contested the validity of art. 5(2) of the *Gender Goods and Services Directive*. Whereas art. 5(1) prohibits "the use of sex as a factor in the calculation of [...] individuals' premium and benefits", article 5(2) permitted member states to create legal provisions derogating this prohibition when sex is a "determining factor" and when the risk assessment is "based on relevant and accurate actuarial and statistical data." The ECJ declared the derogation of article 5(2) incompatible with gender equality and invalid with effect from 21 December 2012. The decision caused an enormous stir in the insurance sector. Possibly, the decision will lead to the use of proxy factors (such as profession, education, lifestyle, etc.) in assessing risk, which in turn might raise questions of indirect discrimination.

Though interesting, investigating these issues in more detail is beyond the scope of this chapter. However, what is relevant for us to note here is that in the *Test-Achats* case the proceedings were completely based on anti-discrimination law, and do not relate to data protection at all. This can be explained by the facts that the claimant was a consumer organisation and not an individual data subject, and that the case did not concern an individual instance of differential treatment but posed a direct challenge to a piece of *AD* legislation. Looking at the *Test-Achats* case it is interesting to speculate whether the data processing related to the gendered differentiation of insurance fees, had it been contested, would have been considered legitimate from a data protection perspective. First, it is not even crystal clear that DP can apply to this type of situation, since the question of whether this type of data qualifies as "personal" in the meaning of Directive 95/46/EC is controverted. It will qualify as such if one refers to the aforementioned opinion of the *WP29*. However, this opinion is not uncontested, and in any event not binding. Second, provided this insurance contract can be considered as a legitimate aim to be pursued (art. 7), this would depend on whether such processing is necessary to the performance of the insurance contract (art. 6 *DP*).

Would the applicability of *DP* be of any help? Often statistical discrimination will not concern any of the protected grounds, rather, attributes such as income, postal code, browsing behaviour, type of car, etc., or complex algorithmic combinations of several attributes. *AD* could be eventually be resorted to if it could be shown that any of these attributes, or algorithmic combinations of these attributes, were used as proxies for any protected ground (indirect discrimination). However, were this is not to be the case then, once more, the "art. 6 *DP* + general equality"-route could prove to be a useful tool to supplement the limitative list of protected grounds in *AD*.

c. Overlaps between DP and AD: many questions left to answer

It would also be interesting to compare the proportionality test in *DP* with the one in *AD* law, but at the moment there is too little case law to say anything conclusive about this issue. Moreover, because of the scattered scope of *AD* law it will be difficult to say whether these considerations are applicable to *AD* law in general, or relate to a specific field, such as nationality based discrimination in *Huber*.

With regard to statistical profiling we can conclude that both data protection and anti-discrimination are struggling to address some of the challenges raised by the spread of this data technique. In the context of data protection, discussions are particularly circled around whether the application of anonymized data to an identifiable person falls within the scope of the Directive.[38] In the context of anti-discrimination, statistical profiling raised the question as to whether the fact that data are accurate and up-to-date exonerates the prohibition of discrimination. Statistical profiling also poses the question whether attributes, and complex algorithmic combinations of attributes, which do *not* belong to any of the specifically protected grounds might bring the concept of indirect discrimination and the "art. 6 *DP* + general equality"-route to the frontline (and as a matter of fact, any difference of treatment that is not based upon the protected grounds).

4.8 Conclusions: Articulating the Two Rights

In the preceding pages, we have attempted to compare data protection and anti-discrimination legislations in the EU legal order.

Beyond differences relating to their respective scope and object, we have observed an increasing convergence in their legal regimes. This convergence, we have argued, can be better understood by tracing back their theoretical underpinnings, and more precisely their nature as human rights embodying the logic of negative freedom as put forth by Berlin, that is, as regulatory human rights.

Because both rights protect the freedom of the individual from the same perspective, it is not excluded that their protection might overlap, as has been shown with the *Huber* and *Test-Achats* cases.

In the light of contemporary practices such as statistical profiling, it seems clear to us that, in the coming years, both rights will increasingly overlap. Therefore, it might be interesting to give some thoughts on precisely *how to best articulate* these rights.

As a matter of fact, we would like to make the point that the protection offered by these two rights is complementary. Hence it is very unlikely that their articulation would lead to clashes or antagonistic results, although some have made the point that this could be the case. In his article, Strahilevitz (2008) argues that having one's data publicly available in a database is actually advantageous and "will reduce the prevalence of distasteful statistical discrimination." (p. 364) Illegitimate, distasteful discrimination is here understood as a heuristic used in

---

[38] And around the right to access the data and the logic involved in statistical profiling (art. 12 *DP*), and the right not to be subjected to a decision based solely on automated processing (art. 15(1) *DP*). However, this is beyond the scope of our discussion.

situations where proper information is lacking (e.g., an employer uses skin colour as a proxy for criminal records – however, if these records would be publicly available the employer would not be forced to take recourse to racist heuristics.) In other words, the more information a person knows the more enlightened his/her choices will be, and thus the chances of undertaking a decision that bears discriminatory consequences will be the lowest possible.

Such a position can probably be traced back to the views developed by Posner in his seminal article *The Right of Privacy* (Posner, 1978), which argues that the efficiency of economic transactions is enhanced by full disclosure of all available information in order to avoid distasteful discrimination. When information is concealed through privacy rights we are more likely to make the 'wrong' choices: e.g. hire an employee who is an ex-convict or has a serious health problem. One could therefore argue for full disclosure of as much information as possible.

This argument is, according to us, flawed. It sets the debate in the wrong terms, as it seems to leave the choice between either total transparency or total privacy. At this point it seems useful to remind that in the EU legal order, data protection and privacy are two different rights, though very much interrelated. Whereas the latter is about the intimacy of the individual and his/her self-determination (Gellert & Gutwirth, 2012, Gutwirth, 2002), the former involves fairness, transparency and legitimacy of the processing of personal data.[39] By default, data protection allows for the processing of personal data, but only at certain conditions. These conditions have been explained in the previous section: in addition to pursuing a legitimate aim, the processing must be necessary and proportional to this aim. Therefore, the point is more about determining the necessity and the proportionality of a processing in view of the legitimate aim that consists in taking a decision that bears no illegitimate discriminatory consequences. In this respect one could eventually argue that the clash between the two rights could shift from "privacy vs transparency" to a clash between two conceptions of necessity and proportionality: a *DP* conception and an *AD* conception. However, this possibility seems highly theoretical and improbable to us, not least because we have shown in the *Huber* case that in order to determine the necessity of a processing, data protection takes anti-discrimination issues into consideration. Therefore, it is difficult for us to see how the rights would clash. On the contrary, it seems to us, that the protection they afford to the individual is complementary: if the protection afforded by one right is not sufficient, the individual can still seek for a protection from the perspective of the other right. This could be the case in the future for discriminations stemming from statistical profiling and thus based on no grounds protected by anti-discrimination: in these cases, the discrimination could still be tackled from the "art. 6 *DP* + general equality"-route.

[39] Some of the data categorized as *sensitive* in art. 8(1) *DP* (race or ethnic origin, political opinions, and religion or belief) overlap with the grounds prohibited by EU anti-discrimination law. However, the provisions regarding sensitive data are *no* exception to the rule of thumb that *DP* does concern the *process* and not the consequences of processing. The only difference between the provisions on ordinary personal data and sensitive ones is that the requirements legitimizing the processing of the latter are somewhat stricter (art. 8(2) *DP*). For an additional discussion on sensitive data, see Annex.

All in all, this necessary complementarity between the two rights stems from their shared nature as regulatory human rights. As such, they are each the materialisation of a specific aspect of negative freedom. Therefore, they protect different dimensions of this negative freedom, and that is the reason why their combination will lead to a protection of negative freedom that is as comprehensive as possible.

Given that the protection of the individual will benefit from the complementarity of *DP* and *AD*, it might be interesting to think about the skilful use that can be made of the specificities of each legal regime. As noted in *supra*, 4.5, *DP* features a bundle of fully-fledged subjective rights, whereas *AD* puts the emphasis on the efficiency of the judicial framework. Therefore, in seeking the best possible protection, the individual could follow a two-step approach that builds upon the strengths of each right. Following this approach, the individual would first use *DP* to ask for access to, and erasure of the data. In case the results were not satisfying, he/she could then go to court with the aid of the subjective rights granted by *AD*.

Recent developments of data processing practices such as automated decision-making in databases lead us to think that issues of discrimination will increasingly come to the fore. It is therefore crucial to have a good understanding of the way in which both data protection and anti-discrimination operate, so as to grant the best possible protection to the individual. The foregoing lays some first elements of reflexion. However, this is work that needs to be further continued.

References

Legislation

a. European Union

Directive 79/7/EEC, on the progressive implementation of the principle of equal treatment for men and women in matters of social security. Official Journal L006/24 (January 10, 1979)

Directive 92/85/EEC, on the introduction of measures to encourage improvements in the safety and health at work of pregnant workers and workers who have recently given birth or are breastfeeding. Official Journal L348/01 (November 28, 1992)

Directive 95/46/EC of the European Parliament and of the Council of 24 October 1995 on the protection of individuals with regard to the processing of personal data and on the free movement of such data. Official Journal L281/31 (November 23, 1995)

Treaty Establishing the European Community (TEC). Official Journal C340/01 (November 10, 1997)

Council Directive 2000/43/EC of 29 June 2000 implementing the principle of equal treatment between persons irrespective of racial or ethnic origin. Official Journal L180/22 (July 19, 2000)

Council Directive 2000/78/EC of 27 November 2000 establishing a general framework for equal treatment in employment and occupation. Official Journal L303/16 (December 2, 2000)

Charter of Fundamental Rights of the European Union (EUCFR) of 7 December 2000. Official Journal C 364/01 (December 18, 2000)
Regulation 45/2001 of the European Parliament and of the Council of 18 December 2000 on the protection of individuals with regard to the processing of personal data by the Community institutions and bodies and on the free movement of such data. Official Journal L8/1 (January 12, 2001)
Directive 2002/58/EC of the European Parliament and of the Council of 31 July 2002 on privacy and electronic communications. Official Journal L201/37 (July 31, 2002)
Directive 2002/73/EC of the European Parliament and of the Council of 23 September 2002 amending Council Directive 76/207/EEC on the implementation of the principle of equal treatment for men and women as regards access to employment, vocational training and promotion, and working conditions. Official Journal L269/15 (October 5, 2002)
Council Directive 2004/113/EC of 13 December 2004 implementing the principle of equal treatment between men and women in the access to and supply of goods and services. Official Journal L373/37 (December 21, 2004)
Directive 2006/54/EC of the European Parliament and of the Council of 5 July 2006 on the implementation of the principle of equal opportunities and equal treatment of men and women in matters of employment and occupation (recast of Directive 2002/73/EC). Official Journal L204/23 (July 26, 2006)
Consolidated version of the Treaty on the Functioning of the European Union (TFEU) of 13 December 2007. Official Journal C115/01 (May 9, 2008)
Proposal for a Council Directive of 2 July 2008 on implementing the principle of equal treatment between persons irrespective of religion or belief, disability, age or sexual orientation. COM, 426 (2008)
Council Framework Decision 2008/877/JHA of 27 November 2008 on the protection of personal data processed in the framework of police and judicial cooperation in criminal matters. Official Journal L350/60 (December 30, 2008)
Council Framework Decision 2008/913/JHA of 28 November 2008 on combating certain forms and expressions of racism and xenophobia by means of criminal law. Official Journal L328/55 (December 6, 2008)
Proposal for a Regulation of 25 January 2012 on the protection of individuals with regard to the processing of personal data and on the free movement of such data (General Data Protection regulation) COM (2012), 11 final

b. National

German Federal Government, AZR-Gesetz [AZR Law] of 2 September 1994 (BGBl. IS. 2265), adjusted by Article 5 of the Law of 22 November 2011 (BGBl. I S. 2258)
Belgian Federal Government, Act Combating Certain Forms of Discrimination [Equality Act] of 10 May 2007, p. 29016. Moniteur Belge/Belgisch Staatsblad (May 30, 2007)
Belgian Federal Government, Act Amending the Act of 30 July 1981 on the Punishment of Certain Acts Motivated by Racism or Xenophobia [Anti-Racism Act] of 10 May 2007, p. 29031. Moniteur Belge/Belgisch Staatsblad (May 30, 2007)
Belgian Federal Government, Act Combating Discrimination Between Women and Men [Gender Discrimination Act] of 10 May 2007, p. 29044. Moniteur Belge/Belgisch Staatsblad (May 30, 2007)

Belgian Federal Government, Act on the protection of privacy regarding the processing of personal data of 8 December 1992, p. 5801. Moniteur Belge/Belgisch Staatsblad (March 18, 1993)

c. International

Council of Europe, European Convention for the Protection of Human Rights and Fundamental Freedoms (signed November 4, 1950)
Organisation for Economic Co-operation and Development (OECD), Guidelines on the Protection of Privacy and Transborder Flow of Personal Data (C(80)58/FINAL) (adopted on September 23, 1980)
Council of Europe, Convention of the Council of Europe for the protection of individuals with regard to automatic processing of personal data, European Treaty Series No. 108 (adopted January 28, 1981)
Proposal for a Regulation of the European Parliament and of the Council on the protection of individuals with regard to the processing of personal data and on the free movement of such data (General Data Protection Regulation), COM, 11 final (January 25, 2012)
United Nations (UN) General Assembly, Guidelines for the Regulation of Computerized Personal Data Files (adopted December 14, 1990)
United Nations (UN) General Assembly, Convention on the Rights of People with Disabilities (adopted December 13, 2006)
Proposal for a Directive on the use of Passenger Name Record (PNR) data for the prevention, detection, investigation and prosecution of terrorist offences and serious crime, COM, 32 final (February 2, 2011)

Case Law

a. European Court of Justice

Lindquist v. Sweden C-101/01 (November 6, 2003)
Lindorfer v. Council, C-227/04P (September 11, 2007)
Opinion of Advocate General Poiares Maduro regarding Huber v. Germany (Case C- 524/06, 16 December 2008) (April 3, 2008)
Huber v. Germany, C-524/06 (Judgment of December 16, 2008)
Arcelor v. Prime Minister, C-127/07 (Judgment of December 16, 2008)
Test Achats v. Council, C-236/09 (Judgment of March 1, 2011)

b. European Court of Human Rights

S and Marper v UnitedKingdom, application nos. 30562/04 and 30566/04 (Judgment of December 4, 2008)

C. National

Rasterfahndung ["trawling"], Federal Constitutional Court of Germany ("Bundesverfassungsgericht"), 1 BvR. 518/02 (Judgment of April 4, 2006)
Huber, Higher Administrative Court for the State of North Rhine-Westphalia ("Oberverwaltungsgericht NRW"), Case 17 A 805/03 (Judgment of June 24, 2009)

Literature

Agre, P.E., Rotenberg, M.: Technology and Privacy: The New Landscape. MIT, Cambridge (2001)

Article 29 Working Party, Opinion 4/2007 on the concept of personal data 4/2007. WP 13 (adopted on June 20, 2007)

Article 29 Working Party, Joint contribution to the Consultation of the European Commission on the legal framework for the fundamental right to protection of personal data. WP 169 (adopted on December 01, 2009)

Article 29 Working Party, Opinion 3/2010 on the principle of accountability. WP 173 (adopted on July 13, 2010)

Berlin, I.: Four Essays on Liberty. Oxford University Press, Oxford (1969)

Bribosia, E.: La lutte contre les discriminations dans l'Union Européenne: une mosaïque de sources dessinant une approche différenciée. In: Bayart, C., Sottiaux, S., Van Drooghenbroeck, S. (eds.) Les Nouvelles Lois Luttant Contre Les Discriminations – De Nieuwe Federale Antidiscriminatiewetten, pp. 31–62. Die Keure, La Charte, Bruges, Bruxelles

Caracciolo di Torella, E.: The Goods and Services Directive: Limitations and Opportunities. Feminist Legal Studies 13(3), 337–347 (2005)

Closset-Marchal, G., Van Drooghenbroeck, J.-F.: L'action en cessation en matière de discriminations. In: Bayart, C., Sottiaux, S., Van Drooghenbroeck, S. (eds.) Les Nouvelles Lois Luttant Contre Les Discriminations – De Nieuwe Federale Antidiscriminatiewetten, pp. 316–414. Die Keure/ La Charte, Bruges/Bruxelles (2008)

Dabin, J.: Le droit subjectif. Dalloz, Paris (1952)

De Hert, P., Gutwirth, S.: Privacy, Data Protection and Law Enforcement. Opacity of the Individual and Transparency of Power. In: Claes, E., Duff, A., Gutwirth, S. (eds.) Privacy and the Criminal Law, pp. 61–104. Intersentia, Antwerps (2006)

De Hert, P., Gutwirth, S.: Regulating profiling in a democratic constitutional state. In: Hildebrandt, M., Gutwirth, S. (eds.) Profiling the European Citizen. Cross Disciplinary Perspectives, pp. 271–291. Springer, Dordrecht (2008)

De Hert, P., Hildebrandt, M., Gutwirth, S., Saelens, R.: De WBP na de Dexia-uitspraken. Privacy & Informatie 10(4), 147–157 (2007)

De Hert, P., Ashiagbor, D. (eds.): Comparative study on access to justice in gender equality and anti-discrimination law. Synthesis report. European Commission, Directorate-General for Employment, Social Affairs and Equal Opportunities (2011)

European Union Agency for Fundamental Rights (FRA), Opinion 1/2011 of the European Union Agency for Fundamental Rights on the Proposal for a Directive on the use of Passenger Name Record (PNR) data for the prevention, detection, investigation and prosecution of terrorist offences and serious crime, COM, 32 final (2011)

Fredman, S.: Discrimination law. Clarendon, Oxford (2002)

Fredman, S.: Changing the norm: positive duties in equal treatment legislation. Maastricht Journal of European Equality Law 12(4), 369–376 (2005)

Fredman, S.: Transformation or Dilution: Fundamental Rights in the EU Social Space. European Law Journal 12(1), 41–60 (2006)

Fredman, S.: Making Equality Effective: The Role of Proactive Measures. Oxford Legal Studies Research Paper No. 53/2010. European Commission, Directorate-General for Employment. Social Affairs and Equal Opportunities, Unit EMPL/G/2 (2009, 2010)

Fribergh, E., Kjaerum, M.: Handbook on European non-discrimination law. Publication Office of the European Union, Luxembourg (2011)

Gandy, O.: Engaging Rational Discrimination. Paper presented at the Ethics, Technology and Identity, TU Delft, June 18-20 (2008)

Gandy, O.: Engaging rational discrimination: exploring reasons for placing regulatory constraints on decision support systems. Ethics and Information Technology 12(1), 29–42 (2009)

Gellert, R., De Hert, P.: La non discrimination comme réalité effective en Europe? Réflexions sur la procéduralisation du droit de l'égalité européen. Revue Belge de Droit Constitutionnel 1 (forthcoming, 2012)

Gellert, R., Gutwirth, S.: Beyond accountability, the return to privacy? In: Guagnin, D., Hempel, L., Ilten, C., Kroener, I., Neyland, D., Postigo, H. (eds.) Managing Privacy through Accountability. Palgrave Macmillan (forthcoming, 2012)

González Fuster, G., De Hert, P., Ellyne, E., Gutwirth, S.: Huber, Marper and Others: Throwing new light on the shadows of suspicion. Justice and Home Affairs, IN:EX Policy Briefs (11) (2010)

Gutwirth, S.: Waarheidsaanspraken in recht en wetenschap. Een onderzoek naar de verhouding tussen recht en wetenschap met bijzondere illustraties uit het informaticarecht. VUBPress/Maklu, Brussel/Antwerpen (1993)

Gutwirth, S.: De polyfonie van de democratische rechtsstaat' [The polyphony of the democratic constitutional state]. In: Elchardus, M. (ed.) Wantrouwen en Onbehagen [Distrust and Uneasiness], vol. 14, Balans. VUBPress, Brussels (1998)

Gutwirth, S.: Privacy and the information age. Rowman & Littlefield, Oxford (2002)

Mayer-Schönberger, V.: Generational Development of Data Protection in Europe. In: Agre, P.E., Rotenberg, M. (eds.) Technology and Privacy: The New Landscape, pp. 219–242. MIT, Cambridge (2001)

McCrudden, C., Prechal, S.: The Concepts of Equality and Non-Discrimination in Europe: A Practical Approach. European Commission, Directorate-General for Employment, Social Affairs and Equal Opportunities, Unit G.2 (2009), `http://ec.europa.eu/social/BlobServlet?docId=4553&langId=en`

Meenan, H.: Equality law in an enlarged European Union: understanding the Article 13 Directives. Cambridge University Press, Cambridge (2007)

More, G.: The principle of equal treatment: From market unifier to fundamental right? In: Craig, P., de Búrca, G. (eds.) The Evolution of EU Law, pp. 517–553. Oxford University Press, Oxford (1999)

Posner, R.A.: The Right of Privacy. Georgia Law Review 2(3), 393–422 (1978)

Poullet, Y., Gutwirth, S.: The contribution of the Article 29 Working Party to the construction of a harmonised European data protection system: an illustration of 'reflexive governance'? In: Perez Asinari, M.V., Palazzi, P. (eds.) Défis du droit à la Protection de la vie Privée – Challenges of Privacy and Data Protection Law, pp. 570–610. Bruylant, Brussels (2008)

Rigaux, F.: La protection de la vie privée et des autres biens de la personnalité. Bruylant, Bruxelles (1990)

Ringelheim, J.: L'évolution contemporaine du droite de la non discrimination. In: Herman, G., Leonard, E., Reman, P. (eds.) Travail, Inégalités et Responsabilité, pp. 163–168. Presses Universitaires de Louvain, Louvain (2010)

Rüegger, M.: La discrimination statistique entre pertinence et arbitraire. Revue de Philosophie Économique 8(1), 73–94 (2007)

Solove, D.J., Rotenberg, M., Schwartz, P.M. (eds.): Information Privacy Law. Aspen, New York (2006)

Strahilevitz, L.J.: Privacy versus Antidiscrimination. University of Chicago Law Review 75(1), 363–382 (2008)

Van Drooghenbroeck, S., Lemmens, K.: Les nouvelles initiatives législatives européennes dans le domaine de la lutte contre la discrimination. Un corpus juris en quête de cohérence. In: Bayart, C., Sottiaux, S., Van Drooghenbroeck, S. (eds.) Actuele Topics Discriminatierecht/Actualités du Droit de la Lutte Contre la Discrimination, pp. 79–125. Die Keure, Brugge (2010)

Weichert, T.: Ausländererfassung in der Bundesrepublik. Die informationelle Sonderbehandlung von Immigrantinnen und Flüchtlingen. Bürgerrechte & Polizei/CILIP 45(2), 30–37 (1993)

Annex: Additional Considerations on Sensitive Data from *DP* and *AD* Perspectives

In the following overview the **bold** categories are the ones that overlap, the *italic* ones partly overlap, and the underlined ones are new additions in the proposal for a Regulation (2012) which would replace Directive 95/46.

DP	Art 8(1), 95/46/EC	Member States shall prohibit the processing of personal data revealing: **racial or ethnic origin**, **political opinions, religious or** *philosophical* **beliefs**, trade-union membership, and the processing of data concerning *health* or *sex life*
DP	Art 9 (1), *Proposal (2012) for a Regulation revising 95/46/EC*	The processing of personal data, revealing: **race or ethnic origin**, **political opinions, religion or beliefs,** trade-union membership, and the processing of ***genetic data*** or data concerning *health* or *sex life* or criminal convictions or related security measures shall be prohibited.
AD	Art. 21 EU Fundamental Rights Charter (EUCFR)	(1) Any discrimination based on any ground such as: sex, **race**, colour, **ethnic** or social **origin**, ***genetic features***, language, **religion or belief**, **political** or any other **opinion**, membership of a national minority, property, birth, *disability*, age or *sexual orientation* shall be prohibited. (2) Within the scope of application of the Treaty [...] any discrimination on grounds of nationality shall be prohibited.

The table clarifies that there is a significant discrepancy between the categories of sensitive data and the prohibited grounds for discrimination. Recently the *European Union Agency for Fundamental Rights* (FRA) suggested in its Opinion *(1/2011)* on the proposed PNR-profiling Directive *(COM(2011) 32 final)*that this discrepancy should be dissolved by classifying all data related to the prohibited grounds of art. 21 as sensitive, because the prohibition of processing such data would help to pre-empt direct discrimination. The Commission has expressed its approval of this suggestion (Computers, Privacy and Data Protection Conference, Brussels, 27 January 2012). However, this could lead to quite absurd results: categories as sex, age, birth, nationality and language probably belong to the most frequently processed personal data. Subjecting the processing of such ubiquitous data to very stringent requirements, merely to reduce the risk that they could be used as the basis for a prohibited unequal treatment, seems a disproportionate measure that mixes up the processing (DP) with the possible outcome (AD).

Chapter 5
The Discovery of Discrimination

Dino Pedreschi, Salvatore Ruggieri, and Franco Turini

Abstract. Discrimination discovery from data consists in the extraction of discriminatory situations and practices hidden in a large amount of historical decision records. We discuss the challenging problems in discrimination discovery, and present, in a unified form, a framework based on classification rules extraction and filtering on the basis of legally-grounded interestingness measures. The framework is implemented in the publicly available DCUBE tool. As a running example, we use a public dataset on credit scoring.

5.1 Introduction

Human right laws (European Union Legislation, 2011; United Nations Legislation, 2011; U.S. Federal Legislation, 2011) prohibit discrimination against protected groups on the grounds of race, color, religion, nationality, sex, marital status, age and pregnancy; and in a number of settings, including credit and insurance; sale, rental, and financing of housing; personnel selection and wages; access to public accommodations, education, nursing homes, adoptions, and health care. Several authorities (regulation boards, consumer advisory councils, commissions) monitor and report on discrimination compliances. For instance, the European Commission publishes an annual report on the progress in implementing the Equal Treatment Directives by the member states (see Chopin & Do, 2010); and in the US the Attorney General reports to the Congress on the annual referrals to the Equal Credit Opportunity Act.

Given the current state of the art of decision support systems (DSS), socially sensitive decisions may be taken by automatic systems, e.g., for screening or ranking applicants to a job position, to a loan, to school admission and so on. Classical approaches adopted in legal cases (Finkelstein & Levin, 2001) are limited to the verification of an hypothesis of possible discrimination by means of statistical

Dino Pedreschi · Salvatore Ruggieri · Franco Turini
Dipartimento di Informatica, Università di Pisa, Italy
e-mail: {pedre,ruggieri,turini}@di.unipi.it

B. Custers et al. (Eds.): Discrimination & Privacy in the Information Society, SAPERE 3, pp. 91–108.
springerlink.com

analysis of past decision records. However, they reveals to be inadequate to cope with the problem of *searching for* niches of discriminatory decisions hidden in a large dataset of decisions.

Discrimination discovery from data consists in the actual discovery of discriminatory situations and practices hidden in a large amount of historical decision records. The aim is to *extract contexts* of possible discrimination supported by *legally-grounded* measures of the degree of discrimination suffered by protected-by-law groups in such contexts. Reasoning on the extracted contexts can support all the actors in an argument about possible discriminatory behaviors. The DSS owner can use them both to prevent incurring in future discriminatory decisions, and as a means to argument against allegations of discriminatory behavior. A complainant in a case can use them to find specific situations in which there is a *prima facie* evidence of discrimination against groups she belongs to. Control authorities can base the fight against discrimination on a formalized process of intelligent data analysis.

However, discrimination discovery from data may reveal itself an extremely difficult task. The reason is twofold. First, personal data in decision records are typically highly dimensional: as a consequence, a huge number of possible contexts may, or may not, be the theater for discrimination. To see this point, consider the case of gender discrimination in credit approval: although an analyst may observe that no discrimination occurs in general, it may turn out that foreign worker women obtain loans to buy a new car only rarely. Many small or large niches may exist, that conceal discrimination, and therefore all possible specific situations should be considered as candidates, consisting of all possible combinations of variables and variable values: personal data, demographics, social, economic and cultural indicators, etc. The anti-discrimination analyst is thus faced with a combinatorial explosion of possibilities, which make her work hard: albeit the task of checking some known suspicious situations can be conducted using available statistical methods and known stigmatized groups, the task of discovering niches of discrimination in the data is unsupported. The second source of complexity is *indirect discrimination* (see e.g., Tobler, 2008), namely apparently neutral practices that take into account personal attributes correlated with indicators of race, gender, and other protected grounds and that result in discriminatory effects on such protected groups. Even when the race of a credit applicant is not directly recorded in the data, racial discrimination may occur, e.g., as in the practice of *redlining*: people living in a certain neighborhood are frequently denied credit; while not explicitly mentioning race, this fact can be an indicator of discrimination, if from demographic data we can learn that most of people living in that neighborhood belong to the same ethnic minority. Once again, the anti-discrimination analyst is faced with a large space of possibly discriminatory situations: how can she highlight all interesting discriminatory situations that emerge from the data, both directly and in combination with further background knowledge in her possession (e.g., census data)?

We present a classification rule mining approach for the discrimination discovery problem, based on the following ideas. Decision policies are induced from past decision records as classification rules of the form: PREMISES $\rightarrow$ DECISION, where each rule comes with a confidence measure, stating the probability of the decision given the premises of the rule; for instance, the rule RACE=BLACK, CITY=NYC $\rightarrow$ CLASS=BAD with confidence 0.75 states that black people from NYC are assigned bad credit with a 75% probability.

Three kinds of facts (items) are used in decision rules: (potentially) discriminatory items, such as RACE=BLACK, (potentially) non-discriminatory items, such as CITY=NYC, and decision items, such as CLASS=BAD. The potentially discriminatory items are specified by a reference legal framework, to denote some designated groups of people protected by the anti-discrimination laws. The non-discriminatory items define the context where a discriminatory decision may take place - here, the set of applicants from the city of NYC.

Given an historical dataset of decision records, the decision rules hidden in the dataset can be found using *association rule mining*, which allows to extract all the classification rules of the desired form that, in the source dataset, are supported by a specified minimum number of decisions. Continuing the example, the rule RACE=BLACK, CITY=NYC $\rightarrow$ CLASS=BAD is automatically found by association rule mining, if the number of black people in NYC receiving the bad credit is above a minimum threshold value. Such a threshold, known as the minimum support, is meaningful from a legal viewpoint, since it accounts for a minimum number of possibly discriminated persons.

In which circumstances does an extracted rule reveal a (possibly unintentional) discriminatory decision strategy? The idea here is to weight the discrimination of a rule by the gain of confidence due to the presence of the potentially discriminatory items in the premise of the rule. In the example, we compare the 0.75 confidence of the rule RACE=BLACK, CITY=NYC $\rightarrow$ CLASS=BAD with the confidence of the rule obtained removing the first item, i.e., CITY=NYC $\rightarrow$ CLASS=BAD. If, e.g., the confidence of the latter rule is 0.25, then we conclude that black people in NYC have a probability of being assigned bad credit which is 3 times larger than that of the general population of NYC. In this case, a measure called *elift* is used to quantify discrimination risk, which is defined as the ratio of the confidence of the two rules above (with and without the discriminatory item). Whether the rule in the example is to be considered discriminatory or not can now be assessed by thresholding the *elift* measure - possibly according to a value specified in the reference legislation, that limits the acceptable disproportion of treatment. While we use *elift* to illustrate examples throughout the chapter, it is worth noting that several other measures of discrimination (see Section 5.2.2) have been considered in the legal and economic literature, none of which is superior to the others. Actually, our approach is parametric in the definition of a reference measure.

By considering all classification rules with a value of the *elift* higher than the threshold, we can find all the contexts where a discriminatory decision has been taken: in the example, by enumerating *all rules* of the form RACE=BLACK, **B** $\rightarrow$ CLASS=BAD an anti-discrimination analyst discovers all situations **B** where black people suffered a discriminatory credit decision, whatever the complexity of the context **B** and in compliance with the reference legal framework.

So far, we have assumed that discriminatory items are recorded in the source data. This is not always the case, e.g., race may be not available or even collectable. What if the discriminatory variables are not directly available? In this case, indirect discrimination may occur. Consider the rule ZIP=10451, CITY=NYC $\rightarrow$ CLASS=BAD, with confidence 0.95, stating that the residents of a given neighborhood of NYC are assigned bad credit with a 95% chance. Apparently, this rule does not unveil any discriminatory practice. However, assume that the following other rule can be coded from available information, such as census data: ZIP=10451, CITY=NYC $\rightarrow$ RACE=BLACK, with confidence 0.80, stating that 80% of residents of that particular neighborhood of NYC are black. Then it is possible to prove a theoretical lower bound of 0.94 for the confidence of the combined rule ZIP=10451, CITY=NYC, RACE=BLACK $\rightarrow$ CLASS=BAD, stating that 94% of black people in that neighborhood are assigned bad credit, around 3.7 times the general population of NYC. This reasoning shows that the original rule unveils a case of redlining.

Different measures of the discrimination power of the mined decision rules can be defined, according to the provision of different anti-discrimination regulations: e.g., the EU Directives (European Union Legislation, 2011) state that discrimination on a given attribute occurs when “a higher proportion of people without the attribute comply or are able to comply” (which we will code as the *risk ratio* measure), while the US Equal Pay Act (U.S. Federal Legislation, 2011) states that: “a selection rate for any race, sex, or ethnic group which is less than four-fifths of the rate for the group with the highest rate will generally be regarded as evidence of adverse impact” (which we will code as the *selection ratio* measure).

Our discrimination discovery approach opens a promising avenue for research, based on an apparently paradoxical idea: data mining, which is typically used to create potentially discriminatory profiles and classifications, can also be used the other way round, as a powerful aid to the anti-discrimination analyst, capable of automatically discovering the patterns of discrimination that emerge from the available data with the strongest *prima facie* evidence. The preliminary experiments on a dataset of credit decisions operated by a German bank show that this method is able to pinpoint evidence of discrimination: the cited highly discriminatory rule that “foreign worker women are assigned bad credit among those who intend to buy a new car” is actually discovered from such a database.

The rest of the chapter is organized as follows. Section 5.2 introduces the technicalities of classification rules and measures of discrimination defined over them. Using those tools, we show how the anti-discrimination analyst can go through the analysis of direct discrimination (Section 5.3), indirect discrimination (Section 5.4), respondent argumentation (Section 5.5), and affirmative actions (Section 5.6). Some details on the analytical tool DCUBE, which supports the discrimination discovery process, are provided in Section 5.7. Finally, we summarize the approach and discuss some challenging lines for future research.

Table 5.1 The German credit case study: attributes (top) and an excerpt of the dataset (bottom)

Attributes

on personal properties: checking account status, duration, savings status, property magnitude, type of housing
on credits: credit history, credit request purpose, credit request amount, installment commitment, existing credits, other parties, other payment
on employment: job type, employment since, number of dependents, own telephone
on personal status: personal status and gender, age, resident since, foreign worker

Decision

CLASS, with values GOOD (grant credit) and BAD (deny credit)

Potentially discriminatory (PD) items

PERSONAL_STATUS=FEMALE *(female)*
AGE=GT_52 *(senior people)*
FOREIGN_WORKER=YES *(foreign workers)*

PERS_STATUS	AGE	JOB	PURPOSE	CREDIT_AMNT	HOUSING	...	CLASS
female	gt_52	self_emp	new_car	lt_38_k	rent	...	bad
male married	30_to_41	unemp	used_car	39k_to_75_k	own	...	good
male single	42_to_51	skilled	business	75k_to_111k	for_free	...	good
female	gt_52	unemp	furniture	lt_38_k	own	...	bad
...	...	...	...	...	...	...	...

5.2 Classification Rules for Discrimination Discovery

As a running example throughout the chapter, we refer to the public domain German credit dataset, publicly available from the UCI repository of machine learning datasets (Newman, Hettich, Blake, & Merz, 1998). The dataset consists of 1000 records over bank account holders. It includes 20 nominal (or discretized) attributes as shown in Table 5.1. The decision attribute takes values representing the good/bad creditor classification of the bank account holder.

5.2.1 Classification Rules

Given a relation with n attributes, we refer to an *item* as an expression $a = v$, where a is an attribute and v one of its possible values. For example PERSONAL_STATUS = MALE SINGLE is an item for the German credit dataset. One of the attributes is taken as the class attribute, i.e., the attribute referring to the decision. In our running example, the class is named CLASS and the two possible items are CLASS = GOOD, that is credit is granted, and CLASS = BAD, that is credit is denied.

A *transaction* T is a set of items, one for each attribute of the relation. Intuitively, a transaction is the set of items corresponding to a row of a table. By an *itemset* $\mathbf{X}$ we mean a set of items, and we say that a transaction T *supports* an itemset $\mathbf{X}$ if every item in $\mathbf{X}$ belongs to T as well, in symbols $\mathbf{X} \subseteq T$. As an example, the transaction corresponding to the first row in Table 5.1 supports the itemset PERSONAL_STATUS

= FEMALE, AGE = GT_52 but not PERSONAL_STATUS = MALE SINGLE, AGE = GT_52. A dataset $\mathscr{D}$ is a set of transactions. Intuitively, it corresponds to the transactions built from a table.

The support of an itemset **X** w.r.t. $\mathscr{D}$ is the proportion of transactions in $\mathscr{D}$ supporting **X**: $supp(\mathbf{X}) = |\{ T \in \mathscr{D} \mid \mathbf{X} \subseteq T \}|/|\mathscr{D}|$, where $|\,|$ is the cardinality operator.

An association rule is an expression $\mathbf{X} \rightarrow \mathbf{Y}$, where **X** and **Y** are disjoint itemsets. **X** is called the *premise* and **Y** is called the *consequence* of the association rule. We say that $\mathbf{X} \rightarrow \mathbf{Y}$ is a *classification rule* if **Y** is a class item. As an example, PERSONAL_STATUS = FEMALE, AGE = GT_52 $\rightarrow$ CLASS = BAD is a classification rule for the German credit dataset.

The support of $\mathbf{X} \rightarrow \mathbf{Y}$ is the support of the itemset obtained by the union of **X** and **Y**, in symbols $supp(\mathbf{X}, \mathbf{Y})$, where $\mathbf{X}, \mathbf{Y}$ is the union of **X** and **Y**. Intuitively, the support of a rule states how often the rule is satisfied in the dataset. A support of 0.1 for the rule PERSONAL_STATUS = FEMALE, AGE = GT_52 $\rightarrow$ CLASS = BAD means that 10% of the transactions support both the premise and the consequence of the rule, i.e., support PERSONAL_STATUS = FEMALE, AGE = GT_52, CLASS = BAD. The confidence of $\mathbf{X} \rightarrow \mathbf{Y}$, defined when $supp(\mathbf{X}) > 0$, is:

$$conf(\mathbf{X} \rightarrow \mathbf{Y}) = supp(\mathbf{X}, \mathbf{Y})/supp(\mathbf{X}).$$

Confidence states the proportion of transactions supporting **Y** among those supporting **X**. A confidence of 0.7 for the rule above means that 70% of the transactions supporting PERSONAL_STATUS = FEMALE, AGE = GT_52 also support CLASS = BAD. Support and confidence range over $[0, 1]$. Since the seminal paper by (Agrawal & Srikant, 1994), many well explored algorithms have been designed for extracting the set of *frequent* itemsets, i.e., itemsets with a specified minimum support. A survey on frequent pattern mining is due to (Han et al. , 2007); a survey on interestingness measures for association rules is reported by (Geng & Hamilton, 2006); a repository of implementations is maintained by (Goethals, 2010).

5.2.2 Measures of Discrimination

A critical problem in the analysis of discrimination is precisely to quantify the degree of discrimination suffered by a given group (say, an ethnic group) in a given context (say, a geographic area and/or an income range) with respect to a decision (say, credit denial). We rephrase this problem in a rule based setting: if **A** is the condition (i.e., the itemset) that characterizes the group which is suspected of being discriminated against, **B** is the itemset that chacterizes the context, and **C** is the decision (class) item, then the analysis of discrimination is pursued by studying the rule $\mathbf{A}, \mathbf{B} \rightarrow \mathbf{C}$, together with its confidence with respect to the underlying decision dataset - namely, how often such a rule is true in the dataset itself.

Civil rights laws explicitly identify the groups to be protected against discrimination, e.g., women or black people. With our syntax, those groups can be represented as items, e.g., SEX=FEMALE or RACE=BLACK. Therefore, we can assume that the laws provide us with a set of items, which we call potentially discriminatory (PD) items, denoting groups of people that could be potentially discriminated. Given a classification rule SEX=FEMALE, CAR=OWN $\rightarrow$ CREDIT=NO, it is straightforward to separate in its premise SEX=FEMALE from CAR=OWN, in order to reason about potential discrimination against women with respect to people owning a car.

However, discrimination typically occurs for subgroups rather than for the whole group (the US courts coined the term "gender-plus allegations" to describe conducts breaching the law on the ground of sex-plus-something-else), or it may occur for multiple causes (called *multiple discrimination* in ENAR, 2007). For instance, we could be interested in discrimination against older women. With our syntax, this group would be represented as the itemset SEX=FEMALE, AGE=OLDER. The intersection of two disadvantaged minorities (here, SEX=FEMALE and AGE=OLDER) is a, possibly empty, smaller (even more disadvantaged) minority as well. As a consequence, we generalize the notion of potentially discriminatory *item* to the one of potentially discriminatory (PD) *itemset*, and assume that the downward closure property holds for PD itemsets (Ruggieri et al., 2010a).

Definition 1. If $\mathbf{A}_1$ and $\mathbf{A}_2$ are PD itemsets, then $\mathbf{A}_1, \mathbf{A}_2$ is a PD itemset as well.

On the technical side, the downward closure property is a sufficient condition for separating PD itemsets in the premise of a classification rule, namely, there is only one way $\mathbf{A}, \mathbf{B}$ of splitting the premise of a rule into a PD itemset $\mathbf{A}$ and a PND itemset $\mathbf{B}$.

Definition 2. A classification rule $\mathbf{A}, \mathbf{B} \rightarrow \mathbf{C}$ is called potentially discriminatory (PD rule) if $\mathbf{A}$ is non-empty, and potentially non-discriminatory (PND rule) otherwise.

PD rules explicitly state conclusions involving potentially discriminated groups. PD rules cannot be extracted from datasets that do not contain potentially discriminatory items. In such a case, PND rules can still indirectly unveil discriminatory practices (see Section 5.4).

Let us consider now how to quantitatively measure the "burden" imposed on such groups and unveiled by a discovered PD rule. Unfortunately, there is no uniformity nor general agreement on a standard quantification of discrimination by legislations. A general principle mentioned by (Knopff, 1986) is to consider group under-representation as a quantitative measure of the qualitative requirement that people in a group are treated "less favorably" (see European Union Legislation, 2011; U.K. Legislation, 2011) than others, or such that "a higher proportion of people without the attribute comply or are able to comply" (see Australian Legislation, 2011) to a qualifying criterium. We recall from (Ruggieri et al., 2010a) the notion of extended lift[1], a measure of the increased confidence in concluding an assertion $\mathbf{C}$ resulting from adding (potentially discriminatory) information $\mathbf{A}$ to a rule $\mathbf{B} \rightarrow \mathbf{C}$ where no PD itemset appears.

---

[1] The term "extended lift" originates from the fact that it conservatively extends the well-known measure of *lift* (or *interest factor*) of an association rule (Tan et al., 2004), which is obtained, as a special case, when $\mathbf{B}$ empty. Conversely, the extended lift of $\mathbf{A}, \mathbf{B} \rightarrow \mathbf{C}$ corresponds to the lift of $\mathbf{A} \rightarrow \mathbf{C}$ over the set of transactions supporting $\mathbf{B}$.

Classification rule $\mathbf{A},\mathbf{B} \to \mathbf{C}$

group	benefit (C) denied	granted	
protected (**A**)	a	b	n_1
unprotected ($\neg$**A**)	c	d	n_2
	m_1	m_2	n (total of **B**)

$$p_1 = a/n_1 \quad p_2 = c/n_2 \quad p = m_1/n$$

$$RD = p_1 - p_2 \quad RR = \frac{p_1}{p_2} \quad RC = \frac{1-p_1}{1-p_2} \quad OR = \frac{RR}{RC} = \frac{a/b}{c/d}$$

$$ED = p_1 - p \quad ER = \frac{p_1}{p} \quad EC = \frac{1-p_1}{1-p}$$

Fig. 5.1 Contingency table and discrimination measures

Definition 3. Let $\mathbf{A},\mathbf{B} \to \mathbf{C}$ be a PD classification rule with $conf(\mathbf{B} \to \mathbf{C}) > 0$. The extended lift of the rule is:

$$elift(\mathbf{A},\mathbf{B} \to \mathbf{C}) = \frac{conf(\mathbf{A},\mathbf{B} \to \mathbf{C})}{conf(\mathbf{B} \to \mathbf{C})}.$$

A rule SEX=FEMALE, CAR=OWN $\to$ CREDIT=NO with an extended lift of 3 means that being a female increases 3 times the probability of being refused credit with respect to the average confidence of people owning a car. While this means that women are discriminated among car owners, notice that we cannot conclude that being a woman is the actual reason of discrimination (see Sect. 5.5 for a discussion). An alternative way, yet equivalent, of defining the extend lift is as the ratio between the proportion of the disadvantaged group **A** in context **B** obtaining the benefit **C** over the overall proportion of **A** in **B**:

$$\frac{conf(\mathbf{B},\mathbf{C} \to \mathbf{A})}{conf(\mathbf{B} \to \mathbf{A})}.$$

This makes it clear how extended lift relates to the principle of group over-representation in benefit denying, or, equivalently, of group under-representation in benefit granting. In addition to extended lift, other measures can be formalized starting from different definitions of discrimination provided by laws. They can be defined over the 2×2 contingency table shown in Figure 5.1, showing the absolute number of transactions in the underlying dataset $\mathscr{D}$ satisfying the itemsets in the X-Y coordinates and the context **B**. Let p_1 (resp., p_2) be the proportion of people in the protected group (resp., not in the protected group) that were not granted a benefit, and let p be the proportion of all people (both protected and not) that were not granted the benefit. The following discrimination measures can be defined:

- *risk difference* (RD $= p_1 - p_2$), also known as *absolute risk reduction*,
- *risk ratio* or *relative risk* (RR $= p_1/p_2$),

- *relative chance* ($RC = (1-p_1)/(1-p_2)$), also known as *selection rate*,
- *odds ratio* ($OR = p_1(1-p_2)/(p_2(1-p_1))$),

and the versions of RD, RR, and RC when the protected group is compared to the average proportion p, rather than to the proportion of the unprotected group:

- *extended difference* ($ED = p_1 - p$);
- *extended ratio* or *extended lift* ($ER = p_1/p$);
- *extended chance* ($EC = (1-p_1)/(1-p)$).

Since one is interested in contexts of higher benefit denial (resp., lower benefit granting) for the protected group compared to the unprotected group or to the average, the values of interest for RR, OR, and ER are those greater than 1; for RD and ED are those greater than 0; and for RC and EC are those lower than 1. Confidence intervals and tests of statistical significance of the above measures are discussed in (Pedreschi et al., 2009; Ruggieri et al., 2010c). Here, we only mention that statistical tests will rank the rules according to how unlikely it is that they would be observed if there was equal treatment, not according to the severity of discrimination. As an example, a mild discrimination among a large population will be ranked higher than a much more severe discrimination in a small community.

From the legal side, different measures are adopted worldwide. UK law (U.K. Legislation, 2011, (a)) mentions risk difference, EU Court of Justice has given more emphasis to the risk ratio (see Schiek et al., 2007, Section 3.5), and US laws and courts mainly refer to the selection rate[2]. Notice that the risk ratio is the ratio of the proportions of *benefit denial* between the protected and unprotected groups, while selection rate is the ratio of the proportions of *benefit granting*. The EU is more concerned about the ratio of denials, while the US is more concerned about the ratio of grants; unfortunately, they do not lead to the same conclusions in discrimination discovery.

Once we are provided with a quantitative measure of discrimination and a threshold between "legal" and "illegal" degree, we are in the position to isolate classification rules whose measure is below/above the threshold (for simplicity, we limit ourselves to the extended lift measure).

Definition 4 (a-protection). We say that a PD classification rule $\mathbf{A}, \mathbf{B} \to \mathbf{C}$ is a-protective if $elift(\mathbf{A}, \mathbf{B} \to \mathbf{C}) < a$. Otherwise, we say that it is a-discriminatory.

Intuitively, a is a fixed threshold stating an acceptable level of discrimination according to laws, regulations, and jurisprudence. Classification rules denying a benefit and with a measure below such a level are considered safe, whilst rules whose measure is greater or equal than such a level can then be considered a *prima facie*[3] evidence

[2] (U.S. Federal Legislation, 2011, (d)) goes further by stating that "a selection rate for any race, sex, or ethnic group which is less than four-fifths (or eighty percent) of the rate for the group with the highest rate will generally be regarded as evidence of adverse impact". This is called the *four-fifths rule*. It turns out to fix a minimum threshold value for *RC* of $4/5 = 0.8$.

[3] *Prima facie* is a Latin term meaning "at first look," or "on its face," and refers to evidence that, unless rebutted, would be sufficient to prove a particular proposition or fact.

of discrimination. While a-protection is defined with reference to *elift*, its definition clearly applies to any measure from Figure 5.1. An extension of a-protection to account for its statistical significance is proposed in (Pedreschi et al., 2009; Ruggieri et al., 2010c). Also, we refer the reader to (Ruggieri et al. 2010a,2010c) for the presentation and experimentation of data mining algorithms able to efficiently extract a-protective classification rules from a large dataset of historical decision records. Finally, (Pedreschi et al., 2012) show that the choice of a reference measure from Figure 5.1 has a critical impact on the ranking imposed over the set of PD classification rules. In other words, selecting a specific discrimination measure is not a neutral choice, in that it implicitly implies a specific moral criterion to evaluate the degree of discrimination in a specific context; i.e., different ways to establish how bad is a discriminatory action. We found it interesting that our quantitative logical framework for discriminatory rules can help understanding the consequences of such choices in law and jurisprudence.

5.3 Direct Discrimination Discovery

From this section on, we formalize various legal concepts in discrimination analysis and discovery as reasonings over the set of extracted classification rules. We start by considering direct discrimination, which, accordingly to (Ellis, 2005), occurs "where one person is treated less favorably than another". For the purposes of making a *prima facie* evidence in a case before the court, it is enough to show that only one individual has been treated unfairly in comparison to another. However, this may be difficult to prove. The complainant may then use aggregate analysis to establish a regular pattern of unfavorable treatment of the disadvantaged group she belongs to. This is also the approach that control authorities and internal auditing may undertake in analysing historical decisions in search of contexts of discrimination against protected-by-law groups. In direct discrimination, we assume that the input dataset contains attributes to denote potentially discriminated groups. This is a reasonable assumption for attributes such as sex and age, or for attributes that can be explicitly added by control authorities, such as pregnancy status. The next section will consider the case of attributes not available at all or not even collectable. Under our assumption, regular patterns of discrimination can then be identified by looking at PD classification rules of the form:

$$\mathbf{A}, \mathbf{B} \rightarrow \text{BENEFIT=DENIED}$$

i.e., where the consequent consists of denying a benefit (a loan, school admission, a job, etc.). Rules of the form above are then screened by selecting/ranking those with a minimum value of a reference discrimination measure. In terms of Def. 4, we are then looking for "a-discrimination of PD classification rules denying benefit".

As an example, consider our running example dataset and fix the PD items as in Table 5.1. By ranking classification rules of the form $\mathbf{A}, \mathbf{B} \rightarrow$ CLASS=BAD accordingly to their extended lift measure, we found near the top positions the following:

PERSONAL_STATUS=FEMALE, FOREIGN_WORKER=YES,

PURPOSE=NEW_CAR $\rightarrow$ CLASS=BAD

with an extended lift of 1.58. The rule can be interpreted as follows: among those applying for loans to buy a new car, female foreign workers had 1.58 times the average chance of being refused the requested credit. The rule above has a confidence of 0.277, meaning that female foreign workers asking a loan to buy a new car had credit denied in 27.7% of cases (precisely, 13 transactions out of 47). The rule for the generality of applicants:

PURPOSE=NEW_CAR $\rightarrow$ CLASS=BAD

has a confidence of 0.175, meaning that people asking a loan to buy a new car had credit denied in 17.5% of cases.

5.4 Indirect Discrimination Discovery

The EU Directives (see European Union Legislation, 2011; Tobler, 2008) provide a broad definition of indirect discrimination (also known as systematic discrimination or disparate impact) as occurring "where an apparently neutral provision, criterion or practice would put persons of a racial or ethnic origin at a particular disadvantage compared with other persons". In other words, the actual result of the apparently neutral provision is the same as an explicitly discriminatory one. In our framework, the "actual result" is modeled by a PD rule $\mathbf{A},\mathbf{B} \rightarrow \mathbf{C}$ that is a-directly discriminatory, while an "apparently neutral provision" is modeled by a potentially non-discriminatory (PND) rule $\mathbf{B} \rightarrow \mathbf{C}$, where PD itemsets do not occur at all. The issue with unveiling indirect discrimination is that the actual result $\mathbf{A},\mathbf{B} \rightarrow \mathbf{C}$ may be unavailable[4], e.g., because the dataset does not contain attributes to denote the potentially discriminated groups. For instance, the information on a person's race is typically not available and, in many countries, not even collectable. In our approach to indirect discrimination, the problem consists then of inferring some PD rule (with a high discrimination measure value) starting from the set of PND rules, and, possibly, from additional background knowledge. The adjective *potentially non-discriminatory* was chosen exactly to underline that, since the rule does not refer to protected groups, it does not unveil any discriminatory practice in a direct way. Nevertheless, it could do that indirectly.

A remarkable example is *redlining*, a form of indirect discrimination that is explicitly banned in the US (U.S. Federal Legislation, 2011, (b)). As sharply pointed out in Figure 5.2, racial segregation very often emerges in most cities characterized by ethnic diversity: the spatial clustering of a city into racially homogeneous areas is observed in reality much more often than the dispersion of races into an integrated structure. We know from Schelling's segregation model (Schelling, 1971) that a natural tendency to spatial segregation emerges, as a collective phenomenon, even if each individual person is relatively tolerant and open-minded: in his famous abstract simulation model, Schelling showed how segregation eventually appears

[4] Otherwise, the technique of Section 5.3 can be adopted to unveil the effects of both direct and indirect discrimination.

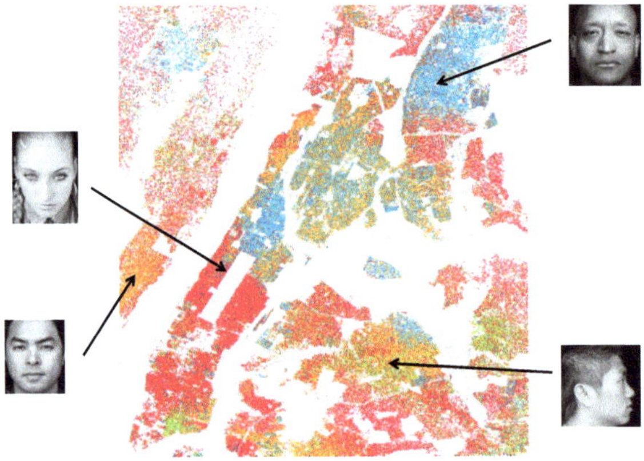

Fig. 5.2 Racial segregation in New York City, based on Census 2000 data (Fischer, 2011). One dot for each 500 residents. Red dots are Whites, blue dots are Blacks, green dots are Asian, orange dots are Hispanic, and yellow dots are other races.

even if each person changes his residence only if less than 30% of his neighbors are of his same race. That's why so many urban territories world-wide, in absence of social restrictions or incentives, developed a structure such that depicted in Figure 5.2; in turn, this explains why denying credit or benefits on the basis of residence – drawing a red line on the border of an urban neighborhood – is often an indirect way to discriminate on the basis of race. Let us consider an example of inference in the context of redlining inspired by the *Hussein vs Saints Complete House Furniture* case reported by (Makkonen, 2006), albeit the numbers reported here are fictious. Assume that a Liverpool furniture store refuses to consider 99% of applicants to a job from a particular postal area ZIP=1234 which had a high rate of unemployment. The extracted classification rule ZIP=1234, CITY=LIVERPOOL $\rightarrow$ APP=NO with confidence $\gamma = 0.99$ is apparently neutral with respect to race discrimination. Assume also that the average refusal rate in the Liverpool area is much lower, say 9%. With our notation, the rule CITY=LIVERPOOL $\rightarrow$ APP=NO has then confidence $p = 0.09$. Assume now to know, e.g., from census background knowledge, that 80% of the population in the postal area ZIP=1234 is black, i.e., that the area is mainly populated by minorities. In formal terms, the association rule ZIP=1234, CITY=LIVERPOOL $\rightarrow$ RACE=BLACK has confidence $\beta = 0.8$. It is now legitimate to ask ourselves whether from such rules, one can conclude a form of redlining, namely the use of ZIP=1234 as a proxy for excluding blacks from a benefit (accepting the side effect of possibly excluding some whites from the same neighborhood). Formally, we want to check whether the extended lift of:

$$(\text{ZIP=1234, RACE=BLACK}),\ \text{CITY=LIVERPOOL} \rightarrow \text{APP=NO} \qquad (\star)$$

is particularly high, where the PD itemset **A** is ZIP=1234, RACE=BLACK, denoting blacks living in the area, and the context **B** is CITY=LIVERPOOL, denoting that the comparison is made against the overall population of that city. The extended lift of such a rule can be read as the ratio of the refusal rate of black people in the ZIP over the mean refusal rate of the whole city. A lower bound for the confidence p_1 of the classification rule ($\star$) can be obtained as $p_1 \geq 1-(1-\gamma)/\beta = 1-0.01/0.8 = 0.9875$ (for details, see Ruggieri et al., 2010a). Intuitively, even in the extreme case that the whole 1% of people in the area who were admitted are blacks, the ratio of un-admitted blacks cannot be lower than 98.75%. By knowing that the average admission rate for the generality of people from Liverpool is 9%, the lower bound for the *elift* measure of ($\star$) is $p_1/p \geq 0.9875/0.09 = 10.97$ – and extremely high ratio stating that black people from that area had at least 10.97 times the average chance (of a Liverpool applicant) of seeing their application refused.

We conclude by mentioning that the redlining inference strategy is one possible inference reasoning for deducing unknown discriminatory effects from observed, apparently non-discriminatory, ones. Additional inference strategies are proposed in (Ruggieri et al., 2010a). In general, an inference strategy consists of deriving lower bounds for a discrimination measure of an unavailable PD rule starting from: assumptions on the form of the premise of the rule; and some background knowledge, which in our framework is coded in the form of association rules. The situation resembles here what occurs in privacy-preserving data mining (Agrawal & Srikant, 2000; Sweeney, 2001), where coupling an anonymized dataset with external knowledge might allow for the inference of the identity of individuals through some attack strategy.

5.5 Argumentation

Consider a PD classification rule denying some benefit:

$$\mathbf{A},\mathbf{B} \rightarrow \text{BENEFIT=DENIED}$$

that has been unveiled, either directly or indirectly. In a case before a court, such a rule supports the complainant position if she belongs to the disadvantaged group **A**, she satisfies the context conditions **B** and the rule is a-directly discriminatory where a is a threshold stated in law, regulations or past sentences. Showing that no rule satisfies those conditions supports the respondent position. However, this is an exceptional case. When one or more such rules exist, the respondent is then required to prove that the "provision, criterion or practice is objectively justified by a legitimate aim and the means of achieving that aim are appropriate and necessary" (see Ellis, 2005). A typical example in the literature is the one of the "genuine occupational requirement", also called "business necessity" by the (U.S. Federal Legislation, 2011, (f)). For instance, assume that the complainant claims for discrimination against women among applicants to a job position. A classification rule SEX=FEMALE, CITY=NYC $\rightarrow$ HIRE=NO with high extended lift supports her position. The respondent might argue that the rule is an instance of a more general rule DRIVE_TRUCK=FALSE, CITY=NYC $\rightarrow$ HIRE=NO. Such a rule is legitimate, since the requirement that prospect workers are able to drive trucks can be considered a genuine occupational requirement (for some specific job). Let us formalize the argumentation of the respondent by saying that a PD classification rule $\mathbf{A},\mathbf{B} \to \mathbf{C}$ is an instance of a PND rule $\mathbf{D},\mathbf{B} \to \mathbf{C}$ when:

- a transaction satisfying $\mathbf{A}$ in context $\mathbf{B}$ satisfies condition $\mathbf{D}$ as well, or, in symbols, $conf(\mathbf{A},\mathbf{B} \to \mathbf{D})$ is close to 1;
- and, the rule $\mathbf{D},\mathbf{B} \to \mathbf{C}$ holds at the same or higher confidence, or, in symbols, $conf(\mathbf{D},\mathbf{B} \to \mathbf{C}) \geq conf(\mathbf{A},\mathbf{B} \to \mathbf{C})$;

A respondent argumenting against discriminatory allegations supported by a PD rule $\mathbf{A},\mathbf{B} \to \mathbf{C}$ must show that the rule is an instance of some PND rule $\mathbf{D},\mathbf{B} \to \mathbf{C}$, and with $\mathbf{D}$ modeling a genuine occupational requirement. On the contrary, a complainant or a control authority can prevent respondent's argumentation by showing that the PD rule $\mathbf{A},\mathbf{B} \to \mathbf{C}$ is not an instance of any PND rule $\mathbf{D},\mathbf{B} \to \mathbf{C}$. In (Ruggieri et al., 2010c), the concept of "instance" has been relaxed to the notion of p-instance, requiring $conf(\mathbf{A},\mathbf{B} \to \mathbf{D}) \geq p$ and $conf(\mathbf{D},\mathbf{B} \to \mathbf{C}) \geq p \cdot conf(\mathbf{A},\mathbf{B} \to \mathbf{C})$. On the experimental side, the vast majority of discriminatory PD rules extracted from the German credit dataset result (p-)instances of some PND rule, thus concluding that it is (fortunately) extremely difficult to characterize *prima facie* evidence of discrimination.

Another defence strategy of the respondent is to resort to the well-known Simpson's paradox. (Bickel, Hammel, & O'Connell, 1975) describes a real case of possible discrimination against women in university admission. Let us rephrase it using our notation. Assume that the rule SEX=FEMALE $\to$ ADMITTED=NO has an high extended lift, so that a possible discrimination is raised. By examining each individual department A of the university, however, it can happen that each rule SEX=FEMALE, DEPT=A $\to$ ADMITTED=NO has a very low extended lift, denoting no discrimination at all. The paradox is that the discrimination observed at university level did not actually occur in any department. If the examination commissions worked at department level, then the department attribute is causal factor, and the standard approach (Pearl, 2009) is to condition probabilities and rules on it. As a consequence, the rules at department level are the correct ones to be looked at, whilst the rule at university level contains confounding factors (the commissions that took decisions).

5.6 Affirmative Actions

Affirmative actions (see ENAR, 2008; Sowell, 2005), sometimes called positive actions or reverse discrimination, are a range of policies to overcome and to compensate for past and present discrimination by providing opportunities to those traditionally denied for. Policies range from the mere encouragement of under-represented groups to quotas in favor of those groups. For instance, US federal contractors are required to identify and set goals for hiring under-utilized minorities and women. Also, universities have voluntarily implemented admission policies that give preferential treatment to women and minority candidates. Affirmative action policies "shall in no case entail as a consequence the maintenance of unequal or separate rights for different racial groups after the objectives for which they were taken have been achieved" (United Nations Legislation, 2011, (a)). It is therefore important to assess and to monitor the application of affirmative actions. In our approach, affirmative actions can be unveiled by proceedings in a similar way as for discriminatory actions. The basic idea is to search, either directly or indirectly, for a-discriminatory PD rules of the form:

$$\mathbf{A},\mathbf{B} \rightarrow \text{BENEFIT=GRANTED}$$

i.e., where the consequent consists of granting a benefit (a loan, a school admission, a job, etc.). Rules of this form with a value of the discrimination measure greater than a fixed threshold highlight contexts **B** where the disadvantaged group **A** was actually favored.

Once again, consider our running example dataset. By ranking classification rules of the form $\mathbf{A},\mathbf{B} \rightarrow$ CLASS=GOOD accordingly to their extended lift measure, we found near the top positions the following:

$$\text{AGE} = \text{GT\_52},\ \text{JOB} = \text{UNEMPLOYED} \rightarrow \text{CLASS=GOOD}$$

with an extended lift of 1.39. The rule can be interpreted as follows: among those unemployed, people older than 52 had 1.39 times the average chance of being granted the requested credit. This could be the case, for instance, of some affirmative actions supporting economic initiatives of unemployed older people.

5.7 The DCUBE Tool

The various concepts and analyses so far discussed, originally implemented as stand-alone programs for achieving the best performances, have been re-designed around an Oracle database, used to store extracted rules, and a collection of functions, procedures and snippets of SQL queries that implement the various legal reasonings for discrimination analysis. The resulting implementation, called DCUBE (Discrimination Discovery in Databases) (Ruggieri et al., 2010), can be accessed and exploited by a wider audience if compared to a stand-alone monolithic application. In fact, SQL is the dominant query language for relational data, with database administrators already mastering issues such as data storage, query optimization, and import/export towards other formats. Discrimination discovery is an interactive and iterative process, where analyses assume the form of deductive reasoning over extracted rules. An appropriately designed database, with optimized indexes, functions and SQL query snippets, can be welcome by a large audience of users, including owners of socially-sensitive decision data, government anti-discrimination analysts, technical consultants in legal cases, researchers in social sciences, economics and law. Typical discrimination discovery questions that DCUBE is able to answer include:

Direct discrimination discovery: *"How much have women been under-represented in obtaining the loan?"* or *"List under which conditions blacks were suffering an extended lift higher than 1.8 in our recruitment data"*. DCUBE comes with all of the legally-grounded measures from Figure 5.1 predefined. The user can adopt any of them or, even, she can easily define new measures over a 4-fold contingency table by adding methods to an Oracle user defined data type.

Indirect discrimination discovery, such as the following redlining question *"I don't have the race attribute in my data, but have the ZIP of residence. By adding background knowledge on the distribution of race over ZIP codes, infer cases where ZIP actually disguises race discrimination."*

Affirmative actions and favoritism: *"List cases where our university admission policies actually favored blacks"*, and *"Under which conditions white males are given the best mortgage rate in comparison to the average?"*

On-line documentation, demo, and download of the DCUBE system can be accessed from `http://kdd.di.unipi.it/dcube`.

5.8 Conclusions

We presented a data mining approach for the analysis and discovery of discrimination in a dataset of socially-sensitive decisions. The approach consists first of extracting frequent classification rules, and then screening/ranking them on the basis of quantitative measures of discrimination. The key legal concepts of protected-by-law groups, direct discrimination, indirect discrimination, genuine occupational requirement, and affirmative actions are formalized as reasonings over the set of extracted rules and, possibly, additional background knowledge. The approach has been implemented in the DCUBE tool and made publicly available. Chapter 13 builds on our approach for the purpose of designing data mining classifiers that do not learn to discriminate, an issue known as discrimination prevention.

As future work, we aim to achieve two goals: on one hand, to improve the methods and the technologies for discovering discrimination, especially looking at data mining methods such at classification and clustering, driven by constraints over specific application contexts (racial profiling, labor market, credit scoring, etc.); on the other hand, to further interact with legal experts both to find out new measures and rules that we may support with our tools and to influence their design and interpretation of legislation. Finally, we are looking at other fields of application, other than credit scoring. An interesting one is discovering possible discrimination (with respect to sex, nationality, etc.) in funding research projects.

References

Agrawal, R., Srikant, R.: Fast algorithms for mining association rules in large databases. In: Proc. of Int. Conf. on Very Large Data Bases (VLDB 1994), pp. 487–499. Morgan Kaufmann (1994)

Agrawal, R., Srikant, R.: Privacy-preserving data mining. In: Proc. of the ACM SIGMOD Int. Conf. on Management of Data (SIGMOD 2000), pp. 439–450. ACM (2000)

Australian Legislation: (a) Age Discrimination Act, 2004; (b) Australian Human Rights Commission Act, 1986; (c) Disability Discrimination Act, 1992; (d) Racial Discrimination Act, 1975; (e) Sex Discrimination Act, 1984; (f) Victoria Equal Opportunity Act, 1995; (g) Queensland Anti Discrimination Act, 1991 (2011), `http://www.hreoc.gov.au`

Bickel, P., Hammel, E., O'Connell, J.: Sex bias in graduate admissions: Data from berkeley. Science 187(4175), 398–404 (1975)
Chopin, I., Do, T.U.: Developing anti-discrimination law in europe. European Network of Legal Experts in Anti-Discrimination (2010), http://ec.europa.eu
Ellis, E.: Eu anti-discrimination law. Oxford University Press (2005)
ENAR. European network against racism, fact sheet 33: Multiple discrimination (2007), http://www.enar-eu.org
ENAR. European network against racism, fact sheet 35: Positive actions (2008), http://www.enar-eu.org
European Union Legislation: (a) European Convention on Human Rights, 1950; (b) Racial Equality Directive, 2000; (c) Employment Equality Directive, 2000; (d) Gender Goods and Services Directive, 2004; (e) Gender Employment Directive, 2006; (f) Equal Treatment Directive (proposal), 2008 (2011), http://eur-lex.europa.eu
Finkelstein, M.O., Levin, B. (eds.): Statistics for lawyers, 2nd edn. Springer (2001)
Fischer, E.: Distribution of race and ethnicity in US major cities (2011), published on line at http://www.flickr.com/photos/walkingsf/sets/72157624812674967 under Creative Commons licence, CC BY-SA 2.0
Geng, L., Hamilton, H.J.: Interestingness measures for data mining: A survey. ACM Computing Surveys 38(3) (2006)
Goethals, B.: Frequent itemset mining implementations repository (2010), http://fimi.cs.helsinki.fi
Han, J., Cheng, H., Xin, D., Yan, X.: Frequent pattern mining: Current status and future directions. Data Mining and Knowledge Discovery 15(1), 55–86 (2007)
Knopff, R.: On proving discrimination: Statistical methods and unfolding policy logics. Canadian Public Policy 12(4), 573–583 (1986)
Makkonen, T.: Measuring discrimination: Data collection and the EU equality law. European Network of Legal Experts in Anti-Discrimination (2006), http://www.migpolgroup.com
Newman, D., Hettich, S., Blake, C., Merz, C.: UCI repository of machine learning databases (1998), http://archive.ics.uci.edu
Pearl, J.: Causality: Models, reasoning, and inference, 2nd edn. Cambridge University Press, New York (2009)
Pedreschi, D., Ruggieri, S., Turini, F.: Measuring discrimination in socially-sensitive decision records. In: Proc. of the SIAM Int. Conf. on Data Mining (SDM 2009), pp. 581–592. SIAM (2009)
Pedreschi, D., Ruggieri, S., Turini, F.: A study of top-k measures for discrimination discovery. In: Proc. of ACM Int. Symposium On Applied Computing (SAC 2012), pp. 126–131. ACM (2012)
Ruggieri, S., Pedreschi, D., Turini, F.: Data mining for discrimination discovery. ACM Trans. on Knowledge Discovery from Data 4(2), 1–40 (2010a)
Ruggieri, S., Pedreschi, D., Turini, F.: DCUBE: Discrimination discovery in databases. In: Proc. of the ACM SIGMOD Int. Conf. on Management of Data (SIGMOD 2010), pp. 1127–1130. ACM (2010b)
Ruggieri, S., Pedreschi, D., Turini, F.: Integrating induction and deduction for finding evidence of discrimination. Artificial Intelligence and Law 18(1), 1–43 (2010c)
Schelling, T.C.: Dynamic models of segregation. Journal of Mathematical Sociology 1, 143–186 (1971)
Schiek, D., Waddington, L., Bell, M. (eds.): Cases, materials and text on national, supranational and international non-discrimination law. Hart Publishing (2007)

Sowell, T. (ed.): Affirmative action around the world: An empirical analysis. Yale University Press (2005)
Sweeney, L.: Computational disclosure control: A primer on data privacy protection. Unpublished doctoral dissertation, MIT, Cambridge, MA (2001)
Tan, P.-N., Kumar, V., Srivastava, J.: Selecting the right objective measure for association analysis. Information Systems 29(4), 293–313 (2004)
Tobler, C.: Limits and potential of the concept of indirect discrimination. European Network of Legal Experts in Anti-Discrimination (2008), http://www.migpolgroup.com
U.K. Legislation: (a) Sex Discrimination Act, 1975; (b) Race Relation Act, 1976 (2011), http://www.statutelaw.gov.uk
United Nations Legislation: (a) Convention on the Elimination of All forms of Racial Discrimination, 1966; (b) Convention on the Elimination of All forms of Discrimination Against Women, 1979 (2011), http://www.ohchr.org
U.S. Federal Legislation: (a) Equal Credit Opportunity Act, 1974; (b) Fair Housing Act, 1968; (c) Employment Act, 1967; (d) Equal Pay Act, 1963; (e) Pregnancy Discrimination Act, 1978; (f) Civil Right Act, 1964, 1991 (2011), http://www.eeoc.gov

Chapter 6
Discrimination Data Analysis: A Multi-disciplinary Bibliography

Andrea Romei and Salvatore Ruggieri

Abstract. Discrimination data analysis has been investigated for the last fifty years in a large body of social, legal, and economic studies. Recently, discrimination discovery and prevention has become a blooming research topic in the knowledge discovery community. This chapter provides a multi-disciplinary annotated bibliography of the literature on discrimination data analysis, with the intended objective to provide a common basis to researchers from a multi-disciplinary perspective. We cover legal, sociological, economic and computer science references.

6.1 Introduction

Discrimination refers to an unjustified distinction of treatment on the basis of any physical or cultural trait, such as gender, race, religion or sexual orientation. The problems of assessing the presence, the extent, the nature, and the trend of discrimination are then of primary importance. In the last fifty years, such problems have been investigated from social, legal, economic, and, recently, from a computer science perspective. The issues of data collection and data analysis are persistent, unifying themes along all the perspectives. We present an annotated multi-disciplinary bibliography specifically focusing on "data-driven", or empirical, or analytical, approaches. The ease of data storage and retention, the ever increasing computing power, the development of intelligent data analysis and mining techniques make it possible to apply "in-the-large" and to improve over classical statistical and econometric techniques. The reference literature, however, is abundant and spread over publications of many disciplines, as witnessed by our references: social sciences, psychology, economics, finance, health research, housing and urban development, statistics, biometrics, econometrics and data mining.

Andrea Romei · Salvatore Ruggieri
Dipartimento di Informatica, Università di Pisa, Italy
e-mail: {romei,ruggieri}@di.unipi.it

B. Custers et al. (Eds.): Discrimination & Privacy in the Information Society, SAPERE 3, pp. 109–135.
springerlink.com

A complete bibliography would be an utopian goal. Our priority is to provide to interested reader with references to survey, comparison, and overview papers as well as with recent works on the subject. The chapter is structured as follows. After introducing the relevant concepts and references from social and legal perspectives in Section 6.2, we concentrate on the vast research on economic models of labour discrimination in Section 6.3. The approaches for collecting and analyzing controlled data using (quasi-)experimental scientific methodologies are presented in Section 6.4. Section 6.5 discusses discrimination in profiling and scoring, and, finally, Section 6.6 reports on recent work on using data mining for discrimination discovery and prevention.

6.2 Sociological and Legal Perspectives

From a sociological perspective, there are three main causes of discrimination: prejudice, rational racism, and unintentional discrimination. *Prejudice* leads to discrimination when it concerns unfairly or unreasonably formed negative attitudes against a protected[1] group. The vicious cycle of discrimination (Newman, 2008) starts from a situation where prejudice causes a protected group to be socially disadvantaged. This is interpreted as evidence that the group is inferior, which, in turn, creates renewed prejudice by increasing social distance, by reinforcing negative stereotypes, and by legitimating negative feelings. Psychologists have investigated situations of anxiety or concerns, called stereotype threats (Steele & Aronson, 1995), where persons have the potential to confirm a negative stereotype of their social group, which results in reduced performances of individuals. *Rational racism* is the result of rational thinking. A form of rational racism is statistical discrimination, occurring when the lack of knowledge about the skills of an individual is compensated by a prior knowledge of the average performances of the group or category the individual belongs to. Another example of rational thinking occurs when an employer foresee a negative impact on his business due to the prejudice of his customers against employers belonging to a protected group. Finally, *unintentional discrimination* occurs not because of malevolent decisions, but due to the lack of awareness on the effects of a decision. This is the case of indifference, incorrect (execution of) procedures or practices, lack of planning and analysis of the decision outcomes. Also, a form of unconscious or implicit discrimination has been considered in the literature (Bertrand et al., 2005; Greenwald & Krieger, 2006; Kang & Banaji, 2010). Together with the concept of indirect discrimination (see later on), unintentional discrimination poses considerable problems for the data analyst to carefully take into account the effects of decisions from the point of view of different protected groups. We refer to (R. Brown, 2010; Newman, 2008) for a sociological overview of prejudice, to (Whitley & Kite, 2009) for a psychological discussion[2], to (Quillian, 2006) for a

[1] We use the term "protected group" for any social group protected by anti-discrimination laws.

[2] See http://www.understandingprejudice.org for links to prejudice-related resources.

review of racial prejudice, and finally, to (Harford, 2008) for a discussion of rational racism. (Yamagishi et al., 1999) review social theories of in-group favoritism.

In the legal context, provisions on equality or non-discrimination[3] are firmly embedded within the key human rights treaties of the United Nations Legislation (United Nations Legislation, 2011). Anti-discrimination laws, however, have evolved differently in common law countries compared to civil law ones. The United States (US) Federal Legislation (U.S. Federal Legislation, 2011), the U.K. Legislation (U.K. Legislation, 2011) and the Australian Legislation (Australian Legislation, 2011) follow the common law characteristic of "the absence of systematisation, or a desire thereof" (Schiek et al., 2007, Introductory Chapter), with the result that laws have been developed ground-by-ground and with reference to specific contexts, possibly with different ruling from one case to another. The European Union (EU) Legislation (European Union Legislation, 2011) and the EU Member States follow a principled approach, resulting in laws covering a (long) list of grounds of discrimination. For a deeper legal discussion and comparison of national and international laws, we refer the reader to books on international group rights (N. Lerner, 2003; Schiek et al., 2007), on EU laws (Ellis, 2005; E.U. Agency for Fundamental Rights, 2011), and on US laws (Bamforth et al., 2008). Several independent authorities (equality enforcement bodies, regulation boards, consumer advisory councils, commissions) provide advice, monitor, and report on discrimination compliances. For instance, the EU Commission[4] publishes an annual report on the progress in implementing the Equal Treatment Directives by the Member States (Chopin & Do, 2010); and in the US Attorney General reports to the Congress about the annual referrals to the Equal Credit Opportunity Act. A general legal principle is to consider *group under-representation* in obtaining a benefit as a quantitative measure of (indirect) discrimination against a protected group. Data collection and statistical data analysis are recognized as fundamental both in the common law and in the civil law countries (R. M. Blank et al., 2004; Makkonen, 2006, 2007). It is commonly agreed, however, that the statistical conclusions establish *a prima facie* evidence of discrimination, which may be rebutted by the respondent using further arguments (e.g., a genuine occupational requirement or an objective justification). We refer to (Wingate & Thornton, 2000; Finkelstein & Levin, 2001) for a review of statistical methods in discrimination litigations. The book edited by (Kaye & Aickin, 1992) contains a collection of papers on the subject. A continuously updated book on statistical methods and case laws is maintained by (Paetzold et al., 1994). Finally, the interdisciplinary economic-legal survey by (Donohue, 2007) provides an overview of the connections between economic models and empirical findings from the one side, and the US anti-discrimination laws on the other side. A related legal concept that is worth mentioning is the one of *affirmative actions*, sometimes called *positive actions*, which are a range of policies to overcome and to compensate for past

[3] The term "non-discrimination law" recalls a set of negative obligations, while "equality law" recalls, in addition, a set of positive obligations to reach the ideal of equal treatment (Bell, 2002).

[4] See also the European Network of Legal Experts http://www.non-discrimination.net, and the Migration Policy Group http://www.migpolgroup.com.

and present discrimination by providing opportunities to those traditionally denied for (ENAR, 2008; Holzer & Neumark, 2004; Sowell, 2005). They range from the mere encouragement of under-represented groups to preferential treatment or quotas in favor of those groups (see e.g., Holzer & Neumark, 2006; R. Lerner & Nagai, 2000).

Since discrimination can arise only through the application of different rules or practices to comparable situations or of the same rule or practice to different situations, a relevant legal distinction is between direct and indirect discrimination. When such rules or practices explicitly treat one person less favorably on a forbidden ground than another is, has been or would be treated in a comparable situation, we have *direct discrimination*, sometimes called *systematic* discrimination or *disparate treatment*. When an apparently neutral provision, criterion or practice results in an unfair treatment of a protected group, we have *indirect discrimination*, sometimes called *adverse impact* (Tobler, 2008). While direct discrimination is intentional and "directed" towards individuals, typically on the basis of their visible traits, such as ethnic origin, race, sex and age, indirect discrimination is concerned with avoiding the circumvention of the prohibition to discriminate, and to enforce such a prohibition substantively, even in the case of unintentionality.

6.3 Labour Economic Perspective

In the labor market, different treatments among groups of workers can be measured in terms of their wages (*wage differentials*), in the degree of participation in the labor force (*employment differentials*), or in the degree of segregation in specific occupations or industries (*segregation differentials*). Public surveys routinely collect data on demographic characteristics and attitudes of residents (e.g., in the US, the General Social Survey - GSS), on the distribution of labor forces in the labor market (e.g., Current Population Survey - GPS), and so on. Empirical research techniques have applied statistical inference to collected data either with the purpose of testing the consequences predicted by a theoretical economic model, or to assess the contribution of different types of discrimination to the overall different treatments in the labor market. The main data analysis techniques adopted include statistical tests on rates and proportions (Agresti, 2002; Fleiss et al., 2003; Sheskin, 2004), (generalized) linear regression models (Dobson & Barnett, 2008; Hardin & Hilbe, 2007; McCullagh & Nelder, 1989), and econometric models (Greene, 2008).

Two major theoretical models of discrimination have been considered in the economic literature. *Taste-based discrimination*, originally proposed by (Becker, 1971), has no rational or economic basis, but only a prejudiced personal taste against protected groups. Wage differentials are due to an additional psychological cost for employing minority workers. Differently, *statistical discrimination*, originated by (Arrow, 1971) and (Phelps, 1972) and systematized by (Aigner & Cain, 1977), starts from the assumption that employers cannot perfectly assess worker productivity at the time of hiring. This market imperfection gives them an incentive to use easily observable characteristics, such as sex and race, as proxies for the expected productivity, estimated by their prior knowledge on the average productivity of the group the worker belongs to. Wages are then set on the basis of the expected productivity of the group, not on the basis of the person's productivity. We refer the reader to (Altonji & Blank, 1999) for a comprehensive mathematical introduction to both theories of economic discrimination, as well as for past empirical approaches to show direct evidence. More recent or comprehensive reviews of theories and empirics in labor market are available in (Cain, 1987; Charles & Guryan, 2011; Kunze, 2008; Lang & Lehmann, 2011). (Weichselbaumer & Winter-Ebmer, 2005) conduct a meta regression analysis of the works on gender wage differentials, where each point of data is not an individual but a research study. (Neal & Johnson, 1996) observed that, after controlling for the ability of a worker, the racial wage gap greatly reduces. In such a study, ability was measured through the controversial *Armed Forces Qualifying Test* (AFQT) score, a test of cognitive skills taken by male adolescents and available from the National Longitudinal Survey of Youth. In the following, we briefly review the most recent lines of research and extensions of the two economic models.

Approaches on taste-based discrimination. The additional cost of minority workers in presence of taste-based discrimination leads to an equilibrium wage differential and to segregation of minority workers in less discriminating firms or for specific occupations. Lower earning for discriminatory firms implies that discrimination occurs mainly in low competitive markets. This is known as the *static implication* of the Becker's model[5]. Influential papers are (Charles & Guryan, 2008), which combine GSS data (to measure racial prejudice) with CPS data (to measure differences in wages), and (Hellerstein et al., 2002), which relate firm profitability to the proportion of female workers both in low competition and high competition markets. Recent approaches using survey data include (Sano, 2009; Tsao & Pearlman, 2010; Zhang & Dong, 2008). On the basis of the identity of the discriminator, Becker's model distinguishes *employer discrimination* (taste in hiring), *customer discrimination* (taste in buying), and *co-employee discrimination* (taste in co-operating). Recent analyses of consumer discrimination have been conducted on data from restaurants (Parrett, 2011; Myers, 2007), contact jobs (Combes et al., 2011), retail stores (Leonard et al., 2010), Major League Baseball (Coyne et al., 2010) and taxicab drivers (Ayres et al., 2005). Evidence of correlation between the predominant race of customers and the race of the marginal hired worker has been shown in (Holzer & Ihlanfeldt, 1998).

[5] The *dynamic implication* of the Becker's model predicts that non-discriminating employers earn higher profits by hiring members of the protected group, and, in the long run and in a competitive market, discriminatory firms will be driven out of the market. The dynamic implication has been investigated in the context of banking deregulation (Black & Strahan, 2001; Levine et al., 2008), globalization (Black & Brainerd, 2004; Neumayer & Soysa, 2007; Oostendorp, 2009) and in the adoption of equality laws worldwide (Weichselbaumer & Winter-Ebmer, 2007).

The working context of *professional sports*, such as baseball, basketball, football, and soccer, offers an unusually good opportunity of studying discrimination. The problem of estimating the productivity of workers is here substantially solved by extensive, publicly available (from online sport almanacs), measures of the performances of players and coaches. Research has covered discrimination in hiring, in retaining (along seasons), in segregating (to specific game roles), and in salary of players, as well as customer discrimination. The last topic is also known as *fan discrimination*, typically measured using TV audience (Aldrich et al., 2005), game attendance (Foley & Smith, 2007; Hersch, 2009; Wilson & Ying, 2003), the trading value of sport cards (Broyles & Keen, 2010; Primm et al., 2011), the votes for best player awards (Jewell et al., 2002). As far as salary discrimination in professional sports is concerned with, there is an extensive literature on the subject. We mention only a few recent papers (Berri & Simmons, 2009; Holmes, 2011; Frick & Deutscher, 2009; Goddard & Wilson, 2009; Palmer & King, 2006; Yang & Lin, 2010), and refer the reader to the surveys (Kahn, 1991b, 2000, 2009).

Extensions of taste-based discrimination, called *search models* (Altonji & Blank, 1999; Lang & Lehmann, 2011), take into account the costs for workers of searching jobs by interacting with prejudiced and non-prejudiced firms, and, for consumers, the costs of searching sellers of their same racial group (Flabbi, 2010; Kuhn & Shen, 2009; Sulis, 2007; Usui, 2009). Finally, a line of studies, initiated by (Hamermesh & Biddle, 1994), investigates the “beauty premium” in labor market. As a recent work, we mention (Cipriani & Zago, 2011), who study favoritism to attractive students in taking exams at University. The effectiveness of blind decisions in reducing gender discrimination has been evaluated for orchestra auctions in (Goldin & Rouse, 2000).

Approaches on statistical discrimination. Some extensions of the statistical discrimination model deal with what happens as the employer’s information on workers’ productivity changes, e.g., at the selection time or over the course of the job. These dynamic extensions, contrasted to a static model, are known as *employer learning* models. (Farber & Gibbons, 1996) propose a dynamic model of learning about worker ability in a competitive labor market. Altonji and Pierret provide a first important strand literature on learning models (Altonji & Pierret, 2001). We complement the studies surveyed in the recent paper (Lang & Lehmann, 2011) by mentioning: (Cheung, 2010), in testing whether parental education is used as a proxy for the ability of workers; and (Wang, 2010), in considering height as an easily observable characteristic.

Also, the differential observability or learnability of worker’s productivity among groups has been taken into account by *screening discrimination* models, originally introduced in (Lang, 1986). Such differences are due, e.g., to miscommunication problems or weak interactions among groups. As an example, (Grogger, 2011) analyzes audio data from telephone interviews to understand the role that speech may play in explaining racial wage differences, and (Pinkston, 2006) shows that the level of education has a large impact on wages. Similar work emerges from the health literature, when testing whether miscommunication problems influence a diagnosis

(Balsa et al., 2005; Mcguire et al., 2008) or whether "expert" patients obtain a more favorable treatment (Grytten et al., 2011). Another strand of statistical discrimination models studies how negative rational stereotypes of employers differentiates firms' hirings and wages, and workers' investments, e.g., in education. (Lang & Lehmann, 2011) call this class as *rational stereotyping* models. Finally, we refer to the survey (Fang & Moro, 2010) for a theoretical discussion of models of statistical discrimination and affirmative actions.

6.4 (Quasi-)Experimental Perspective

A recurring problem in discrimination analysis is the collection of controlled data, as opposed to observational data, for which the results of analytical and statistical techniques can be interpreted without any concern for external or confounding factors. This has been tackled through quasi-experimental and experimental methods, that we review in the next two subsections.

6.4.1 Auditing

Auditing, also known as *field experiments*, follows a quasi-experimental approach to investigate for the presence of discrimination by controlling the factors that may influence decision outcomes. The basic idea consists of using pairs of *testers* (also called *auditors*), who have been matched to be similar on all characteristics that may influence the outcome except race, gender, or other grounds of possible discrimination. The tester pairs are then sent into one or more situations in which discrimination is suspected, e.g., to rent an apartment or to apply for a job, and the decision outcome is recorded. The difference in the outcomes among the paired groups provides then a measure of discrimination. A summary of recent audit studies in employment discrimination is due to (Pager, 2007). (Riach & Rich, 2002) review and compare the statistical significance of field experiments on racial, sex, and disability discrimination in employment, and on discrimination on housing sale and rental. Criticism of the conclusions drawn from audit methods is discussed in (Heckman & Siegelman, 1993) and (Heckman, 1998), while (Riach & Rich, 2004) comment on ethical implications of such methods. (Quillian, 2006) discusses how the measurement of discrimination through audit methods should incorporate recent advancement in psychological theories of prejudice.

We categorize three different approaches in detecting discrimination by auditing.

Situation testing occurs when the testers come in contact with the decision maker. This is the case, for instance, of job interviews involving human testers, who are selected and trained in advance to act similar each other (Bendick et al., 2010; Moreno et al., 2004; Pager & Quillian, 2005; Pager et al., 2009; Turner & Ross, 2005; Turner et al., 2002). A strong point in favor of situation testing is that testers can record the cause of discrimination, such as prejudice or stereotypes, hence allowing for a causal analysis of the discrimination cases. A limitation of situation testing is that the phase of data collection is expensive. In addition, situation testing cannot be applied at all in some contexts, e.g., in wage rising discrimination, or in disparate application of contractual terms, e.g., in house lending (Roscigno et al., 2009). (Bendick, 2007) reviews more than 30 situation testing studies in employment discrimination in the US, while (Rorive, 2009) covers the EU Member States context.

In *correspondence testing*, the data scarcity problem is mitigated by designing paired ad-hoc fake resumes or application forms to be sent to advertised vacancies, and by assigning to each of them a typical white American name or an African-American sounding name (Arai et al., 2008; Banerjee et al., 2009; Bertrand & Mullainathan, 2004; Carlsson & Rooth, 2007; Kaas & Manger, 2010; Neumark, 2010). Other grounds of discrimination have been covered with a correspondence testing approach in job applications, including sex (Riach & Rich, 2006; Booth & Leigh, 2010), obesity (Rooth, 2009), sexual orientation (Drydakis, 2009), ethnicity (McGinnity et al., 2009).

Larger opportunities for data collection are offered by emerging Internet job advertisement services, known as *e-recruiting* (Booth et al., 2010; Edin & Lagerstrøm, 2006). The synthetic generation of resumes is tackled in (Lahey & Beasley, 2009) by a parametric tool that mitigates the bias that is present in manually generated CVs. The legal implications of possible discrimination in e-recruiting, as compared to classical means of recruiting, are discussed in (Hogler et al., 1998). In addition, contexts other than employment can be covered, such as discrimination in product advertising in internet marketing (Doleac & Stein, 2010; Nunley et al., 2010), and in on-line rental housing (Ahmed & Hammarstedt, 2008; Bosch et al., 2010; Friedman et al., 2010; Hanson & Hawley, 2011; Taylor, 2010).

6.4.2 Controlled Experiments

Field experiments construct control groups by matching similar persons and then observing the outcome of a quasi-experiment in a *natural* environment, e.g., in a job selection procedure. Empirical data from field experiments reflect a variety of environmental factors: disentangling these factors may be difficult if not impossible. Controlled experiments are conducted in an *artificial* environment, such as a laboratory, under tightly controlled conditions, including selection of treatment and control groups and strict rules on their behavior and actions. On the one hand, the impact of a specific factor can be evaluated by systematically varying it. On the other hand, confounding variables and other extraneous stimuli can be minimized. Controlled experiments are very useful to test the predictions of some theoretical model or to pre-test the impact of some ruling or laws before their application. Also, controlled experiments are repeatable, by definition, and less expensive than field experiments. The main criticism against controlled experiments is that they suffer of lack of realism, also called *external validity*. (Harrison & List, 2004) propose a taxonomy of experiments. We refer to (Charness & Kuhn, 2011; Levitt & List, 2007) and (R. M. Blank et al., 2004, Chapter 6) for an in-deep discussion on methodological strengths and on the limits of generalizing results obtained from experiments.

We distinguish here two classes of controlled experiments, namely laboratory experiments and natural experiments.

(Levitt & List, 2007) review five classes of games used in the economic literature to measure social preferences through *laboratory experiments*, including fairness, trust, and conditional reciprocity. The reviewed games include dictator and ultimatum games, public goods games, trust and gift exchange games. As an example, (Fershtman & Gneezy, 2001) adopt trust games, dictator and ultimatum games to test for ethnic discrimination. The trust game assumes a "player A", who is given a fixed amount of money and asked to transfer a certain amount to "player B". The transferred amount is triplicated. Then, "player B" can choose to transfer any part of the received amount back to "player A". Players A and B are randomly paired from students of different ethnicity. The lower average amount of money transferred to players of a specific ethnicity, compared to others ethnicities, is considered evidence of discrimination. Recent controlled experiments can be found in the context of sports card market (J. List, 2004), employment (Feltovich & Papageorgiou, 2004; Falk et al., 2008) and wages differentials (Güth et al., 2010; Dickinson & Oaxaca, 2009), beauty and speech differences (Andreoni & Petrie, 2008; Rödin & Özcan, 2011). Moreover, gender (Slonim & Guillen, 2010), racial (Castillo & Petrie, 2010) and district-based (Falk & Zehnder, 2007) differences have been studied in the context of in-group discrimination and favoritism.

Natural experiments occur in real life (yet, controlled) situations. The experimenter only observes the behavior of participants, who typically are not aware of the experiment. Television game shows are a typical example, where discriminatory choices of participants can be studied in a controlled environment. Discrimination analysis has been reported in (Antonovics et al., 2005, 2009; Bagues & Villadoniga, 2008; Levitt, 2004), with data gathered from the *Weakest Link* game show, in (Lee, 2009) with data from *American Idol* TV contest show, and in (J. A. List, 2006) with data from *Friend or Foe?*. Sources of favoritism to attractive people by analysing data from a TV game show based on the prisoner's dilemma are studied in (Belot et al., 2008). In addition to the criticism of external validity, natural experiments have also the problem that not all factors are under control, e.g., the selection of participants to a TV game show.

6.5 Profiling Perspective

Profiles consists of patterns, rules, or any other form of knowledge that can be used to screen people when searching for those with a certain behavior. They occur in many context, from criminal investigation to marketing, from genetic screening to web site personalization, from fraud prevention to location-based services. Profiling is the process of extracting profiles, either by manually eliciting them from domain experts or by automatically inferring them from historical data using increasingly sophisticated machine learning and data mining techniques. The process of profiling also concerns the application of profiles to screen individuals, e.g., as in the case of credit risk scoring and in the identification of security risks – which are covered in the next two subsections. We refer to (Hildebrandt & Gutwirth, 2008) for a cross-disciplinary perspective of automated profiling.

6.5.1 Racial Profiling

Profiling is an illegal practice as soon as its application results in direct or indirect discrimination against protected groups. In this section, we concentrate on *racial profiling*, defined as "the practice of subjecting citizens to increased surveillance or scrutiny based on racial or ethnic factors rather than reasonable suspicion" (J. Chan, 2011). Among several possible contexts of racial profiling, vehicle stops have attracted the vast majority of studies[6]. Numerous data collection efforts have been initiated by law enforcement agencies, often as a result of litigation or of legislation, for the purpose of understanding the vehicle stop practices of its officers. Attributes collected concern the stop (time, date, location, reason, duration), driver (race, gender, age), vehicle (make, model), officer (age, gender, race, education, experience), and the outcome of the stop (e.g., warning, citation, arrest, search, seizure of contraband). The objective of data analysis is to identify racial patterns of disparity. One of the early surveys on racial profiling is due to (Engel et al., 2002). More recent papers include (Farrell & McDevitt, 2010; Tillyer et al., 2010), reviewing vehicle stops approaches. The adequacy of statistical analysis of racial profiling in addressing legal issues is also discussed in (Tillyer et al., 2008). For a legal comparison of US and EU laws, see (Baker & Phillipson, 2011).

(Tillyer et al., 2010) categorize existing approaches depending on whether they deal with the initial decision or with the outcome of a stop.

In *initial stop* studies, the actual rate of stops by drivers' race is compared with benchmark data providing the expected rate of stops assuming no police bias. The outermost difficulty of the approach consists of identifying accurate benchmarks of the expected driver population at risk of being stopped. (Engel & Calnon, 2004), and (R. M. Blank et al., 2004, Chapter 9) outline strengths and limitations of six primary data sources and their use in the design of benchmark data: census data, observations of roadway usage, official accident data, assessments of traffic violating behavior, citizen surveys, and internal departmental comparisons. Alternative means for collecting benchmark data are proposed in (Alpert et al., 2004; Jobard & Lévy, 2011; Quintanar, 2009; Ridgeway & MacDonald, 2009; Gelman et al., 2007).

Post-stop outcome studies focus on the identification of racial disparities in a specific outcome of the stop by taking as reference population the whole set of stops. An example of post-stop outcome analysis consists of checking whether the search for drugs among stopped vehicles is biased against the driver's race. In this respect, starting from the influential paper proposed in (Knowles et al., 2001), several extensions and critiques have been presented (Antonovics & Knight, 2009; Anwar & Fang, 2006; Gardner, 2009; Rowe, 2008; Sanga, 2009). We refer to the surveys

[6] Other contexts include profiling in airport security (Gabbidon et al., 2011; Persico & Todd, 2005), fraud investigators (Leopold & Meints, 2008), capital sentences (Alesina & Ferrara, 2011), and consumer profiling (Gabbidon et al., 2008; Schreurs et al., 2008).

(Tillyer et al., 2010; Engel, 2008) for extensive references. Recent additional approaches include (Anbarci & Lee, 2008; Blalock et al., 2007; Pickerill et al., 2009; Ridgeway, 2006).

6.5.2 Credit Markets

Discrimination in the lending process may occur at several steps, from advertising, to pre-application enquires, to loan approval/denial, up to loan administration (Turner & Skidmore, 1999). Among the various credit markets, mortgage lending has received most of the interest. In all cases, however, the main challenge is in the difficulty of estimating the risk of granting a loan to an applicant on the basis of her financial capacity and her personal characteristics.

In the US, the *Home Mortgage Disclosure Act* (HMDA) requires lenders to gather and to make available census data about their mortgage applications. Since 1990, the HMDA has been integrated with information on discrimination grounds of applicants. One of the first relevant contribution is due to researchers at the Federal Reserve Bank of Boston in the research work known as *Boston Fed Study* (Munnell et al., 1996). They supplemented the original census HMDA data for Boston with additional information on the credit history of more than 3,000 individual applicants, including data from more than one hundred financial institutes. Several criticisms of the Boston Fed study appeared in the literature (Ross & Yinger, 2002, Chapter 5), (Longhofer & Peters, 1999), (Turner & Skidmore, 1999, Chapter 3). Among the problems highlighted, we mention data errors, misclassification problems, endogenous explanatory variables and the omitted variables bias (e.g., loan amount and indicator of cosigner were missing). A theoretical and empirical survey on racial disparities in mortgage lending markets in the context of the fair housing legislation is provided in (LaCour-Little, 1999). (G. Dymski, 2006) describes the state-of-the-art on discrimination in housing and credit markets both from a legal and an economic perspective. A recent review has been proposed in (Yezer, 2010), which devises three approaches of testing disparities in loan approval decisions: mortgage rejection, pricing and defaults.

In *mortgage rejection*, the disproportionate rate of rejected decisions between racial groups of applicants is considered *prima facie* evidence of discrimination. Empirical studies (Clarke et al., 2009; Dietrich, 2009; Dietrich & Johannsson, 2005; Goenner, 2010; Sanandaji, 2009) include the analysis of HMDA data at *bank level* (i.e., a model for each bank under analysis) or at a *market level* (i.e., a single model aggregating variables for several banks). An experimental comparison of the two approaches is reported in (Blackburn & Vermilyea, 2006). Other sources of data range from micro-lending data (Agier & Szafarz, 2010) to on-line data derived from a peer-to-peer lending site (Pope & Sydnor, 2011).

Mortgage pricing concentrates on the dataset of approved loans, by considering whether a minority group is systematically charged with the highest interest rates.

Recent mortgage pricing studies consider gender and racial discrimination in consumer credit (Edelberg, 2007), such as credit cards and education loans, in private firm credit (Albareto & Mistrulli, 2011; Blanchard et al., 2008; Blanchflower et al., 2003; Cavalluzzo et al., 2002; Muravyev et al., 2009), in subprime home loans (Bocian et al., 2008; Reid & Laderman, 2009), in household credit (Weller, 2008). Using survey data, (P. Cheng et al., 2009) found that women pay higher rates because they do not search for best-rate loans as much as men do.

Mortgage default studies adopt the percentage of mortgage defaults as a measure of discrimination. Intuitively, if different default rates are observed for equally creditworthy groups that differ in some discrimination ground, this is considered *prima facie* evidence of discrimination. Recent contributions on the subject include (C. L. Brown & Simpson, 2010; S. Chan et al., 2010; Yezer, 2010). A discussion of the limitations of data on mortgage defaults, including unobserved variables and sample-selection bias, can be found in (Turner & Skidmore, 1999, Chapter 5).

Discrimination in mortgage rejection and pricing has often occurred indirectly, through the practice of *redlining* (Hillier, 2003),(Turner & Skidmore, 1999, Chapter 4), which consists of denying credit or of applying higher interest rates to people living in some specific neighborhood. The use of geographic attributes may hide (intentionally or not) the fact that such a neighborhood is populated mainly by people of a specific race or minority. US cities, in particular, show a very high racial divide. The percentage of individuals of a protected group in a neighborhood is often used as a measure of the level of segregation (James & Tauber, 1985; Reardon & Firebaugh, 2002). Empirical works combine HMDA data with census data (Silverman, 2005; E. Blank et al., 2005; Blackburn & Vermilyea, 2007; Ding et al., 2008; Ezeala-Harrison et al., 2008; Wyly et al., 2008; Rugh & Massey, 2010; Squires et al., 2009; Vicki et al., 2009; G. A. Dymski et al., 2011) to test for such a form of indirect discrimination. As an alternative, (Campbell et al., 2008) use proprietary data on unsecured debt. Other studies on redlining use house market data (Aalbers, 2007; Ezeala-Harrison et al., 2008), consumer credit card data (Brevoort, 2011; Cohen-Cole, 2009), and insurance data (Ong & Stoll, 2007; Ross & Tootell, 2004).

Finally, in the related context of consumer markets, *price discrimination* is the practice of a retailer, wholesaler, or manufacturer of selling the same product, with the same marginal cost, at different prices based on buyers' willingness to pay (Armstrong, 2006). Differential pricing discriminating racial minorities has been observed in the car sales market (Ayres, 1995; Ayres & Siegelman, 1995; Goldberg, 1996).

6.6 Knowledge Discovery Perspective

The issue of discrimination analysis has been considered from a knowledge discovery, also known as data mining, perspective along two directions: discrimination discovery and prevention.

Discrimination discovery from data consists in the actual discovery of discriminatory situations and practices hidden in a large amount of historical decision records. The aim is to unveil contexts of possible discrimination on the basis of *legally-grounded* measures of the degree of discrimination suffered by protected-by-law groups in such contexts. The legal principle of under-representation has inspired existing approaches for discrimination discovery based on pattern mining. Starting from a dataset of historical decision records, (Pedreschi et al., 2008; Ruggieri et al., 2010a) propose to extract classification rules such as RACE=BLACK, PURPOSE=NEW_CAR → CREDIT=NO, called *potentially discriminatory* (PD) rules, to unveil contexts (here, people asking for a loan to buy a new car) where the protected group (here, black people) suffered from under-representation with respect to the decision (here, credit denial). The approach has been implemented on top of an Oracle database by relying on tools for frequent itemset mining (Ruggieri et al., 2010b), and extended in (Pedreschi et al., 2009; Ruggieri et al., 2010c; Luong, 2011). The main limitation of the approach is that there is no control of the characteristics (e.g., capacity to repay the loan) of the protected group, versus, or as opposed to others in this context.

This results in an overly large number of PD rules that need to be further screened. (Luong et al., 2011) exploit the idea of situation testing. For each member of the protected group with a negative decision outcome, testers with similar characteristics are searched for in a dataset of historical decision records. If one can observe significantly different decision outcomes between the testers of the protected group and the testers of the unprotected group, one can ascribe the negative decision to a bias against the protected group, thus labeling the individual as discriminated. The approaches so far described assume that the dataset under analysis contains items to denote protected groups. This may be not the case when such items are not available, or not even collectable at micro-data level, e.g., as in the case of the loan applicant's race. (Ruggieri et al., 2010a, 2010c) adopt a form of rule inference to cope with the indirect discovery of (either direct or indirect) discrimination.

Discrimination prevention in data mining and machine learning consists of extracting models (typically, classifiers) that trade off accuracy for non-discrimination. In fact, mining from historical data may mean to discover traditional prejudices that are endemic in reality (i.e., taste-based discrimination), or to discover patterns of lower performances, skills or capacities of protected-by-law groups (i.e., statistical discrimination). Mining algorithms may then assign to such discriminatory practices the status of general rules, which are subsequently used for automatic decision making in socially sensitive tasks (see e.g., (N. Cheng et al., 2011; Chien & Chen, 2008; Yap et al., 2011)).

Discrimination prevention has been recognized as an issue in the tutorial (Clifton, 2003, Slide 19), where the danger of building classifiers capable of redlining discrimination in home loans has been put forward. In predictive statistics, the same issue has been raised by (Pope & Sydnor, 2007). The naïve approach of deleting attributes that denote protected groups from the original dataset does not prevent a classifier to indirectly learn discriminatory decisions, since other attributes strongly correlated with them could be used as a proxy by the model extraction algorithm.

This issue has been observed in (Pope & Sydnor, 2007; Ruggieri et al., 2010a). We categorize three non mutually-exclusive strategies toward discrimination prevention: (i) a controlled distortion of the training set (a pre-processing approach) (Kamiran & Calders, 2009; Zliobaite et al., 2011; Luong et al., 2011; Hajian et al., 2011); (ii) a modification of the classification learning algorithm (an in-processing approach), by integrating anti-discrimination criteria within it (Calders & Verwer, 2010; Kamiran et al., 2010; Kamishima et al., 2011); (iii) a post-processing of the classification model, once it has been extracted, to correct its decision criteria (Pedreschi et al., 2009; Calders & Verwer, 2010).

6.7 Conclusions

The collection and analysis of observational and experimental data is the main tool for assessing the presence, the extent, the nature, and the trend of discrimination phenomena. In this chapter, we provided an annotated bibliography of the main references and of recent works on discrimination data analysis from a multi-disciplinary perspective. Our intended objective was to provide a guidance through the abundant literature to researchers and anti-discrimination analysts that are faced with data analysis problems. Substantively, the reader is referred to works on sociological causes, legal norms, economic models, empirical studies, data collection approaches, profiling methods, discrimination discovery techniques, and discrimination prevention algorithms in data mining. The bibliography section includes 262 references, half of which appeared in the last five years (2007-2011). This demonstrates a never-ending interest on the topic of discrimination data analysis.

References

Aalbers, M.: What types of neighbourhoods are redlined? Journal of Housing and the Built Environment 22(2), 177–198 (2007)

Agier, I., Szafarz, A.: Microfinance and gender: Is there a glass ceiling in loan size? In: CEB Working Paper No. 10-047, Université Libre de Bruxelles (2010), http://ssrn.com

Agresti, A.: Categorical data analysis, 2nd edn. Wiley-Interscience (2002)

Ahmed, A.M., Hammarstedt, M.: Discrimination in the rental housing market: A field experiment on the internet. Journal of Urban Economics 64(2), 362–372 (2008)

Aigner, D.J., Cain, G.G.: Statistical theories of discrimination in labor markets. Industrial and Labor Relations Review 30, 175–187 (1977)

Albareto, G., Mistrulli, P.: Bridging the gap between migrants and the banking system (Economic working paper No. 794). Economic Research Department, Bank of Italy (2011), http://www.bancaditalia.it

Aldrich, E.M., Arcidiacono, P.S., Vigdor, J.L.: Do people value racial diversity? Evidence from Nielsen ratings. Journal of Economic Analysis & Policy 5(1), Art. 4 (2005)

Alesina, A.F., Ferrara, E.L.: A test of racial bias in capital sentencing (Working Paper No. 16981). National Bureau of Economic Research (2011), http://www.nber.org

Alpert, G.P., Smith, M.R., Dunham, R.G.: Toward a better benchmark: Assessing the utility of not-at-fault traffic crash data in racial profiling research. Justice Research and Policy 6(1), 43–70 (2004)

Altonji, J.G., Blank, R.M.: Race and gender in the labor market. In: Ashenfelter, O., Card, D. (eds.) Handbook of Labor Economics-Part C, vol. 3, pp. 3143–3259. Elsevier (1999)

Altonji, J.G., Pierret, C.R.: Employer learning and statistical discrimination. The Quarterly Journal of Economics 116(1), 313–350 (2001)

Anbarci, N., Lee, J.: Speed discounting and racial disparities: Evidence from speeding tickets in Boston (Discussion Paper No. 3903). Institute for the Study of Labor, IZA (2008), http://ftp.iza.org

Andreoni, J., Petrie, R.: Beauty, gender and stereotypes: Evidence from laboratory experiments. Journal of Economic Psychology 29(1), 73–93 (2008)

Antonovics, K.L., Arcidiacono, P., Walsh, R.: Games and discrimination: Lessons from the Weakest Link. Journal of Human Resources 40(4), 918–947 (2005)

Antonovics, K.L., Arcidiacono, P., Walsh, R.: The effects of gender interactions in the lab and in the field. Review of Economics and Statistics 91(1), 152–162 (2009)

Antonovics, K.L., Knight, B.G.: A new look at racial profiling: Evidence from the Boston Police Department (Working Paper No. 10634). National Bureau of Economic Research (2009), http://www.nber.org

Anwar, S., Fang, H.: An alternative test of racial prejudice in motor vehicle searches: Theory and evidence. American Economic Review 96(1), 127–151 (2006)

Arai, M., Moa Bursell, M., Nekby, L.: Between meritocracy and ethnic discrimination: The gender difference (Research Papers in Economics No. 2008:4). Institute for the Study of Labor, IZA (2008), http://ftp.iza.org

Armstrong, M.: Recent developments in the economics of price discrimination. In: Blundell, R., Newey, W.K., Persson, T. (eds.) Proc. of World Congress on Advances in Economics and Econometrics Theory and Applications, vol. 2, pp. 1–46. Cambridge University Press (2006)

Arrow, K.J.: The theory of discrimination. In: Ashenfelter, O., Rees, A. (eds.) Discrimination in Labor Markets, pp. 3–33. Princeton University Press (1971)

Australian Legislation: (a) Age Discrimination Act, 2004; (b) Australian Human Rights Commission Act, 1986; (c) Disability Discrimination Act, 1992; (d) Racial Discrimination Act, 1975; (e) Sex Discrimination Act, 1984; (f) Victoria Equal Opportunity Act, 1995, (g) Queensland Anti Discrimination Act, 1991 (2011), http://www.hreoc.gov.au

Ayres, I.: Further evidence of discrimination in new car negotiations and estimates of its cause. Michigan Law Review 94(1), 109–147 (1995)

Ayres, I., Siegelman, P.: Race and gender discrimination in bargaining for a new car. The American Economic Review 85(3), 304–321 (1995)

Ayres, I., Vars, F.E., Zakariya, N.: To insure prejudice: Racial disparities in taxicab tipping. Yale Law Journal 114(7), 1613–1674 (2005)

Bagues, M.F., Villadoniga, M.J.P.: Why do i like people like me? (Working Paper No. 2008-06). Universidad Carlos III, Departamento de Economía de la Empresa (2008), http://papers.ssrn.com

Baker, A., Phillipson, G.: Policing, profiling and discrimination law: US and European approaches compared. Journal of Global Ethics 7(1), 105–124 (2011)

Balsa, A.I., Mcguire, T.G., Meredith, L.S.: Testing for statistical discrimination in health care. Health Services Research 40(1), 227–252 (2005)

Bamforth, N., Malik, M., O'Cinneide, C.: Discrimination law: Theory & context, text and materials, 1st edn. Sweet & Maxwell (2008)

Banerjee, A., Bertrand, M., Datta, S., Mullainathan, S.: Labor market discrimination in Delhi: Evidence from a field experiment. Journal of Comparative Economics 37(1), 14–27 (2009)
Becker, G.S.: The economics of discrimination (economic research studies), 2nd edn. University of Chicago Press, Chicago (1971)
Bell, M.R.: Anti-discrimination law and the European Union. Oxford University Press (2002)
Belot, M., Bhaskar, V., van de Ven, J.: Beauty and the sources of discrimination (Working Paper No. 241). ESRC Centre for Economic Learning and Social Evolution (2008), http://eprints.ucl.ac.uk
Bendick, M.: Situation testing for employment discrimination in the United States of America. Horizons Stratégiques 3(5), 17–39 (2007)
Bendick, M., Rodriguez, R.E., Jayaraman, S.: Employment discrimination in upscale restaurants: Evidence from matched pair testing. Social Science Journal 47(4), 802–818 (2010)
Berri, D., Simmons, R.: Race and the evaluation of signal callers in the National Football League. Journal of Sports Econonomics 10(1), 23–43 (2009)
Bertrand, M., Chugh, D., Mullainathan, S.: Implicit discrimination. American Economic Review 95(2), 94–98 (2005)
Bertrand, M., Mullainathan, S.: Are Emily and Greg more employable than Lakisha and Jamal? A field experiment on labor market discrimination. American Economic Review 94(4), 991–1013 (2004)
Black, S.E., Brainerd, E.: Importing equality? The impact of globalization on gender discrimination. Industrial and Labor Relations Review 57(4), 540–559 (2004)
Black, S.E., Strahan, P.E.: The division of spoils: Rent-sharing and discrimination in a regulated industry. American Economic Review 91(4), 814–831 (2001)
Blackburn, M.L., Vermilyea, T.: A comparison of unexplained racial disparities in bank-level and market-level models of mortgage lending. Journal of Financial Services Research 29(2), 125–147 (2006)
Blackburn, M.L., Vermilyea, T.: The role of information externalities and scale economies in home mortgage lending decisions. Journal of Urban Economics 61(1), 71–85 (2007)
Blalock, G., DeVaro, J.L., Leventhal, S., Simon, D.H.: Gender bias in power relationships: Evidence from police traffic stops (Working Paper). School of Industrial and Labor Relations, Cornell University (2007), http://ssrn.com
Blanchard, L., Zhao, B., Yinger, J.: Do lenders discriminate against minority and woman entrepreneurs? Journal of Urban Economics 63(2), 467–497 (2008)
Blanchflower, D.G., Levine, P.B., Zimmerman, D.J.: Discrimination in the small-business credit market. The Review of Economics and Statistics 85(4), 930–943 (2003)
Blank, E., Venkatachalam, P., McNeil, L., Green, R.: Racial discrimination in mortgage lending in Washington, D.C.: A mixed methods approach. The Review of Black Political Economy 33(2), 9–30 (2005)
Blank, R.M., Dabady, M., Citro, C.F. (eds.): Measuring racial discrimination - panel on methods for assessing discrimination. National Academies Press (2004)
Bocian, D.G., Ernst, K.S., Li, W.: Race, ethnicity and subprime home loan pricing. Journal of Economics and Business 60(1-2), 110–124 (2008)
Booth, A.L., Leigh, A.: Do employers discriminate by gender? A field experiment in female-dominated occupations. Economics Letters 107(2), 236–238 (2010)
Booth, A.L., Leigh, A., Varganova, E.: Does racial and ethnicdiscrimination vary across minority groups? Evidence from a field experiment (Discussion paper No. 4947). Institute for the Study of Labor, IZA (2010), http://ftp.iza.org

Bosch, M., Carnero, M.A., Farr, L.: Information and discrimination in the rental housing market: Evidence from a field experiment. Regional Science and Urban Economics 40(1), 11–19 (2010)

Brevoort, K.P.: Credit card redlining revisited. Review of Economics and Statistics 93(2), 714–724 (2011)

Brown, C.L., Simpson, W.G.: An analysis of alternative methodologies and interpretations of mortgage discrimination research using simulated data. Academy of Banking Studies Journal 9(2), 65–75 (2010)

Brown, R.: Prejudice: Its social psychology, 2nd edn. Wiley-Blackwell (2010)

Broyles, P., Keen, B.: Consumer discrimination in the NBA: An examination of the effect of race on the value of basketball trading cards. The Social Science Journal 47(1), 162–171 (2010)

Cain, G.G.: The economic analysis of labor market discrimination: A survey. In: Ashenfelter, O., Layard, R. (eds.) Handbook of Labor Economics, vol. 1, pp. 693–781. Elsevier (1987)

Calders, T., Verwer, S.: Three naive bayes approaches for discriminationfree classification. Data Mining & Knowledge Discovery 21(2), 277–292 (2010)

Campbell, R., Roberts, B., Rogers, K.: An evaluation of lender redlining in allocation of unsecured consumer credit. Urban Studies 45(5), 1243–1254 (2008)

Carlsson, M., Rooth, D.-O.: Evidence of ethnic discrimination in the Swedish labor market using experimental data. Labour Economics 14(4), 716–729 (2007)

Castillo, M., Petrie, R.: Discrimination in the lab: Does information trump appearance? Games and Economic Behavior 68(1), 50–59 (2010)

Cavalluzzo, K., Cavalluzzo, L., Wolken, J.: Competition, small business financing, and discrimination: Evidence from a new survey. Journal of Business 75(4), 641–680 (2002)

Chan, J.: Racial profiling and police subculture. Canadian Journal of Criminology and Criminal Justice 53(1), 75–78 (2011)

Chan, S., Gedal, M., Been, V., Haughwout, A.F.: The role of neighborhood characteristics in mortgage default risk: Evidence from New York City (Working Paper No. 4551). NYUWagner School and Furman Center for Real Estate & Urban Policy (2010), `http://www.ssrn.com`

Charles, K.K., Guryan, J.: Prejudice and wages: An empirical assessment of becker's the economics of discrimination. Journal of Political Economy 116(5), 773–809 (2008)

Charles, K.K., Guryan, J.: Studying discrimination: Fundamental challenges and recent progress. Annual Review of Economics 3, 479–511 (2011)

Charness, G., Kuhn, P.: Lab labor: What can labor economists learn from the lab? In: Ashenfelter, O., Card, D. (eds.) Handbook of Labor Economics, vol. 4, pp. 229–330. Elsevier (2011)

Cheng, N., Chandramouli, R., Subbalakshmi, K.: Author gender identification from text. Digital Investigation 8(1), 78–88 (2011)

Cheng, P., Lin, Z., Liu, Y.: Do women pay more for mortgages? The Journal of Real Estate Finance and Economics, 1–18 (2009) (published online: November 17, 2009)

Cheung, S.: A test of employer learning in the labour market for young Australians. Applied Economics Letters 17(1), 93–98 (2010)

Chien, C.-F., Chen, L.: Data mining to improve personnel selection and enhance human capital: A case study in high-technology industry. Expert Systems with Applications 34(1), 280–290 (2008)

Chopin, I., Do, T.U.: Developing anti-discrimination law in europe. European Network of Legal Experts in Anti-Discrimination (2010), `http://ec.europa.eu`

Cipriani, G.P., Zago, A.: Productivity or discrimination? beauty and the exams. Oxford Bulletin of Economics and Statistics 73(3), 428–447 (2011)

Clarke, J., Roy, N., Courchane, M.: On the robustness of racial discrimination findings in mortgage lending studies. Applied Economics 41(18), 2279–2297 (2009)

Clifton, C.: Privacy preserving data mining: How do we mine data when we aren't allowed to see it? In: Proc. of the ACM SIGKDD Int. Conf. on Knowledge Discovery and Data Mining (KDD 2003), Tutorial (2003), http://www.cs.purdue.edu

Cohen-Cole, E.: Credit card redlining (Working Paper No. QAU08-1). Federal Reserve Bank of Boston (2009), http://www.bos.frb.org

Combes, P.-P., Decreuse, B., Laouenan, M., Trannoy, A.: A test of customer discrimination: Theory and evidence from the french labor market (2011), preliminary version, http://econ.sciences-po.fr

Coyne, C.J., Isaacs, J., Schwartz, J.: Entrepreneurship and the taste for discrimination. Journal of Evolutionary Economics 20(4), 609–627 (2010)

Dickinson, D.L., Oaxaca, R.L.: Statistical discrimination in labor markets: An experimental analysis. Southern Economic Journal 76(1), 16–31 (2009)

Dietrich, J.: Does multiple jeopardy exist in mortgage markets? (Economics Working Paper No. 2009-3). Office of the Comptroller of the Currency (2009), http://www.occ.treas.gov

Dietrich, J., Johannsson, H.: Searching for age and gender discrimination in mortgage lending (Economics Working Paper No. 2005-2). Office of the Comptroller of the Currency (2005), http://www.occ.treas.gov

Ding, L., Ratcliffe, J., Stegman, M., Quercia, R.: Neighborhood patterns of high-cost lending: The case of Atlanta. Journal of Affordable Housing 17(3), 194–211 (2008)

Dobson, A.J., Barnett, A.G.: Introduction to generalized linear models, 3rd edn. Chapman and Hall/CRC (2008)

Doleac, J.L., Stein, L.C.: The visible hand: Race and online market outcomes (Discussion Paper No. 09-015). Stanford Instititute for Economic Policy Research (2010), http://www.stanford.edu

Donohue, J.J.: The law and economics of antidiscrimination law. In: Polinsky, A.M., Shavell, S. (eds.) Handbook of Law and Economics, pp. 1387–1472. Elsevier (2007)

Drydakis, N.: Sexual orientation discrimination in the labour market. Labour Economics 16(4), 364–372 (2009)

Dymski, G.: Discrimination in the credit and housing markets: Findings and challenges. In: Rodgers, W. (ed.) Handbook on the Economics of Discrimination, 3rd edn., Edward Elgar Publishing (2006)

Dymski, G.A., Hernandez, J., Mohanty, L.: Race, power, and the subprime/foreclosure crisis: A mesoanalysis (Economics Working Paper No. 669). Levy Economics Institute of Bard College (2011), http://ssrn.com

Edelberg, W.: Racial dispersion in consumer credit interest rates (Finance and Economics Discussion Series Nos. 2007–2028). Board of Governors of the Federal Reserve System, U.S. (2007)

Edin, P.-A., Lagerstrøm, J.: Blind dates: quasi-experimental evidence on discrimination (Working Paper No. 4). Institute for Labour Market Policy Evaluation (2006), http://www.ifau.se

Ellis, E.: Eu anti-discrimination law. Oxford University Press (2005)

ENAR. European network against racism, fact sheet 35: Positive actions (2008), http://www.enar-eu.org

Engel, R.S.: A critique of the "outcome test" in racial profiling research. Justice Quarterly 25(1), 1–36 (2008)
Engel, R.S., Calnon, J.M.: Comparing benchmark methodologies for police-citizen contacts: Traffic stop data collection for the Pennsylvania State Police. Police Quarterly 7(1), 97–125 (2004)
Engel, R.S., Calnon, J.M., Bernard, T.J.: Theory and racial profiling: Shortcomings and future directions in research. Justice Quarterly 19(2), 1–36 (2002)
E.U. Agency for Fundamental Rights. Handbook on European nondiscrimination law. European Court of Human Rights (2011), http://fra.europa.eu
European Union Legislation: (a) European Convention on Human Rights, 1950; (b) Racial Equality Directive, 2000; (c) Employment Equality Directive, 2000; (d) Gender Goods and Services Directive, 2004; (e) Gender Employment Directive, 2006; (f) Equal Treatment Directive (proposal), 2008 (2011), http://eur-lex.europa.eu
Ezeala-Harrison, F., Glover, G., Shaw-Jackson, J.: Housing loan patterns toward minority borrowers in Mississippi: Analysis of some micro data evidence of redlining. The Review of Black Political Economy 35(1), 43–54 (2008)
Falk, A., Walkowitz, G., Wirth, W.: Do ex-offenders face discrimination in the labor market? Because of expected inferior reciprocity (2008) (unpublished manuscript), http://www.uni-graz.at
Falk, A., Zehnder, C.: Discrimination and in-group favoritism in a citywide trust experiment (Working Paper No. iewwp318). Institute for Empirical Research in Economics, University of Zurich (2007), http://www.iew.uzh.ch
Fang, H., Moro, A.: Theories of statistical discrimination and affirmative action: A survey. In: Benhabib, J., Jackson, M., Bisin, A. (eds.) Handbook of Social Economics, vol. 1A, pp. 134–200. Elsevier (2010)
Farber, H.S., Gibbons, R.: Learning and wage dynamics. The Quarterly Journal of Economics 111(4), 1007–1047 (1996)
Farrell, A., McDevitt, J.: Identifying and measuring racial profiling by the police. Sociology Compass 4(1), 77–88 (2010)
Feltovich, N., Papageorgiou, C.: An experimental study of statistical discrimination by employers. Southern Economic Journal 70(4), 837–849 (2004)
Fershtman, C., Gneezy, U.: Discrimination in a segmented society: An experimental approach. The Quarterly Journal of Economics 116(1), 351–377 (2001)
Finkelstein, M.O., Levin, B. (eds.): Statistics for lawyers, vol. 2. Springer (2001)
Flabbi, L.: Prejudice and gender differentials in the US labor market in the last twenty years. Journal of Econometrics 156(1), 190–200 (2010)
Fleiss, J.L., Levin, B., Paik, M.C.: Statistical methods for rates and proportions, 3rd edn. Wiley (2003)
Foley, M., Smith, F.H.: Consumer discrimination in professional sports: New evidence from Major League Baseball. Applied Economics Letters 14(13), 951–955 (2007)
Frick, B., Deutscher, C.: Salary determination in the Ge rman "Bundesliga": A panel study (IASE Conference Paper No. 0811). International Association of Sports Economists (2009)
Friedman, S., Squires, G.D., Galvan, C.: Cybersegregation in Boston and Dallas: Is Neil a more desirable tenant than Tyrone or Jorge? (Presented at the Population Association of America 2010 Annual Meeting (2010), http://paa2010.princeton.edu
Gabbidon, S.L., Craig, R., Okafo, N., Marzette, L.N., Peterson, S.A.: The consumer racial profiling experiences of Black students at historically Black colleges and universities: An exploratory study. Journal of Criminal Justice 36(4), 354–361 (2008)

Gabbidon, S.L., Higgins, G.E., Nelson, M.: Public support for racial profiling in airports: Results from a statewide poll. Criminal Justice Policy Review (2011) (published online: March 14, 2011)

Gardner, J.: Deterrence externalities and racial bias in law enforcement (Working Paper No. 88). Carnegie Mellon University (2009), http://www.heinz.cmu.edu

Gelman, A., Fagan, J., Kiss, A.: An analysis of the new york city police departments "stop-and-frisk" policy in the context of claims of racial bias. Journal of the American Statistical Association 102(479), 813–823 (2007)

Goddard, J., Wilson, J.O.S.: Racial discrimination in English professional football: Evidence from an empirical analysis of players' career progression. Cambridge Journal of Economics 33(2), 295–316 (2009)

Goenner, C.: Discrimination and mortgage lending in Boston: The effects of model uncertainty. Journal of Real Estate Finance and Economics 40(3), 260–285 (2010)

Goldberg, P.K.: Dealer price discrimination in new car purchases: Evidence from the consumer expenditure survey. Journal of Political Economy 104(3), 622–654 (1996)

Goldin, C., Rouse, C.: Orchestrating impartiality: The impact of "blind" auditions on female musicians. American Economic Review 90(4), 715–741 (2000)

Greene, W.H.: Econometric analysis, 7th edn. Prentice-Hall (2008)

Greenwald, A.G., Krieger, L.H.: Implicit bias: Scientific foundations. California Law Review 94(4), 945–967 (2006)

Grogger, J.T.: Speech patterns and racial wage inequality. Journal of Human Resources 46(1), 1–25 (2011)

Grytten, J., Skau, I., Sørensen, R.: Do expert patients get better treatment than others? Agency discrimination and statistical discrimination in obstetrics. Journal of Health Economics 30(1), 163–180 (2011)

Güth, W., Kocher, M.G., Popova, V.: Co-employment of permanently and temporarily employed agents (Jena Economic Research Paper Nos. 2010–016). Friedrich-Schiller-University Jena, Max-Planck-Institute of Economics (2010), http://www.econ.mpg.de

Hajian, S., Domingo-Ferrer, J., Martínez-Ballesté, A.: Rule Protection for Indirect Discrimination Prevention in Data Mining. In: Torra, V., Narakawa, Y., Yin, J., Long, J. (eds.) MDAI 2011. LNCS, vol. 6820, pp. 211–222. Springer, Heidelberg (2011)

Hamermesh, D., Biddle, J.: Beauty and the labor market. The American Economic Review 84(5), 1174–1194 (1994)

Hanson, A., Hawley, Z.: Do landlords discriminate in the rental housing market? Evidence from an Internet field experiment in U.S. cities (Working Paper No. 2011-05). Experimental Economics Center, Andrew Young School of Policy Studies, Georgia State University (2011), http://excen.gsu.edu

Hardin, J., Hilbe, J.: Generalized linear models and extensions, 2nd edn. Stata Press (2007)

Harford, T.: The logic of life. The Random House Publishing Group (2008)

Harrison, G.W., List, J.A.: Field experiments. American Economic Literature 42(4), 1009–1055 (2004)

Heckman, J.: Detecting discrimination. The Journal of Economic Perspectives 12(2), 101–116 (1998)

Heckman, J., Siegelman, P.: The Urban Institute audit studies: Their methods and findings. In: Fix, M., Struyk, R. (eds.) Clear and Convincing Evidence: Measures of Discrimination in America, pp. 187–248. The Urban Institute Press (1993)

Hellerstein, J.K., Neumark, D., Troske, K.R.: Market forces and sex discrimination. Journal of Human Resources 37(2), 353–380 (2002)

Hersch, P.L.: Customer discrimination against black Major League Baseball pitchers reconsidered. Applied Economics Letters 17(2), 205–208 (2009)

Hildebrandt, M., Gutwirth, S. (eds.): Profiling the European citizen: Cross-disciplinary perspectives. Springer, Heidelberg (2008)

Hillier, A.E.: Spatial analysis of historical redlining: A methodological explanation. Journal of Housing Research 14(1), 137–168 (2003)

Hogler, R.L., Henle, C., Bemus, C.: Internet recruiting and employment discrimination: A legal perspective. Human Resource Management Review 8(2), 149–164 (1998)

Holmes, P.: New evidence of salary discrimination in Major League Baseball. Labour Economics 18(3), 320–331 (2011)

Holzer, H.J., Ihlanfeldt, K.R.: Customer discrimination and employment outcomes for minority workers. The Quarterly Journal of Economics 113(3), 835–867 (1998)

Holzer, H.J., Neumark, D.: The economics of affirmative action. Edward Elgar, Cheltenham (2004)

Holzer, H.J., Neumark, D.: Affirmative action: What do we know? Journal of Policy Analysis and Management 25(2), 463–490 (2006)

James, D.R., Tauber, K.E.: Measures of segregation. Sociological Methodology 13, 1–32 (1985)

Jewell, R.T., Brown, R.W., Miles, S.E.: Measuring discrimination in Major League Baseball: Evidence from the baseball hall of fame. Applied Economics 34(2), 167–177 (2002)

Jobard, F., Lévy, R.: Racial profiling: The Parisian police experience. Canadian Journal of Criminology and Criminal Justice 53(1), 87–93 (2011)

Kaas, L., Manger, C.: Ethnic discrimination in Germany's labour market: A field experiment (Discussion Paper No. 4741). Institute for the Study of Labor, IZA (2010), `http://ftp.iza.org`

Kahn, L.M.: Customer discrimination and affirmative action. Economic Inquiry 29(3), 555–571 (1991a)

Kahn, L.M.: Discrimination in professional sports: A survey of the literature. Industrial and Labor Relations Review 44(3), 395–418 (1991b)

Kahn, L.M.: The sports business as a labor market laboratory. Journal of Economic Perspectives 14(3), 75–94 (2000)

Kahn, L.M.: The economics of discrimination: Evidence from basketball (Discussion Paper No. 3987). Institute for the Study of Labor, IZA (2009), `http://ftp.iza.org`

Kamiran, F., Calders, T.: Classification without discrimination. In: Proc. of the 2nd Int. Conf. on Computer, Control & Communication (IEEE-IC4 2009). IEEE Press (2009)

Kamiran, F., Calders, T., Pechenizkiy, M.: Discrimination aware decision tree learning. In: Proc. of the 10th IEEE Int. Conf. on Data Mining (ICDM 2010), pp. 869–874. IEEE Computer Society (2010)

Kamishima, T., Akaho, S., Sakuma, J.: Fairness-aware learning through regularization approach. In: Proc. of the IEEE Int. Workshop on Privacy Aspects of Data Mining (PADM 2011), pp. 643–650. IEEE Computer Society (2011)

Kang, J., Banaji, M.R.: Fair measures: A behavioral realist revision of "affirmative action". UCLA Law Review 58(2), 465–520 (2010)

Kaye, D., Aickin, M. (eds.): Statistical methods in discrimination litigation. Marcel Dekker, Inc. (1992)

Knowles, J., Persico, N., Todd, P.: Racial bias in motor vehicle searches: Theory and evidence. Journal of Political Economy 109(1), 203–229 (2001)

Kuhn, P.J., Shen, K.: Employers' preferences for gender, age, height and beauty: Direct evidence (Working Paper No. 15564). National Bureau of Economic Research (2009), http://www.nber.org
Kunze, A.: Gender wage gap studies: Consistency and decomposition. Empirical Economics 35(1), 63–76 (2008)
LaCour-Little, M.: Discrimination in mortgage lending: A critical review of the literature. Journal of Real Estate Literature 7(1), 15–49 (1999)
Lahey, J.N., Beasley, R.A.: Computerizing audit studies. Journal of Economic Behavior & Organization 70(3), 508–514 (2009)
Lang, K.: A language theory of discrimination. The Quarterly Journal of Economics 101(2), 363–382 (1986)
Lang, K., Lehmann, J.-Y.K.: Racial discrimination in the labor market: Theory and empirics (2011) (unpublished manuscript, Boston University)
Lee, J.: American Idol: Evidence on same-race preferences. The Berkeley Electronic Journal of Economic Analysis & Policy 9(1), Article 28 (2009)
Leonard, J.S., Levine, D.I., Giuliano, L.: Customer discrimination. The Review of Economics and Statistics 92(3), 670–678 (2010)
Leopold, N., Meints, M.: Profiling in employment situations (fraud). In: Hildebrandt, M., Gutwirth, S. (eds.) Profiling the European Citizen: Crossdisciplinary Perspectives, pp. 236–257. Springer (2008)
Lerner, N.: Group rights and discrimination in international law, 2nd edn. Martinus Nijhoff Publishers (2003)
Lerner, R., Nagai, A.K.: Reverse discrimination by the numbers. Journal Academic Questions 13(3), 71–84 (2000)
Levine, R., Levkov, A., Rubinstein, Y.: Racial discrimination and competition (Working Paper No. 14273). National Bureau of Economic Research (2008), http://www.nber.org
Levitt, S.D.: Testing theories of discrimination: Evidence from Weakest Link. Journal of Law & Economics 47(2), 431–452 (2004)
Levitt, S.D., List, J.A.: What do laboratory experiments measuring social preferences reveal about the real world? Journal of Economic Perspectives 21(2), 153–174 (2007)
List, J.: The nature and extent of discrimination in the marketplace: Evidence from the field. The Quarterly Journal of Economics 119(1), 49–89 (2004)
List, J.A.: Friend or foe? A natural experiment of the prisoner's dilemma. Review of Economics and Statistics 88(3), 463–471 (2006)
Longhofer, S.D., Peters, S.R.: Why is mortgage discrimination illegal? A fresh look at the mortgage discrimination debate. Regulation, Cato Institute 22(4), 28–36 (1999)
Luong, B.T.: Generalized discrimination discovery on semi-structured datasupported by ontology. Unpublished Doctoral Dissertation, IMT Institute for Advanced Studies, Lucca, Italy (2011)
Luong, B.T., Ruggieri, S., Turini, F.: k-NN as an implementation of situation testing for discrimination discovery and prevention. In: Proc. of the ACM SIGKDD Int. Conf. on Knowledge Discovery and Data Mining (KDD 2011), pp. 502–510. ACM (2011)
Makkonen, T.: Measuring discrimination: Data collection and the EU equality law. European Network of Legal Experts in Anti-Discrimination (2006), http://www.migpolgroup.com
Makkonen, T.: European handbook on equality data. European Network of Legal Experts in Anti-Discrimination (2007), http://ec.europa.eu
McCullagh, P., Nelder, J.A.: Generalized linear models, 2nd edn. Chapman and Hall (1989)

McGinnity, F., Nelson, J., Lunn, P., Quinn, E.: Discrimination in recruitment - evidence from a field experiment (Report). The Equality Authority and The Economic and Social Research Institute (2009), http://www.equality.ie

Mcguire, T.G., Ayanian, J.Z., Ford, D.E., Henke, R.E.M., Rost, K.M., Zaslavsky, A.M.: Testing for statistical discrimination by race/ethnicity in panel data for depression treatment in primary care. Health Research and Educational Trust 43(2), 531–551 (2008)

Moreno, M., Ñopo, H., Saavedra, J., Torero, M.: Gender and racial discrimination in hiring: A pseudo audit study for three selected occupations in metropolitan Lima (Working Paper No. 321). Econometric Society (2004), http://econpapers.repec.org

Munnell, A.H., Tootell, G.M.B., Browne, L.E., McEneaney, J.: Mortgage lending in Boston: Interpreting HMDA data. American Economic Review 86(1), 25–53 (1996)

Muravyev, A., Talavera, O., Schfer, D.: Entrepreneurs' gender and financial constraints: Evidence from international data. Journal of Comparative Economics 37(2), 270–286 (2009)

Myers, C.K.: Ladies first? A field study of discrimination in Coffee Shops (Working Paper No. 711). Middlebury College, Department of Economics (2007), http://community.middlebury.edu

Neal, D.A., Johnson, W.R.: The role of premarket factors in black-white wage differences. Journal of Political Economy 104(5), 869–895 (1996)

Neumark, D.: Detecting discrimination in audit and correspondence studies (Working Paper No. 16448). National Bureau of Economic Research (2010), http://www.nber.org

Neumayer, E., de Soysa, I.: Globalisation, women's economic rights and forced labour. The World Economy 30(10), 1510–1535 (2007)

Newman, D.M.: Sociology: Exploring the architecture of everyday life, 7th edn. Pine Forge Press (2008)

Nunley, J.M., Owens, M.F., Howard, R.S.: The effects of competition and information on racial discrimination: Evidence from a field experiment (Working Paper No. 201007). Middle Tennessee State University, Department of Economics and Finance (2010), http://frank.mtsu.edu

Ong, P.M., Stoll, M.A.: Redlining or risk? A spatial analysis of auto insurance rates in Los Angeles. Journal of Policy Analysis and Management 26(4), 811–830 (2007)

Oostendorp, R.: Globalization and the gender wage gap. World Bank Economic Review 23(1), 141–161 (2009)

Paetzold, R.L., Willborn, S.L., Baldus, D.C.: The statistics of discrimination: Using statistical evidence in discrimination cases. Shepard's/McGraw-Hill (1994)

Pager, D.: The use of field experiments for studies of employment discrimination: Contributions, critiques, and directions for the future. The ANNALS of the American Academy of Political and Social Science 609(1), 104–133 (2007)

Pager, D., Quillian, L.: Walking the talk? What employers say versus what they do. American Sociological Review 70(3), 355–380 (2005)

Pager, D., Western, B., Bonikowski, B.: Discrimination in a low-wage labor market: A field experiment (Discussion Paper No. 4469). Institute for the Study of Labor, IZA (2009), http://ftp.iza.org

Palmer, M., King, R.: Has salary discrimination really disappeared from Major League Baseball? Eastern Economic Journal 32(2), 285–297 (2006)

Parrett, M.: Customer discrimination in restaurants: Dining frequency matters. Journal of Labor Research 32(2), 87–112 (2011)

Pedreschi, D., Ruggieri, S., Turini, F.: Discrimination-aware data mining. In: Proc. of the ACM SIGKDD Int. Conf. on Knowledge Discovery and Data Mining (KDD 2008), pp. 560–568. ACM (2008)

Pedreschi, D., Ruggieri, S., Turini, F.: Measuring discrimination in socially-sensitive decision records. In: Proc. of the SIAM Int. Conf. on Data Mining (SDM 2009), pp. 581–592. SIAM (2009)

Persico, N., Todd, P.: Passenger profiling, imperfect screening, and airport security. American Economic Review 95(2), 127–131 (2005)

Phelps, E.S.: The statistical theory of racism and sexism. American Economic Review 62(4), 659–661 (1972)

Pickerill, J.M., Mosher, C., Pratt, T.: Search and seizure, racial profiling, and traffic stops: A disparate impact framework. Law & Policy 31(1), 1–30 (2009)

Pinkston, J.C.: A test of screening discrimination with employer learning. Industrial and Labor Relations Review 59(2), 267–284 (2006)

Pope, D.G., Sydnor, J.R.: Implicit statistical discrimination in predictive models (Working Paper No. 2007-09-11). Risk Management and Decision Processes Center. The Wharton School of the University of Pennsylvania (2007), `http://opim.wharton.upenn.edu`

Pope, D.G., Sydnor, J.R.: Whats in a picture? Evidence of discrimination from Prosper.com. Journal of Human Resources 46(1), 53–92 (2011)

Primm, E., Piquero, N.L., Piquero, A.R., Regoli, R.M.: Investigating customer racial discrimination in the secondary baseball card market. Sociological Inquiry 81(1), 110–132 (2011)

Quillian, L.: New approaches to understanding racial prejudice and discrimination. Annual Review of Sociology 32(1), 299–328 (2006)

Quintanar, S.M.: Man vs. machine: An investigation of speeding ticket disparities based on gender and race (Departmental Working Paper No. 2009-16). Department of Economics, Louisiana State University (2009), `http://bus.lsu.edu`

Reardon, S.F., Firebaugh, G.: Measures of multigroup segregation. Sociological Methodology 32(1), 33–67 (2002)

Reid, C., Laderman, E.: The untold costs of subprime lending: Examining the links among higher-priced lending, foreclosures and race in california (Working Paper). Paper presented at the Institute for Assets and Social Policy, April 15. Brandeis University (2009), `http://iasp.brandeis.edu`

Riach, P.A., Rich, J.: Field experiments of discrimination in the market place. The Economic Journal 112(483), 480–518 (2002)

Riach, P.A., Rich, J.: Deceptive field experiments of discrimination: Are they ethical? Kyklos 57(3), 457–470 (2004)

Riach, P.A., Rich, J.: An experimental investigation of sexual discrimination in hiring in the english labor market. The B.E. Journal of Economic Analysis & Policy 6(2), Art. 1 (2006)

Ridgeway, G.: Assessing the effect of race bias in post-traffic stop outcomes using propensity scores. Journal of Quantitative Criminology 22(1), 1–29 (2006)

Ridgeway, G., MacDonald, J.M.: Doubly robust internal benchmarking and false discovery rates for detecting racial bias in police stops. Journal of the American Statistical Association 104(486), 661–668 (2009)

Rödin, M., Özcan, G.: Is it how you look or speak that matters? An experimental study exploring the mechanisms of ethnic discrimination (Research Papers in Economics No. 2011:12). Stockholm University, Department of Economics (2011), `http://www2.ne.su.se`

Rooth, D.-O.: Obesity, attractiveness, and differential treatment in hiring: A field experiment. Journal of Human Resources 44(3), 710–735 (2009)

Rorive, I.: Proving Discrimination Cases - the Role of Situation Testing. Centre For Equal Rights & Migration Policy Group (2009), `http://www.migpolgroup.com`

Roscigno, V., Karafin, D.L., Tester, G.: The complexities and processes of racial housing discrimination. Social Problems 56(1), 49–69 (2009)

Ross, S.L., Tootell, G.M.B.: Redlining, the community reinvestment act, and private mortgage insurance. Journal of Urban Economics 55(2), 278–297 (2004)

Ross, S.L., Yinger, J.: The color of credit: Mortgage discrimination, research methodology, and fair-lending enforcement. The MIT Press (2002)

Rowe, B.: Gender bias in the enforcement of traffic laws: Evidence based ona new empirical test (Working Paper No. 3). The Berkeley Electronic Press (2009), `http://www-personal.umich.edu`

Ruggieri, S., Pedreschi, D., Turini, F.: Data mining for discrimination discovery. ACM Trans. on Knowledge Discovery from Data 4(2), 1–40 (2010a)

Ruggieri, S., Pedreschi, D., Turini, F.: DCUBE: Discrimination discovery in databases. In: Proc. of the ACM SIGMOD Int. Conf. on Management of Data (SIGMOD 2010), pp. 1127–1130. ACM (2010b)

Ruggieri, S., Pedreschi, D., Turini, F.: Integrating induction and deduction for finding evidence of discrimination. Artificial Intelligence and Law 18(1), 1–43 (2010c)

Rugh, J.S., Massey, D.S.: Racial segregation and the American foreclosure crisis. American Sociological Review 75(5), 629–651 (2010)

Sanandaji, T.: Reversion to the racial mean and mortgate discrimination (Working Paper No. 811). Research Institute of Industrial Economics (2009), `http://www.ifn.se`

Sanga, S.: Reconsidering racial bias in motor vehicle searches: Theory and evidence. Journal of Political Economy 117(6), 1155–1159 (2009)

Sano, S.: Testing the taste-based discrimination hypothesis: Evidence from data on japanese listed firms. Japan Labor Review 6(1), 36–50 (2009)

Schiek, D., Waddington, L., Bell, M. (eds.): Cases, materials and text on national, supranational and international non-discrimination law. Hart Publishing (2007)

Schreurs, W., Hildebrandt, M., Kindt, E., Vanfleteren, M.: The role of data protection law and non-discrimination law in group profiling in the private sector. In: Hildebrandt, M., Gutwirth, S. (eds.) Profiling the European Citizen: Cross-Disciplinary Perspectives, pp. 258–287. Springer (2008)

Sheskin, D.J. (ed.): Handbook of parametric and non-parametric statistical procedure, 3rd edn. Chapman & Hall/CRC (2004)

Silverman, R.: Redlining in a majority black city?: Mortgage lending and the racial composition of Detroit neighborhoods. The Western Journal of Black Studies 29(1), 531–541 (2005)

Slonim, R., Guillen, P.: Gender selection discrimination: Evidence from a trust game. Journal of Economic Behavior & Organization 76(2), 385–405 (2010)

Sowell, T. (ed.): Affirmative action around the world: An empirical analysis. Yale University Press (2005)

Squires, G.D., Hyra, D.S., Renner, R.N.: Segregation and the subprime lending crisis (Briefing Paper No. 244). Economic Policy Institute (2009), `http://www.epi.org`

Steele, C.M., Aronson, J.: Stereotype threat and the intellectual test performance of african americans. Journal of Personality and Social Psychology 69(5), 797–811 (1995)

Sulis, G.: Gender wage differentials in Italy: A structural estimation approach (Working Paper CRENoS No. 2007-15). Centre for North South Economic Research, University of Cagliari and Sassari, Sardinia (2007), `http://crenos.unica.it`

Taylor, A.: Discrimination in rental housing markets: Evidence from Craigslist audits (2010), `http://aysps.gsu.edu` (unpublished manuscript)

Tillyer, R., Engel, R.S., Cherkauskas, J.C.: Best practices in vehicle stop data collection and analysis. Journal of Police Strategies and Management 33(1), 69–92 (2010)

Tillyer, R., Engel, R.S., Wooldredge, J.: The intersection of racial profiling research and the law. Journal of Criminal Justice 36(2), 138–153 (2008)

Tobler, C.: Limits and potential of the concept of indirect discrimination. European Network of Legal Experts in Anti-Discrimination (2008), http://www.migpolgroup.com

Tsao, T.-Y., Pearlman, A.: Decomposition of the black-white wage differential in the physician market (Economics Working Paper No. wp 588). The Levy Economics Institute (2010), http://www.levyinstitute.org

Turner, M.A., Ross, S.L.: How racial discrimination affects the search for housing. In: de Souza Briggs, X. (ed.) The Geography of Opportunity, pp. 81–100. Brookings Institution Press (2005)

Turner, M.A., Ross, S.L., Galster, G.C., Yinger, J.: Discrimination in metropolitan housing markets: National results from phase i of hds 2000. Urban Institute, Dep. Hous. Urban Dev. (2002), http://www.urban.org

Turner, M.A., Skidmore, F. (eds.): Mortgage lending discrimination: A review of existing evidence. The Urban Institute (1999), http://www.urban.org

U.K. Legislation: (a) Sex Discrimination Act, 1975, (b) Race Relation Act, 1976 (2011), http://www.statutelaw.gov.uk

United Nations Legislation: (a) Universal Declaration of Human Rights, 1948, (b) International Covenant for Civil and Political Rights, 1966, (c) International Covenant on Ecomomic, Social and Cultural Rights, 1966, (d) Convention on the Elimination of All forms of Racial Discrimination, 1966, (e) Convention on the Elimination of All forms of Discrimination Against Women, 1979 (2011), http://www.ohchr.org

U.S. Federal Legislation: (a) Equal Credit Opportunity Act, 1974; (b) Fair Housing Act, 1968; (c) Employment Act, 1967; (d) Equal Pay Act, 1963; (e) Pregnancy Discrimination Act, 1978; (f) Civil Right Act, 1964, 1991 (2011), http://www.eeoc.gov

Usui, E.: Wages, non-wage characteristics, and predominantly male jobs. Labour Economics 16(1), 52–63 (2009)

Vicki, B., Ellen, I., Madar, J.: The high cost of segregation: Exploring racial disparities in high-cost lending. Fordham Urban Law Journal 36(3), 361–393 (2009)

Wang, S.-Y.: Statistical discrimination, productivity and the height of immigrants (Working Paper No. 3344). eSocialSciences (2010), http://www.esocialsciences.org

Weichselbaumer, D., Winter-Ebmer, R.: A meta-analysis of the international gender wage gap. Journal of Economic Surveys 19(3), 479–511 (2005)

Weichselbaumer, D., Winter-Ebmer, R.: The effects of competition and equal treatment laws on gender wage differentials. Economic Policy 22, 235–287 (2007)

Weller, C.E.: Credit access, the costs of credit and credit market discrimination (Working Paper No. 171). Political Economy Research Institute, University of Massachusetts at Amherst (2008), http://www.peri.umass.edu

Whitley, B.E., Kite, M.E.: The psychology of prejudice and discrimination, 2nd edn. Wadsworth Publishing (2009)

Wilson, D.P., Ying, Y.-H.: Nationality preferences for labour in the international football industry. Applied Economics 35(14), 1551–1559 (2003)

Wingate, P.H., Thornton, G.C.: Statistics and employment discrimination law: An interdisciplinary review. In: Martocchio, J., Liao, H., Joshi, A. (eds.) Research in Personnel and Human Resources Management, vol. 19, pp. 295–337. Emerald Group Publishing Ltd. (2000)

Wyly, E.K., Moos, M., Foxcroft, H., Kabahizi, E.: Subprime mortgage segmentation in the American urban system. Journal of Economic and Social Geography 99(1), 3–23 (2008)

Yamagishi, T., Jin, N., Kiyonari, T.: Bounded generalized reciprocity: Ingroup boasting and ingroup favoritism. In: Lawler, E.J., Macy, M.W. (eds.) Advances in Group Processes, vol. 16, pp. 161–197. Jai Press Inc. (1999)

Yang, C.-H., Lin, H.-Y.: Is there salary discrimination by nationality in the NBA? Foreign talent or foreign market. Journal of Sports Economics (2010) (published online: December 27, 2010)

Yap, B.W., Ong, S.H., Husain, N.H.M.: Using data mining to improve assessment of credit worthiness via credit scoring models. Expert Systems with Applications 38(10), 13274–13283 (2011)

Yezer, A.M.: A review of statistical problems in the measurement of mortgage market discrimination and credit risk (Report). Research Institute for Housing America (2010), http://www.housingamerica.org

Zhang, L., Dong, X.-Y.: Male-female wage discrimination in Chinese industry - investigation using firm-level data. Economics of Transition 16(1), 85–112 (2008)

Zliobaite, I., Kamiran, F., Calders, T.: Handling conditional discrimination. In: Proc. of the 11th IEEE Int. Conf. on Data Mining (ICDM 2011), pp. 992–1001. IEEE Computer Society (2011)

Chapter 7
Risks of Profiling and the Limits of Data Protection Law

Bart Schermer

Abstract. Profiling and automated decision-making may pose risks to individuals. Possible risks that flow forth from profiling and automated decision-making include discrimination, de-individualisation and stereotyping. To mitigate these risks, the right to privacy is traditionally invoked. However, given the rapid technological developments in the area of profiling, it is questionable whether the right to informational privacy and data protection law provide an adequate level of protection and are effective in balancing different interests when it comes to profiling. To answer the question as to whether data protection law can adequately protect us against the risks of profiling, I will discuss the role of data protection law in the context of profiling and automated decision-making. First, the specific risks associated with profiling and automated decision-making are explored. From there I examine how data protection law addresses these risks. Next I discuss possible limitations and possible drawbacks of data protection law when it comes to the issue of profiling and automated decision-making. I conclude with several suggestions to for making current data protection law more effective in dealing with the risks of profiling. These include more focus on the actual goals of data processing and 'ethics by design'.

7.1 Introduction

Profiling, the application of profiles to individuate and represent a subject or to identify a subject as a member of a group or category (Hildebrandt 2008), is commonplace in our data-driven information society. While profiling may have many benefits for businesses, the government and citizens themselves, there are also potential risks for data subjects attached to profiling. To mitigate these risks, traditionally the right to (informational) privacy is invoked. However, given the

Bart Schermer
eLaw, Institute for Law in the Information Society, Leiden University, The Netherlands
e-mail: schermer@considerati.com

B. Custers et al. (Eds.): Discrimination & Privacy in the Information Society, SAPERE 3, pp. 137–152.
springerlink.com

rapid technological developments in the area of profiling and automated decision-making, it is questionable whether the right to informational privacy and more specifically data protection law (still) provide an adequate level of protection and whether they balance the interests of the actors involved effectively.

In this chapter I explore the possible risks associated with profiling and examine whether the current legal framework can mitigate these risks effectively.[1] I shall do so by seeking answers to the following questions:

- What risks does profiling pose for individuals (and groups)?
- How are these risks addressed by the current data protection framework?
- Does the current legal framework for data protection provide adequate protection whilst also taking into account the legitimate interest of profilers?

After answering these questions I examine what changes might be necessary in order to mitigate the risks posed by profiling.

7.2 Risks Associated with Profiling

While profiling can be a valuable aid for businesses and governments, profiling may also entail risks. Risks commonly associated with profiling are: discrimination, de-individualisation, stereotyping, information asymmetries, inaccuracy and the abuse of profiles.

7.2.1 Discrimination

Classification and division are at the heart of profiling. As such, discrimination is part and parcel of profiling. However, there are situations where discrimination is considered unethical and even illegal. This can occur for instance when a profiling exercise is focussed on characteristics such as ethnicity, gender, religion or sexual preference. But even without a prior desire to judge people on the basis of particular characteristics, there is the risk of inadvertently discriminating against particular groups or individuals.

7.2.2 De-individualisation

In many cases profiling is in large parts concerned with classification and thus there is the risk that persons are judged on the basis of group characteristics rather than on their own individual characteristics and merits (Vedder 1999). Group profiles usually contain statistics and therefore the characteristics of group profiles may be valid for the group and for individuals as members of that group, though not for individuals as such. For instance, people who live in a particular neighbourhood may have a 20% higher chance of defaulting their loan than the

[1] In discussing data protection legislation, I shall focus exclusively on the EU framework for data protection.

average person. This characteristic goes for the group (i.e., people living in that particular neighbourhood), for the individuals as members of that group (i.e., randomly chosen people living in the neighbourhood), but not necessarily for the individuals as such (i.e., for John, Mary and William who all live in the same neighbourhood). When individuals are judged by group characteristics they do not possess as individuals, this may negatively affect them (Custers 2010).

Group profiling may not only have direct negative effects on individuals, but may also lead to stigmatisation of group members. Moreover, divisions into groups can damage societal cohesion. When group profiles, whether correct or not, become public knowledge, people may start treating each other accordingly. For instance, when people start believing that individuals from a particular neighbourhood default their loans more often, they may conclude that those individuals live in a 'bad' neighbourhood.

7.2.3 Stereotyping

Closely related to the risk of de-individualisation and stigmatisation is that of stereotyping. A profile casts us on the basis of predetermined categories (e.g., 'valuable customer', 'young urban professional', but also 'security risk' or 'dodgy debtor'). For a profiling exercise to remain effective and efficient there are a finite number of general categories. These profiles are, almost by definition, incapable of accurately reflecting all the nuances of our personality. As such, the profile we fit will become a stereotype on the basis of which we are judged. Moreover, these profiles can also make it more difficult for a person to 'escape' the stereotype.

7.2.4 Information Asymmetries

A fourth risk associated with profiling is that it can lead to information asymmetries. In other words, through profiling, the position of the data controller improves with regard to the data at his disposal, whereas that of the data subject remains the same. This is a particular issue when the data subject is unaware of the profiling exercise, or does not have complete information about the profiling exercise. Information asymmetries may lead to an imbalance in the playing field between government and citizens, and between businesses and consumers, upsetting the current balance of power between different parties.

In the context of the relation between government and citizens, information asymmetries can also affect individual autonomy. If data mining indeed yields information the government can act upon, the government will have more power. Moreover, the fear of strong data mining capabilities on the part of the government may 'chill' the willingness of people to engage in political activities, given the fear of being watched. For this fear to materialise, profiling does not even have to be effective (Schermer 2007, p. 137).

In the context of the relation between businesses and consumers, information asymmetries may lead to unfair economic practices and discriminatory pricing. For instance, certain goods or services may be withheld from individuals, solely on the basis of them fitting or not fitting a particular profile. It is also possible to adjust prices of goods and services on the basis of the profile of the individual. Charging different prices on the basis of particular characteristics (e.g., race, sex, or sexual preference) is likely a violation of anti-discrimination legislation.

7.2.5 *Inaccuracy*

A fifth risk associated with profiling is that profiles might be inaccurate. In particular there is the problem of 'false positives' and 'false negatives'. This means that people that in fact do not fit the profile are fitted within it (a false positive), or people that fit the profile are left outside of it (false negative). False positives and false negatives occur for various reasons, for instance because insufficient data is available, or the data is inaccurate. False positives and false negatives are a particular problem in automated decision making since there is no human intervention and it is not an adversarial process where both sides are heard. This is troublesome as it places the burden of proof on the side of the data subject: they must prove that they do or do not fit the profile.

7.2.6 *Abuse*

A final risk associated with profiling is that data controllers or third parties (for instance hackers) abuse profiles and/or the information contained therein. Possibilities for abuse arise in particular when the profile can be linked to an identified individual. A profile could for instance be made public leading to reputational damage for the data subject (e.g., the data subject is exposed as a dodgy debtor), or the (personal) data contained in the profile could be used for fraudulent purposes.

7.3 Privacy and Data Protection in Light of Profiling

To mitigate the risks mentioned in the previous paragraphs, traditionally the right to (informational) privacy is invoked. The right to informational privacy acts as a boundary against the free flow of information and thus ensures the protection of personal information. An important aspect of informational privacy is personal data protection. In particular in the context of the private sector, data protection legislation has become the most important aspect of informational privacy protection. Van den Hoven (2008, p. 311) lists four different moral reasons for protecting personal data. They are: 1) protection against information based-harm, 2) protection against informational inequality, 3) protection against informational injustice and, 4) the protection of moral autonomy.

Information based-harm
Because information can be used to cause harm (e.g., identity theft, fraud) or other serious disadvantages to data subjects, personal data needs to be protected from access by parties who wish to cause harm using personal data. Data protection sets

forth rules on access and security to personal data, thwarting the efforts of those who wish to cause harm.

Informational inequality

A second moral reason for the protection of personal data is that it reduces the negative effects of informational inequality. Since consumers are not always (fully) aware of the economic opportunities their personal data may present, and/or not in a position to trade their identity-relevant information in a fair and transparent market, they may be disadvantaged in the marketplace for identity-relevant information. Constraints on the flow of personal data need to be put in place in order to guarantee economic equality of arms, transparency and fairness (Van den Hoven 2008, p. 313).

While van den Hoven only describes the issue of informational inequality from a private sector perspective, it is also relevant in the context of the relationship between governments and citizens. In this realm, informational inequality is closely associated with personal autonomy. If the government knows a great deal about its citizens, but is not equally transparent, the balance of power is upset.

Informational injustice

A third moral reason for data protection is to avoid informational injustice. Informational injustice occurs when the boundaries of the 'spheres of access' are disrespected. People do not mind when there data are being processed for a legitimate goal (e.g., their medical data being used for their treatment). But if a sphere of access is disrespected (e.g., the medical data is being used in a job application procedure) informational injustice takes place.

Moral autonomy and moral identification

A fourth reason to invoke data protection rules is that they allow us to set a 'distance' between the outside world and ourselves. This distance is crucial for what van den Hoven calls 'shaping our own moral biographies' (Van den Hoven 2008, p. 316). Without the observing gaze of others we can freely develop our thoughts and our identity. Furthermore, it allows us to present ourselves to the outside world as we see fit. When the outside world can readily access personal data across a number of different contexts, the individual's freedom to shape our own moral biography is reduced.

These moral foundations for protecting personal data are also relevant when we observe the possible risks of profiling. For instance, stereotyping and de-individualisation encroach upon our sense of moral autonomy, informational inequality may occur when profiling is surreptitious or when profiles become too rich, and informational injustice may occur when profiles cross the boundaries of spheres of access. Therefore, the right to informational privacy and data protection law are also relevant in the context of profiling.

7.4 Data Protection Law

In Europe there are two main bodies of law that address profiling for purposes other than national security and law enforcement.[2] They are the Data protection directive (1995/46/EC) and the ePrivacy directive (2002/58/EC), which was amended in 2009 by Directive 2009/136/EC. The Data protection directive deals with the use of 'personal data' in general, whereas the ePrivacy directive deals with the use of unique identifiers and tracking technologies that can be used to facilitate profiling (e.g., cookies).

European data protection law has its roots in the OECD principles on privacy protection and the transborder flow of personal data and the Council of Europe treaty on personal data protection.[3] It aims to strike a balance between the (informational) privacy of the data subject and the free flow of information. The Data protection directive does this by providing a harmonised framework for the secure and legitimate exchange of personal data throughout Europe.[4]

The Data protection directive states that personal data must be processed fairly and lawfully and only for specified, explicit and legitimate purposes. To ensure fair and lawful processing the data protection sets a number of rules for the processing of personal data. These include –amongst others- obligations to keep the data secure, ensure its quality, inform the data subject, register the process in a public register, and grant the data subject access to the data.

In order for the provisions of the Data protection directive to be applicable, data must first be qualified as 'personal data'. Personal data is described in article 2(a) as:

"any information relating to an identified or identifiable natural person ('data subject'); an identifiable person is one who can be identified, directly or indirectly, in particular by reference to an identification number or to one or more factors specific to his physical, physiological, mental, economic, cultural or social identity"

An individual is considered 'identified' when that individual can be distinguished from all other members of a group.[5] Identification is commonly achieved through

---

[2] The use of profiling techniques for law enforcement purposes is governed –for the most part- by the law of criminal procedure, which differs from member state to member state. Though they differ from country to country, all laws that govern profiling must be in accordance with the rules set forth in article 8 of the European Charter of Human Rights (ECHR).

[3] Council of Europe Convention for the Protection of Individuals with regard to Automatic Processing of Personal Data (Convention ets. no. 108, Strasbourg 28-1-1981).

[4] Early December 2011, a draft version of a new general Regulation on data Protection prepared by the European Commission leaked (version 56, 29 November 2011). Relevant provisions include more strict rules on profiling (article 18) and the inclusion of online identifiers such as cookies in the definition of personal data. Given the fact that this Regulation is still in the drafting phase it is not discussed further in this chapter.

[5] Opinion N° 4/2007 on the concept of personal data, Article 29 Working Party, p. 12.

the combination of certain 'identifiers' that hold a particularly privileged and close relationship to the individual.[6] Common identifiers are: name, physical appearance, and certain unique numbers such as a social security number. The extent to which certain identifiers are sufficient to achieve identification depends on the context of the particular situation.[7]

Van den Hoven explains that the data used in an identification process are *referential*, meaning that the data refers to a specific person, not just any person (Van den Hoven 2008, p. 309) This means that personal data always need an identity-relevant context. Without such context data have no meaning and are just *attributive*: they would describe a situation or fact without reference to any specific individual. One could argue that attributive data are *conditionally referential*; they can become personal data if another identity-relevant condition occurs, for instance because the raw data are placed in an identity-relevant context or combined with a piece of identity relevant information (Terstegge 2009). So while an individual might not be directly identified on the basis of a unique identifier such as a name, he or she may nonetheless be 'identifiable'. In the context of profiling three situations may occur that would render attributive data referential, personal data. They are: 1) adding identifying data to a profile, 2) spontaneous identification based on the uniqueness of the profile, 3) linking the profile to an individual by means of unique identifiers.

The first situation occurs when referential data (personal data) is added to attributive data. By adding information that is considered uniquely identifying (e.g., full name, date of birth, address) to a profile, all the data in that profile will become personal data.

The second situation occurs when the data contained in a profile leads to the spontaneous identification of the data subject. This is the case when the constellation of data is considered so unique, that the profile can only fit a single person and that person can be identified on the basis of the profile.[8] A well-known example of this is the case of 'AOL searcher 4417749'. In 2006 AOL published an anonymised dataset consisting of search queries for research purposes. But it did not take researchers long to trace back the search queries of an anonymous user (4417749) to her real name: Thelma Arnold (Barbaro and Zeller 2006). The combinations of search queries were so unique for each individual that they were able to trace back the queries to Ms. Arnold.

The third situation occurs when the profile of a data subject is associated with unique identifiers that are closely associated with him. Apart from identification on the basis of unique attributes such as for instance name, address and/or social security number, a profile may also be linked to a natural personal via other means. Most often this will be the case with profiling, since profiling is only effective if the data subject can be somehow be linked to the relevant profile. Apart from using identifiers that are unique to the data subject (e.g., name, address, place and date of birth), other types of identifiers may also be used. A common method is to link a data subject to a profile through the terminal equipment (e.g., mobile

[6] Opinion N° 4/2007 on the concept of personal data, Article 29 Working Party, p. 12.
[7] Opinion N° 4/2007 on the concept of personal data, Article 29 Working Party, p. 12.
[8] Opinion N° 4/2007 on the concept of personal data, Article 29 Working Party, p. 13.

phone, computer) used by the data subject. For instance, a profile may be linked to a unique number associated with the terminal equipment. Identifiers that can function in this manner are IP-addresses, IMEI-numbers and MAC-addresses. Another option is to read from and write information to the terminal equipment, for instance by means of a cookie.

According to the article 29 Working Party when these indirect links are sufficient to single out a person, the associated profile should be considered personal data:

"Without even enquiring about the name and address of the individual it is possible to categorise this person on the basis of socio-economic, psychological, philosophical or other criteria and attribute certain decisions to him or her since the individual's contact point (a computer) no longer necessarily requires the disclosure of his or her identity in the narrow sense."[9]

The applicability of the Data protection directive to profiling in this manner is further confirmed in the Article 29 Working Party opinion on behavioural advertising. In this opinion, the Article 29 Working Party explains why it feels that personal data is processed in the context of behavioural advertising:

"This is due to various reasons: i) behavioural advertising normally involves the collection of IP addresses and the processing of unique identifiers (through the cookie). The use of such devices with a unique identifier allows the tracking of users of a specific computer even when dynamic IP addresses are used. In other words, such devices enable data subjects to be 'singled out', even if their real names are not known. ii) Furthermore, the information collected in the context of behavioural advertising relates to, (i.e. is about) a person's characteristics or behaviour and it is used to influence that particular person."[10]

The key element is that the identifier enables the person to be singled out for a specific treatment on the basis of the associated profile. On the basis of this reasoning by the Article 29 Working Party, most if not all, profiling exercises will fall under the scope of the Data protection directive. In those cases where an identifier is used that reads and/or writes information to terminal equipment, article 5(3) of Directive 2002/58/EC also applies. We may thus conclude that profiling in most if not all current forms, falls within the scope of the Data protection acquis in Europe.

7.5 Drawbacks to the Current Approach to Data Protection in the Context of Profiling

We have established that there are moral reasons for the protection of (personal) data in the context of profiling. If we follow the (broad) interpretation of the

[9] Opinion N° 4/2007 on the concept of personal data, Article 29 Working Party, p. 13.
[10] Opinion N° 2/2010 on online behavioural advertising, Article 29 Working Party, p. 9.

concept of personal data as set forth by the Article 29 Working Party we may also conclude that data protection law apply to many profiling practices.

The goal of profiling is to individualise and give a representation of a subject, or to identify that subject as a member of a group or category (Hildebrandt 2008, p. 17). For profiling to be effective it is unnecessary to know the data subjects actual identity. Furthermore, it is most often unnecessary to distinguish an individual from other members of a group. Rather, profiling calls for categorising an individual, for instance by fitting that individual into a certain predetermined target group. While it might be possible to identify an individual on the basis of a profile, more often than not, the data controller is not interested in actually identifying the data subject.[11] Nevertheless, as signalled by the Article 29 Working Party a user can be singled out on the basis of a profile in combination with a unique identifier such as a cookie or an IP-address. The idea is that because a person can be singled out and the information contained in the profile is used to make decisions about the person, data protection law should apply. While this is understandable from a privacy perspective, it is questionable whether data protection law is always the most effective mechanism for dealing with the risks posed by profiling. This question is important, as there are possible drawbacks to applying the current data protection law to profiling. Below I shall discuss several drawbacks that may prompt us to rethink the current approach to data protection in the light of profiling.[12]

7.5.1 The 'Binary' Nature of Data Protection Law

The first drawback of current data protection law is its binary nature. The Data protection directive only applies to the processing of personal data. While this sounds logical, it leads to difficulties in practice. The difficulty with the binary nature of data protection law is that it is oftentimes difficult to establish when data should be considered personal data. A combination of different pieces of data may all of a sudden become personal data when a referential piece of information is added, or when the different pieces of data by themselves spontaneously identify an individual. Moreover, while the profiling exercise itself may not be aimed at identifying a data subject, the possibility of identification is always present. For instance, pieces of information (oftentimes outside of the control of the data controller) may be linked to the profile, enabling identification. The question then becomes: at what point are all the protection mechanisms and legal obligations of the Data protection directive exactly to come into play. This leads to a great deal of uncertainty for (potential) data controllers. But the binary nature of data protection law is also problematic for data subjects, as it is unclear to whom they should turn for redress when a profile is misused or abused.

---

[11] While the data controller might not be interested in identifying or re-identifying the individual other parties may wish to do so. However, given the limited space available, we shall not address this issue in this chapter.

[12] It is relevant to note that in this discussion the focus is mainly on the use of profiling for commercial purposes.

The solution to this issue as set forth by the Article 29 Working Party seems to be to err on the side of caution, and consider any form of profiling the processing of personal data. The difficulty with this is that once a dataset or a profile is considered personal data, all the rules of the Data protection directive apply. In practice, this leads to a substantial administrative burden for data controllers (see paragraph 7.2). Moreover, it dilutes the effectiveness of enforcement (see paragraph 7.3).

7.5.2 The Procedural Nature of Data Protection Law

The second drawback, which ties in with the binary nature of the current data protection law, is the procedural nature of the law. Data protection legislation in its current form is primarily aimed at the *ex-ante* protection of privacy and personal data. This means that data controllers need to ensure that their processing of data is compliant with all the demands set forth by the Data protection directive. Though this should ensure the privacy of the data subject, in practice the effect of this ex-ante approach is often limited. In practice, privacy protection is for data controllers mainly an issue of compliance and following the procedural rules of the Directive (e.g. registering the processing in a public register, informing the data subject), rather than a discussion on what is considered a sustainable, ethical and responsible (business) process.

For the data subject there are also possible drawbacks. A significant drawback is that the data subject has limited options for redress in case there is a misuse or an abuse of personal data.[13] The reason is that the Data protection authorities are in charge of the enforcement of the law, leaving less room for individual redress.

7.5.3 Inflation of the Personal Sphere

Another issue with the application of the Data protection directive in the context of profiling is that it further expands the scope of the Data protection directive. The risk this brings with it is that as more and more activities fall under the header of personal data protection, the protection the law can provide actually decreases. Zwenne (2010, p. 335) for instance argues that a law that applies to essentially everything applies to effectively nothing. Blok (2002) also warns for this problem, calling the expansion of the concept of personal data 'an inflation of the personal sphere'. The main problem that might arise as a result of this inflation is that key privacy interests get heaped up with less important infringements of privacy, leading to an overstretched enforcement apparatus, confusion on how the law should apply, and possibly a degradation of the importance of privacy as a human right and the underlying values which it aims to protect.

A further problem with the inflation of the personal sphere is that the data protection authority becomes the *de facto* judge of what is considered the ethical use

[13] Whether there are options for individual redress is dependent on the actual implementation of the Data protection directive and associated privacy laws in national law. For the most part though we can say that the Data protection authority is in the lead when it comes to the enforcement of privacy rules, rather than the data subject.

of ICT. As soon as something is considered personal data, the data protection authority can decide whether or not a particular use of technology is acceptable or not. While the data protection authority can provide valuable input for discussion, oftentimes other institutions are more suited to this task. In the context of discrimination for instance, an equal treatment committee is probably more suited to determine when data processing is discriminatory. Moreover, when it comes to balancing different interests, it is important to involve all relevant stakeholders. This is particularly relevant in the context of profiling as most data protection authorities seem more 'privacy oriented' than 'data controller oriented', which could lead to an unfair balancing of different interests.

7.5.4 Data Minimisation

The Data protection directive states that the personal data processed must be adequate with regards to the goal of the processing. In essence, this means that no more data may be processed than is necessary (data minimisation). But the converse is also true: since the data processed must be adequate for the specified goal, processing too little personal data is also undesirable. In theory, the more attributes that are added to a profile, the more accurate a profile will be. So from the perspective of accuracy it could be argued that data *maximisation* rather than data *minimisation* should be a goal.[14] This leads to an interesting paradox when it comes to privacy and profiling. The goal of privacy and data protection legislation is to minimise the amount of data being processed. However, this may lead to profiles being less accurate, which in turn may engender the risks mentioned in paragraph (e.g. false positives/negatives, stereotyping, de-individualisation and discrimination). Furthermore, data minimisation is not necessarily a guarantee against discriminatory effects. Research in this area suggests that even by eschewing sensitive data (race for instance) altogether, discrimination may still occur (Verwer and Calders 2010). For a possible alternative to the current approach of data minimisation see Chapter 15 on data mini*mum*misation.

7.6 Is Data Protection Law an Adequate Solution?

The binary nature of data protection legislation has led to a situation whereby more and more data are considered personal data, in turn leading to an inflation of the personal sphere. As discussed in the previous paragraphs, this inflation is troublesome for several reasons. First of all, it leads to legal uncertainty and unnecessary burdens for data controllers. Second, the inflation of the personal sphere dilutes the effectiveness of enforcement and places too much emphasis on the role of the Data protection authority. Thirdly, the role of privacy and data protection legislation in addressing societal issues associated with profiling will become too big. This last point requires some further explanation. Informational privacy,

[14] Of course in saying this we must take the constraints of computer science into account, since an excessive amount of data may be detrimental for the efficiency and effectiveness of the algorithms used.

while an important human right in itself, is oftentimes more a means than an end. By limiting access and use of data via the right to informational privacy and data protection (the *means*), we limit the possibilities for misuse and abuse of these data, thus protecting interests such as personal autonomy, reputation and equal treatment (the *ends*). For instance, the right to privacy in the context of government surveillance is aimed at protecting personal autonomy: because knowledge is power, less information about citizens means less power for governments. In the context of processing data about an individual's race or religion the primary interest is not privacy protection, but rather equal treatment and/or avoiding discrimination: by not allowing racial information to be processed, it will be impossible to discriminate on the basis of these data. So by regulating the use of personal data, we mitigate possible risks and protect underlying goals (i.e., the moral reasons for data protection).

While this approach has proven useful, it also has its limitations. The binary nature of data protection law (it either applies, or it does not) also means that there are few possibilities to differentiate in the application of data protection legislation. On the one hand this may mean that too strict a regime is applied to 'mundane' privacy issues, while serious issues such as discrimination do not get the attention they deserve and are only treated as data protection issues. Moreover, too strong a focus on data protection may draw away our attention from alternative (legislative) solutions that provide more protection for individuals and groups as well as take into account the interests of the profiler.

7.7 Shifting the Focus in Data Protection Law

We have seen that the EU Data protection directive is quite expansive in its scope because of the broad interpretation of the concept of personal data, which may be troublesome. As such, we may conclude that there are limits to the effectiveness of data protection law in the context of profiling. Nonetheless, privacy and data protection law provide an important barrier against privacy intrusions and there are compelling moral reasons for protecting personal data. Therefore it is worthwhile to explore how we can make privacy and data protection law more effective, particularly in the context of profiling.

7.7.1 Differentiation in Data Protection: Data Centric Approach

A first option to make data protection law more effective is to differentiate in the application of data protection law based upon the data being processed. Depending on factors like the type of data processed, the likelihood of identification, and the scope of the data processing exercise we could set different standards of protection. When it comes to the appropriate (legal) safeguards, we could for instance employ a light regime that focuses on transparency, data quality and data security, possibly linked with stronger *ex-post* protection mechanisms, for data that is not easily identifiable; and a stronger regime that employs all the (ex-ante) safeguards of the data protection directive for data with a clear link to an identified person.

Ohm (2010) proposes a differentiation based on the roles different entities can play in the identification or re-identification process. He argues that because identification or re-identification is made possible (or easier) by combining different data sets from different entities, entities that process large amounts of (personal) data (what Ohm calls 'large entropy reducers'), should have a higher duty of care (e.g., companies like Google, Microsoft and Choicepoint).

Schwarz & Solove (2011) propose a differentiated system based on the difference between 'identified data' and 'identifiable data'. They divide the use of data into three risk categories: identified, identifiable, and non-identifiable. Rather than defining these categories in law, they opt for a more flexible, standards based approach to determine under which circumstances what regime should apply.

7.7.2 Focus on the 'Why' Instead of the 'What': Goal Oriented Approach

While a more fine-grained data centric approach will, to some extent, remedy the issues associated with the binary and procedural nature of data protection legislation, it does not necessarily deal effectively with the possible risks of profiling. Therefore, we should also look towards other mechanisms to function alongside data protection legislation.

An alternative (or an addition) to the data centric approach is a more goal oriented approach. Depending on the actual goal of the data processing and the possible risks involved, the most effective protective measures may be chosen.

Purpose specification and purpose binding already form key elements of the structure of the current Data protection directive. Data controllers need to have a specified, explicit and legitimate purpose for collecting personal data and any further processing may not be incompatible with the specified purpose (see article 6 of the Directive). However, the goal of the data processing does not determine which rules should apply. Rather, the general rules of the Data protection directive apply, regardless whether they are the most effective protective measures.

7.7.3 Revisiting the Moral Reasons for Data Protection

In a goal-oriented approach the type and level of protection would be based primarily on the goal of the profiling exercise and the risks associated with this goal, rather than on the basis of the fact that certain data is considered personal data. By looking more closely at the risks involved with a particular type of processing we can ascertain whether data protection law should apply, and to what extent. A more goal-oriented approach makes data protection rules more context-sensitive, opening up the possibility for other legal protection mechanisms (such as consumer protection, equal treatment, and unfair commercial practices legislation) that might be more effective or suitable.

A goal-oriented approach to data protection and profiling would therefore place more emphasis on the moral grounds for data protection than is currently the case. This may also entail that other types of legislation (anti-discrimination legislation for instance) may come into play in addition to data protection law. In some cases these rules may even supersede data protection legislation. For instance, data minimisation rules and prohibitions on the processing of sensitive data may be overruled if they undermine the accuracy of a profiling exercise, or if they deny us the possibility to detect discrimination in a profiling exercise.

A goal-oriented approach will likely mean less focus on ex-ante protection and more focus on ex-post protection mechanisms. A positive effect of this shift is that it will force data controllers to actually make an assessment of the risks involved in their data processing and profiling activities, rather than reducing privacy and data protection to a mere compliance issue, as is currently often the case.

7.7.4 From 'Privacy by Design' to 'Ethics by Design'

Apart from the application of data protection and the rules associated with it, we should also examine other means of regulation. In particular the 'code as code' solution of *privacy by design* should be taken into consideration (Lessig 2006). 'Privacy by design' refers to the notion that we must incorporate privacy-protecting measures into the architecture of information systems (Borking 2010). In this way we can 'hardwire' the rules into the system. While privacy by design is an important measure when it comes to the protection of the individual in the context of profiling we should also be cognisant of the limitations and drawbacks of such an approach. In particular, we should take into account the limitations of privacy and data protection law in drafting functional and legal requirements for IT systems. Rather than a narrow focus on privacy and data protection we should look towards the actual goal of the profiling exercise and determine whether we can apply appropriate safeguards. This requires a broader focus than privacy, so instead of privacy by design we should focus on *ethics by design*. While closely related to the idea of privacy by design, ethics by design allows for a more focused, context sensitive approach to dealing with the possible risks of profiling.

7.8 Conclusion

Profiling is becoming an increasingly important tool for the public and private sector. While an effective tool, profiling might also entail risks for individuals and groups. These risks include stereotyping, inaccuracies in the application of profiles, stigmatisation, de-individualisation and abuse of profiles.

Currently these risks are addressed mainly through the application of data protection law. However, it is questionable whether this legal framework in its current form and application effectively mitigates the risks of profiling.

By stretching the definition of personal data to include profiles and the identifiers that link these profiles to individuals (e.g., IP-addresses and cookies) the protection mechanisms of the Data protection directive apply. While such an approach is understandable, there are some drawbacks. Because of the binary and procedural nature of data protection law there is no way to differentiate in the application of the Data protection directive. Labelling all data as personal data either because there are moral reasons to have some form of protection, or because there is a risk of identification or re-identification, will lead to an inflation of the personal sphere. In some cases, applying data protection law may be ineffective and even counterproductive.

To counter the drawbacks that currently flow forth from the application of data protection legislation, a first option could be to differentiate between different types of data (identified, identifiable, non-identifiable). While this would make data protection law more flexible and practical it will not necessarily address all the issues associated with profiling, nor remedy all the drawbacks of applying data protection law in the context of profiling. Therefore, in addition to the data centric approach, we should focus more on the actual goal of the profiling exercise and determine on the basis of the actual risks associated with this goal which safeguards should apply. Not only would such an approach likely provide more protection to the individual, it would also allow for a better balancing of the interests of the data subject and those of the data controller.

References

Barbaro, M., Zeller, T.: A face is exposed for AOL searcher no. 44177179. New York Times (2006), online version via: `http://www.nytimes.com/2006/08/09/technology/09aol.html` (last visited: December 28, 2011)

Borking, J.J.F.M.: Privacyrecht is code. Kluwer, Deventer (2010) (in Dutch)

Blok, P.: Het Recht op Privacy. Boom Juridische uitgevers, Den Haag (2002) (in Dutch)

Bygrave, L.: Minding the machine: article 15 of the EC data protection directive and automated profiling. Computer Law & Security Report 17, 17–24 (2001)

Custers, B.H.M.: Data Mining with Discrimination Sensitive and Privacy Sensitive Attributes. In: Proceedings of ISP 2010, International Conference on Information Security and Privacy, Orlando, Florida, July 12/14 (2010)

Hildebrandt, M.: Defining Profiling: A New Type of Knowledge? In: Hildebrandt, M., Gutwirth, S. (eds.) Profiling the European Citizen, Cross-Disciplinary Perspectives. Springer Science (2008)

van den Hoven, J.: Privacy and the Varieties of Informational Wrongdoing. Australian Journal of Professional and Applied Ethics, Special Issue 1 (June 1999)

van den Hoven, J.: Information Technology, Privacy and the Protection of Personal Data. In: van den Hoven, J., Weckert, J., et al. (eds.) Information Technology and Moral Philosophy, pp. 301–332. Cambridge University Press, Cambridge (2008)

Lessig, L.: Code 2.0. Perseus Book Group, New York (2006)

Ohm: Broken Promises of Privacy: Responding to the Surprising Failure of Anonymization. UCLA Law Review 57, 1701–1777 (2010)

Schermer, B.W.: Software Agents, Surveillance, and the Right to Privacy: a Legislative Framework for Agent-enabled Surveillance, PhD. Thesis. Leiden University (2007)

Schwarz, Solove: The PII Problem: Privacy and a new concept of personal identifiable information. New York University Law Review 86, 1814 (2011)

Terstegge, J.: Back to basics: privacy ethics (2009), http://jeroenterstegge.blogspot.com (last visited: December 28, 2011)

Vedder, A.: KDD: The challenge to individualism. Ethics and Information Technology 1(4) (December 1999)

Verwer, Calders: Three Naive Bayes Approaches for Discrimination-Free Classification. In: Data Mining: Special Issue with Selected Papers from ECML-PKDD 2010. Springer (2010)

Zwenne, G.J.: Over persoonsgegevens en IP-adressen, en de toe komst van privacywetgeving. In: Mommers, L., Franken, H., Klaauw, F., van der Herik, H., van der Zwenne., G.-J. (eds.) Het Binnenstebuiten, Liber Amicoru m Aernout Schmidt, pp. 321–341. Universiteit Leiden (2010) (in Dutch)

Part III
Practical Applications

Chapter 8
Explainable and Non-explainable Discrimination in Classification

Faisal Kamiran and Indrė Žliobaitė

Abstract. Nowadays more and more decisions in lending, recruitment, grant or study applications are partially being automated based on computational models (classifiers) premised on historical data. If the historical data was discriminating towards socially and legally protected groups, a model learnt over this data will make discriminatory decisions in the future. As a solution, most of the discrimination-free modeling techniques force the treatment of the sensitive groups to be equal and do not take into account that some differences may be explained by other factors and thus justified. For example, disproportional recruitment rates for males and females may be explainable by the fact that more males have higher education; treating males and females equally will introduce reverse discrimination, which may be undesirable as well. Given that the law or domain experts specify which factors are discriminatory (e.g. gender, marital status) and which can be used for explanation (e.g. education), this chapter presents a methodology how to quantify the tolerable difference in treatment of the sensitive groups. We instruct how to measure, which part of the difference is explainable and present the *local* learning techniques that remove exactly the illegal discrimination, allowing the differences in decisions to be present as long as they are explainable.

8.1 Introduction

Data mining builds computational models from historical data. Classification is a data mining task, where the goal is to learn the relation between given variables in order to apply the learnt model in the future for decision making. Suppose an

Faisal Kamiran
Lahore Leads University, Pakistan
e-mail: `faisal.kamiran@gmail.nl`

Indrė Žliobaitė
Bournemouth University, UK
e-mail: `izliobaite@bournemouth.ac.uk`

B. Custers et al. (Eds.): Discrimination & Privacy in the Information Society, SAPERE 3, pp. 155–170.
springerlink.com

automated CV screening in recruitment. Given education, employment history and qualifications (the input variables called *attributes*) of an individual the task is to decide whether this individual should be selected for an interview (the outcome called *label*). An automated classifier for such decisions can be built using historical data examples where the relations between the attributes and labels are known.

Nowadays more and more decisions in lending, recruitment, grant or study applications are partially being automated based on models trained on historical data. That historical data may be discriminatory[1]; for instance, racial or gender discrimination may have affected the selection of job candidates in the historical data. In such a case classifiers trained on this discriminatory data are likely to learn the discriminatory relation, and, as a result, they will make discriminatory predictions when applied to new data in the future.

Training a discrimination-free model on the historical data that is discriminatory is challenging. Removing the sensitive attribute, e.g., gender, from the training data is not enough to prevent discrimination. If gender is related to some of the remaining attributes, e.g, marital status, the model will capture the discriminatory decisions indirectly. A number of techniques have been developed (Calders et al., 2009; Calders and Verwer, 2010; Kamiran et al., 2010; Kamiran and Calders, 2010) focusing on how to train discrimination-free classifiers over the discriminatory training data. These techniques aim at making the probabilities of positive decision equal across the sensitive groups, e.g., male and female. They do not take into account that some differences in treatment may be explainable by other attributes, such as education level. This chapter presents a methodology how to quantify and measure the explainable and non-explainable parts of discrimination and introduces classification techniques to remove the non-explainable part only. The methodology is referred to as *conditional* non-discrimination.

The studies by (Pedreschi et al., 2008, 2009; Ruggieri et al., 2010) aim at the detection of discrimination from training data and identify the potentially discriminatory classification rules. A central notion in these works on identifying discriminatory rules is that of the *context* of the discrimination. That is, specific regions in the data are identified in which the discrimination is particularly high. These works focus also on the case where the discriminatory attribute is not present in the dataset and background knowledge for the identification of discriminatory guidelines has to be used. However, we assume that the discriminatory data is given but the discrimination should be avoided in future predictions. Our chapter discusses the next steps after detecting discrimination.

The chapter is organized as follows. In Section 8.2 we discuss the problem of explainable and illegal discrimination in classifier design. Section 8.3 analyzes the concept of explainable and non-explainable discrimination, and instructs how to measure the explainable part of the discrimination. In Section 8.4 we present the local modeling techniques for removing illegal discrimination and experimentally illustrate their performance. Section 8.5 concludes the chapter.

[1] Discrimination is the prejudicial treatment of an individual based on their membership in a certain group or category.

8.2 Explainable and Non-explainable Discrimination

It is in the best interest of the decision makers (e.g., banks, consultancies, universities) to ensure that the classifiers they build are discrimination-free even if the historical data is discriminatory.

8.2.1 Discussion of the Legal Aspects

Proving a case as discriminatory in court requires proving that there were no genuine reasons for the biased treatment. As an example, employment practices may be considered discriminatory if they have a disproportionate adverse impact on members of a minority group. The US Equal Pay Act 1963 (Legislation, 1963) requires men and women to be given equal pay for equal work in the same workplace. The jobs need not to be identical, but they must be substantially equal. If the difference in jobs can be justified, the differences in salary are tolerable.

To illustrate the legal context and the difficulty of the task, consider the following case. Recently one of the world largest consultancy firms was accused of discrimination against ethnic minorities in a law suit (Ahearn, 2010) as a result of using criminal records to turn down candidates in pre-employment screening. Not the use of criminal records itself was considered problematic, but the correlation between race and criminality in this particular case. Indirectly, the use of criminal records led to racial discrimination. So, even though the company did not intend to discriminate, the decisions were deemed discriminatory by the court, while having been convicted was deemed to be irrelevant for pre-screening purposes by the court. This example shows that discrimination may occur even if the sensitive information is not used in the model and that such indirect discrimination is as well forbidden. Many attributes can be used only to the extent that they do not lead to indirect discrimination.

8.2.2 Motivation for the Explainable Discrimination

The mainstream solutions to make classifiers discrimination-free (Calders et al., 2009; Calders and Verwer, 2010; Kamiran et al., 2010; Kamiran and Calders, 2010) aim at removing all discrimination present in the data; the probabilities of a positive decision for all subgroups (e.g., male and female) must be equal. Such approaches; however, have a significant limitation, as they do not take into account the fact that a part of the differences in the probability of acceptance for the two groups may be objectively explainable by other attributes.

For instance, in the Adult dataset (Asuncion and Newman, 2007), females on average have a lower annual income than males. However, one can observe that females work fewer hours per week on average; see Table 8.1. If we assume that job requires the attendance of employee for full working hours (e.g., job at information desk), work hours per week gives a good justification for low income. Assume the task is to build a classifier to determine a salary, given an individual.

The previous works would correct the decision making in such a way that males and females would get on average the same income, say \$ 20,000, leading to a reverse discrimination as it would result in male employees being assigned a lower salary than female for the same amount of working hours. In many real world cases, if the difference in the decisions can be justified, it is not considered as illegal discrimination. Moreover, making the probabilities of acceptance equal for both would lead to favoring the group which is being deprived, in this example females.

Table 8.1 Summary statistics of the Adult dataset (Asuncion and Newman, 2007)

	hours per week	annual income (K$)
female	36.4	10.9
male	42.4	30.4
all data	40.4	23.9

8.2.3 Discrimination in Decision Making

To analyze the effects of discrimination and design discrimination-free learning techniques, a model describing how discrimination happens needs to be assumed.

We consider that discrimination happens in the following way in relation to experimental findings reported in (Hart, 2005). The historical data originates from decision making by human experts. First the qualifications of a candidate are evaluated and a preliminary score is obtained. The qualifications are evaluated objectively. Then the score is corrected with a discrimination bias by looking at, e.g., the gender of a candidate and either adding or subtracting a fixed (the same) bias from the qualification score. The final acceptance decision is made by comparing the score to a fixed acceptance threshold. If the score is higher than the threshold then the candidate is accepted.

This discrimination model has two important implications. First, the decision bias is more likely to affect the individuals whose objective score is close to the decision threshold. If an individual has very good qualifications, adding or subtracting the discriminatory bias would not change the acceptance decision.

Second, there may be attributes within the training data, however, that objectively explain the score, but at the same time are correlated with the sensitive attribute. When observing the decisions it would seem due to correlation that the decision is using the sensitive attribute. Next we discuss how to quantify, which part of the difference in decision across the sensitive groups is explainable and which is due to discrimination bias.

It is important to mention here that this discrimination model does not guarantee to cover the all possible scenarios that lead to discrimination, however, it covers the most important and typical scenario.

8.3 Conditional Non-discrimination in Decision Making

Even though the historical data contains discrimination, new classifiers trained on this data should not discriminate in the future decision making. The first solution that comes to mind to address this discrimination-aware classification problem is that we should remove the sensitive attribute from the training data before learning a new classifier. Unfortunately, this approach does not remove discrimination from future decision making if some of the attributes in the training data are correlated with the sensitive attribute. For instance, postal code may be highly correlated with race. A classifier will be able indirectly (internally) predict the race from the postal code and then it will still use race in the acceptance decisions. That is indirect discrimination, known as *redlining*.

To get rid of such discriminatory relations among attributes, one would also need to remove the attributes that are correlated with the sensitive attribute. It is not a good solution if these attributes carry the objective information about the outcome, as the predictions will become less accurate. For instance, a postal code in addition to the racial information may carry information about real estate prices in the neighborhood, which is objectively informative for loan decisions. The aim is to use the objective information, but not the sensitive information of such attributes.

The explanatory attribute is the attribute in the training data that is correlated with the sensitive attribute, and at the same time gives some objective information about the outcome. In general there is no objective truth which attribute is more reasonable to use as the explanation for discrimination. For instance, in case gender is the sensitive attribute, some attributes, such as relationships ('wife' or 'husband') are not a good explanation, as semantically they are closely related to gender. On the other hand, difference in working hours may be an appropriate reason to have different monthly salaries. What is discriminatory and what is legal to use as an explanation of the outcome depends on the law and goals of the anti-discrimination policies. Thus, the law or domain experts define, which attributes are sensitive and which are allowed to be treated as explanatory.

8.3.1 An Example on University Admission

Consider a model of the admission decisions to a fictitious university[2], that will help to analyze the difference between the explainable and illegal discrimination. Note that the model presents a simplified version of reality and is intended to cover the key mechanisms of decision making, and does not cover a full application process. Gender is the sensitive attribute; male (m) and female (f) are the sensitive groups, against which discrimination may occur. There are two programs: medicine (med) and computer science (cs) with potentially different acceptance standards. Program is considered to be the explanatory attribute. In this example, we assume that the differences in acceptance statistics between male and female that can be attributed to

[2] This model does not express our belief how the admission procedures is modeled. We use it for the purpose of illustration only.

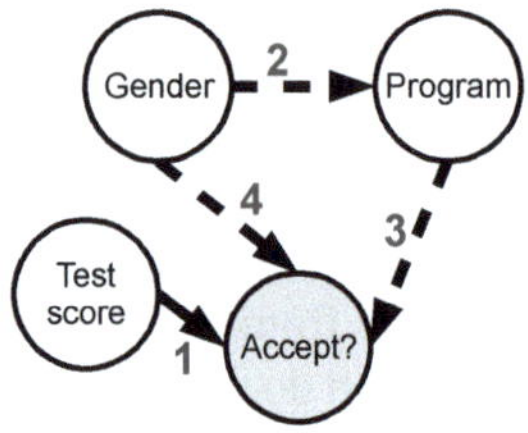

Fig. 8.1 An example illustrating an admission procedure of a fictitious university

different participation rates in the programs are acceptable. All applicants take a test for which their score is recorded (T), which, we assume, is objective. The acceptance (+) decision is made personally for each candidate during the final interview. Figure 8.1 shows the setting. The probabilities that are fixed are shown in the illustration.

There are four relations between variables in this example. Relation (1) shows that the final decision whether to accept partially depends on the test score. Notice that the test scores are assumed to be independent from gender or program. Relation (3) shows that the probability of acceptance depends on the program. For example, the competition for medicine may be higher, thus less applicants are accepted in total. Relation (2) shows that the choice of program depends on gender. For instance, the larger part of the female candidates may apply to medicine, while more males apply to computer science. Relation (4) shows that acceptance also depends on gender, which is a bias in the decision making that is clearly a case of illegal discrimination. The presence of illegal, explainable or both discriminations depends on the strength of the relations (2),(3) and (4), as the following examples will show.

8.3.2 *Measuring Discrimination*

We will present the discrimination measures using the example about university admission procedure, consider a historical dataset as illustrated in Figure 8.2.

In the existing discrimination-aware classification (Calders et al., 2009; Calders and Verwer, 2010; Kamiran et al., 2010; Kamiran and Calders, 2010) the discrimination is considered to be present if the acceptance rate for the favored community (denote m for males) and the deprived community (denote f for

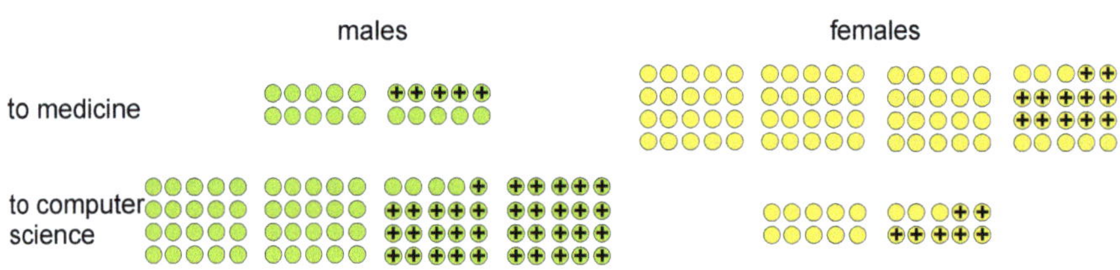

Fig. 8.2 Model dataset, '+' means accepted, empty means rejected

females) were not equal. The acceptance rate for males is the number of males that have been accepted (in our example 41) divided by the total number of male applicants (in our example 100), the acceptance rate is 41%. The acceptance rate for females is calculated in the same way $\frac{19}{100} = 19\%$. Discrimination is measured as the difference between the two acceptance rates $D_{all} = 41\% - 19\% = 22\%$, thus our data contains 22% discrimination in total. This 22% difference may include the results of a discrimination bias in decision making (undesirable part) as well as a part that is objectively explainable by an explanatory attribute (tolerable part).

We can see that in our dataset gender and program are not independent, i.e., medicine was more popular among females, while computer science was more popular among males. It so happened that medicine turned to be more competitive program, only 17 candidates out of 100 were accepted, while for computer science 43 out of 100 candidates were accepted. Our task is to establish whether a lower female acceptance rate is explainable by differences in program acceptance rates or there is a bias towards gender in decision making.

Ideally, we would need pairs of candidates that have all attributes equal except gender to assess the existence of gender discrimination, however, in real data a variety of values for different attributes are possible, thus grouping all the data into identical pairs is not realistic. In order to quantify the explainable and illegal discrimination we need to group the data so that we can treat the individuals as equal within each group. The explanatory attribute serves as such a criterion.

Suppose the program is allowed to be treated as the explanatory attribute. Thus we will require males and females to be treated equally within each program, while the acceptance rates for males and females will be allowed to be different across different programs.

Let us calculate the acceptance rates within each program for our university example, starting from medicine. The acceptance rate for females in medicine is calculated as the number of females accepted to medicine divided by the number of females that applied to medicine, i.e., $\frac{12}{80} = 15\%$. Similarly, the acceptance rate for males to medicine is $\frac{5}{20} = 25\%$. We see that the acceptance rates within medicine are different, thus there is a case of illegal discrimination.

Similarly, the current acceptance rate for females to computer science is $\frac{7}{20} = 35\%$ and the acceptance rate for males to computer science is $\frac{36}{80} = 45\%$.

To calculate the explainable part we first need to find what would have been the correct acceptance rate within each program if there was no discrimination bias in decision making. Following the discrimination model (Section 8.2.3) the unbias acceptance rate to medicine is the (weighted[3]) average of the acceptance rates of males and females to medicine: $(15\% + 25\%)/2 = 20\%$. Similarly, the non-discriminating acceptance rate to computer science should thus be $(35\% + 45\%)/2 = 40\%$.

Now what we know is the 'correct' acceptance rates within each program, we can calculate what would be the total difference between accepted males and females if

[3] The average needs to be weighted by the proportions of the applicants to each program. In our case 100 applicants applied to medicine and 100 to computer science, each weight is 0.5, thus the weighted average is the same as non weighted $0.5 \times 15\% + 0.5 \times 25\% = 20\%$.

this correct rate is applied, i.e., if we accepted 20% of male applicants and 20% of female applicants to medicine and if we accepted 40% of male applicants and 40% of female applicants to computer science. Figure 8.3 illustrates the situation. Observe that the total number of accepted applicants has not changed, 60 applicants were accepted before and 60 are accepted now.

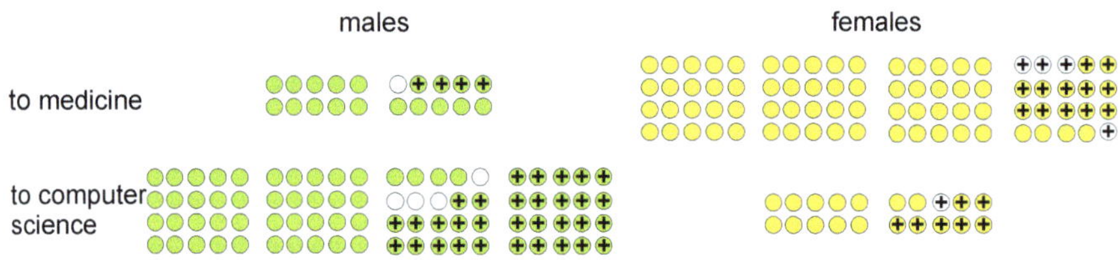

Fig. 8.3 Corrected data, empty dots indicate the corrected decisions

We can see from the figure that the total acceptance rates of males and females are still different, 36% and 24% correspondingly. However, as long as we treat males and females equally within each program, there is no illegal discrimination. Thus all this 12% difference is explainable and tolerable. Recall, that there is 22% over all discrimination in the original data shown in Figure 8.2. Thus, as we see that 12% is explainable and thus the remaining 22% – 10% = 10% is non-explainable, and we should aim to eliminate only this illegal part of discrimination when training a classifier. See (Zliobaite et al., 2011) for further technical information on how to calculate the explainable discrimination.

8.3.3 *Illustration of the Redlining Effect*

Suppose that it is no longer allowed to discriminate females directly, the gender information is kept hidden from the admission committee to avoid the gender discrimination. The committee will treat male and female applicants within medicine and within computer science equally. However, knowing the fact that females prefer to apply to medicine, it is still possible to discriminate indirectly (without knowing the gender of an applicant). A decision maker who wants to discriminate, may reduce the overall acceptance rates to medicine and increase the acceptance rate to computer science. This phenomenon is known as the *redlining* (Tootell, 1996).

Figure 8.4 illustrates the situation with our university example. Recall that in our example 80 females chose to apply to medicine and 20 to computer science, while 20 males chose medicine and 80 chose computer science. Within each program both genders are treated equally. Figure 8.4 plots the situation when an adversary varies the acceptance rates within each program (keeping the total number of accepted people fixed to 60 as in the original example). The black dots illustrate the acceptance rates where all the difference between males and females is explainable (20% for medicine and 40% for computer science as calculated in the previous section). If

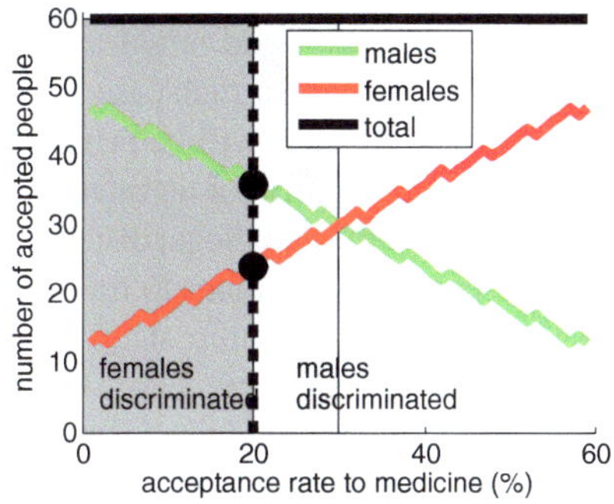

Fig. 8.4 Illustration of redlining

the acceptance rate to medicine is reduced, the illegal discrimination is introduced, females are discriminated (the grey area to the left). If the admission committee chooses to increase the acceptance rates to medicine, then males actually become discriminated, as more females are accepted than the explainable difference. The light grey area illustrates the Simpson's paradox (Simpson, 1951), in which a relation exists in different groups is reversed when the groups are combined. We see that more males are accepted, but in fact males are being discriminated, because females are applying to a more competitive program. Such situation was reported in the Berkeley study (Bickel et al., 1975). The next section presents a more elaborate example of a reverse discrimination.

8.3.4 Illustration of the Reverse Discrimination

We illustrate what happens if we do *not* take into account the explainable aspect of discrimination. For simplicity let us modify our university example. Now there are males and females applying for a job. The candidates are assessed based on their experience. Male applicants happen to have on average longer job experience (in years) than females. The recruitment company selects 6 candidates for an interview, that have the highest experience level.

Figure 8.5 (a) shows the example situation. We see that males and females were treated equally, thus all the difference in acceptance rates (the acceptance rate for males is $\frac{4}{8} = 50\%$, for females $\frac{2}{8} = 25\%$) is explainable and happens due to the fact that the experience attribute is correlated with the gender. On the other hand, if we

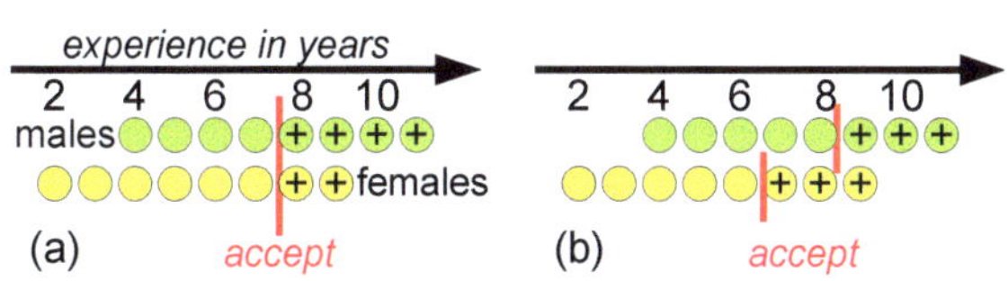

Fig. 8.5 Illustration of reverse discrimination

try to make the acceptance rates for males and females equal, we will *introduce* a reverse discrimination, as illustrated in (b). In such case the acceptance rates will be equal ($\frac{3}{8} = 38\%$ for both); however, females will be privileged, as they will require a lower experience level than male to qualify for an interview.

We discussed the issues of explainable and illegal discrimination in decision making. In the next section we present two techniques to train classifiers with an aim to remove only the illegal discrimination from future decision making.

8.4 Removing the Illegal Discrimination When Training a Classifier

In order to ensure that the built classifier is free from illegal discrimination one needs to control two constraints. Using the university admission example, the two conditional non-discrimination constraints are as follows:

1. acceptance rates of males and females *within each program* need to be equal (although programs may have different acceptance rates, e.g., medicine 20% and computer science 40%);
2. acceptance rates within each program need to be consistent with the original data, i.e., even if the acceptance rate to medicine for males and females become equal, the acceptance rate to medicine should not artificially decreased, e.g., to 10%.

The second condition is necessary to prevent the *redlining effect*, as discussed in the previous section.

8.4.1 Techniques

We present two techniques (Zliobaite et al., 2011) for removing illegal discrimination that modify labels of the historical data so that the historical data is no longer discriminatory, i.e., it satisfies the two conditional non-discrimination constraints.

The techniques work as follows. First, the 'correct' acceptance rates to each program need to be computed. Next, the acceptances of some of the individuals in the historical data are modified in such a way that the acceptance rates within each program are 'correct'. The classifiers trained on the modified data, which does not contain illegal discrimination, are expected to produce decisions that would not contain illegal discrimination. The two presented techniques differ in a way how they modify the historical data.

Local Massaging

The local massaging within every group, defined by a unique value of the explanatory attribute[4], modifies the values of labels until the historical data contains no

[4] E.g., one group will be formed of all students that applied to medicine, the other group will consists of students that applied to computer science.

discrimination. The discrimination model in Section 8.2.3 implies that discrimination is more likely to affect the objects that are closer to the decision boundary. To this end, massaging identifies the instances that are close to the decision boundary and changes the values of their labels to the opposite ones (e.g., positive to negative or negative to positive). Suppose females have been discriminated as in our university admission model and the discrimination is reflected in the historical data. The local massaging identifies a number of females that were almost accepted, and makes their labels positive, and identifies a number of males that were very likely, but have not been rejected, and makes their labels negative. To choose the cases for relabeling, individuals are ordered according to their probability of acceptance using an internal ranker (a classifier that outputs posterior probabilities), learned on the training data for each program separately.

The local massaging uses the same principles as the massaging technique (Kamiran et al., 2010). However, local massaging works on the partitioned data, within each program separately. In addition, it also modifies and controls the number of accepted males and females, to ensure no redlining. The procedure for local massaging is illustrated in Figure 8.6.

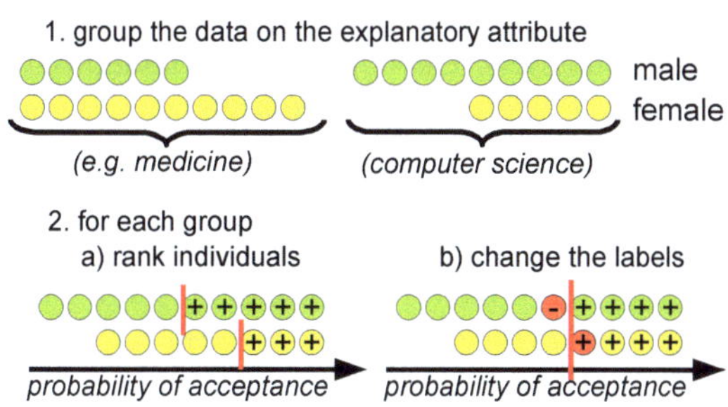

Fig. 8.6 Local massaging

Local Preferential Sampling

The preferential sampling technique does not modify the training instances or labels, instead it modifies the composition of the training set. It deletes and duplicates training instances so that the labels of new training set contain no discrimination.

Following the discrimination model where the discrimination is more likely to affect the individuals that are closer to the decision boundary, the preferential sampling deletes the 'wrong' cases that are close to the decision boundary and duplicates the cases that are 'correct' and close to the boundary. The cases are selected using a ranker learned in the same way as in the local massaging. In the university example the local preferential sampling will delete a number of males that were almost rejected and duplicate the males that were almost accepted. It will also delete a number of females that were almost accepted and duplicate the females that were almost rejected.

The local preferential sampling applies the same principles of preferential sampling (Kamiran and Calders, 2010) but now locally to partitions of the data. It modifies and controls the number of accepted male and female, to ensure no redlining. The procedure for local preferential sampling is presented in Figure 8.7.

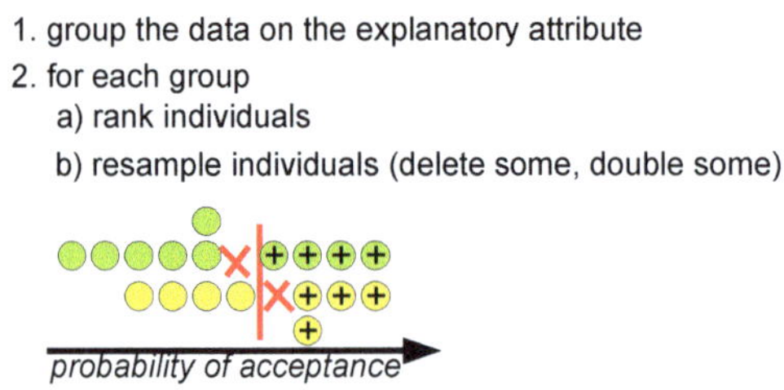

Fig. 8.7 Local preferential sampling

8.4.2 *Computational Experiments*

In this section we demonstrate the performance of the local discrimination handling techniques on real world datasets. The objective is to minimize the absolute value of the *illegal* discrimination while keeping the accuracy as high as possible. It is important not to overshoot and end up with a reverse discrimination.

Data

For our experiments we use two real datasets. The **Adult** dataset comes from UCI (Asuncion and Newman, 2007), the task is to classify individuals into high and low income classes. Our dataset consists of a uniform sample of 15 696 instances, which are described by 13 attributes and a class label. Originally 6 of the 13 attributes were numeric attributes, which we discretized. Gender is the sensitive attribute, income is the label. We repeat our experiments several times, where any of the other attributes in turn is selected as explanatory. Figure 8.8 (left) shows the discrimination in the dataset. The horizontal axis denotes the index of the explanatory attribute.

In the Adult dataset a number of attributes are weakly related with gender (such as workclass, education, occupation, race, capital loss, native country). Therefore, nominating any of those attributes as explanatory will not explain much of the discrimination. For instance, knowledge of biology suggests that race and gender are independent. Thus, race cannot explain the discrimination on gender; that discrimination is either illegal or it is due to some other attributes. Indeed, the plot shows that all the discrimination is illegal, when treating race (attribute #7) as explanatory. On the other hand, we observe that the relationship (attribute #6) explains a great deal of D_{all}. Judging subjectively, the values of this attribute 'wife' and 'husband' clearly capture the gender information and from the data mining perspective, if we are allowed to treat it as acceptable, a large part of the discrimination is explained. Age, and working hours per week are other examples of explanatory attributes that

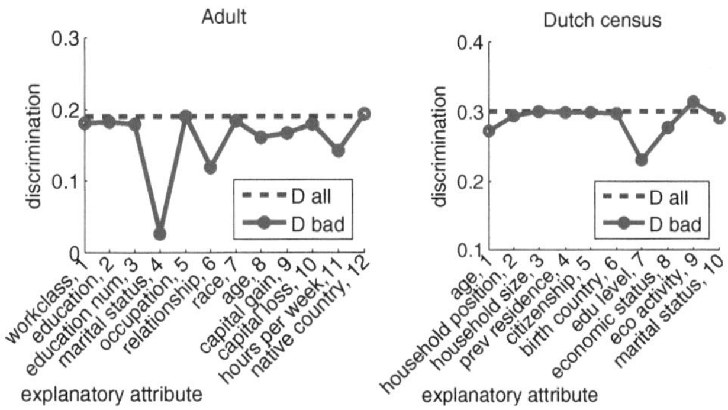

Fig. 8.8 Discrimination in the datasets. We label the over all discrimination as D_{all} and the illegal discrimination as D_{bad}.

justify some discrimination. Whether relationship is an acceptable argument to justify differences in income is to be determined by law.

Another dataset that we use is the **Dutch Census of 2001** (Dutch Central Bureau for Statistics, 2001), that represents aggregated groups of inhabitants of the Netherlands. We formulate a binary classification task to classify the individuals into 'high income' (prestigious) and 'low income' professions, using occupation as the class label. Individuals are described by 11 categorical attributes. After removing the records of under-aged people, several professions in the middle level and people with unknown professions our dataset consists of 60420 instances. Gender is treated as the sensitive attribute.

Figure 8.8 (right) presents the discrimination contained in this data. The difference between the all and the illegal discrimination is much less than in the Adult data. Here many attributes are not that strongly correlated with gender. Simply removing the sensitive attribute should therefore perform reasonably well. Nevertheless, education level, age and economic activity present cases for conditional non-discrimination, thus we explore this dataset in our experiments.

Non-discrimination Using Local Techniques

Let us analyze how the local techniques handle discrimination[5]. We expect them to remove exactly the illegal discrimination and nothing more. For comparison we add a technique that does not use any discrimination handling strategies (blank) and two local techniques (that, as we discussed, risk to introduce reverse discrimination).

Figure 8.9 shows the resulting discrimination after applying the local massaging and the local preferential sampling. Both local techniques perform well on the Adult data. Illegal discrimination is reduced to nearly zero, except for relationship as explanatory attribute when massaging is applied to the Adult dataset. The techniques also do not produce the reverse discrimination as, e.g., global massaging does.

[5] The performance is tested with decision trees J48 via 10-fold cross validation.

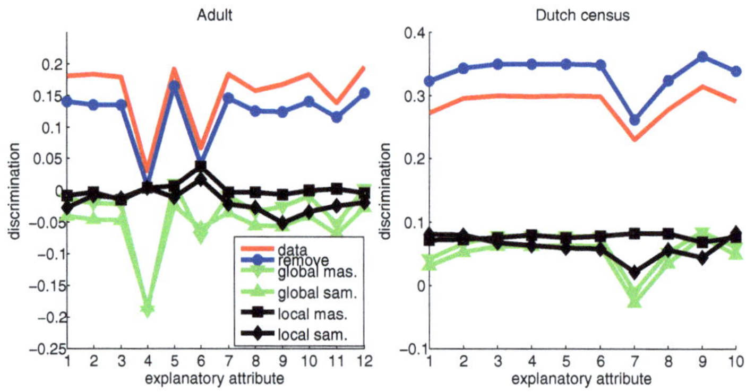

Fig. 8.9 Discrimination with *the local* techniques

The approaches do not perform that well with the Dutch census data, as the sensitive attribute is not very strongly correlated with any other attribute in the dataset. The local techniques are primarily designed to handle high correlations with the sensitive attribute that induce *redlining*.

When classifiers become discrimination-free, they may lose some accuracy. Figure 8.10 presents the resulting accuracies scores of the results of our experiments. We observe that the local methods lose some accuracy, but work more accurately than the global methods. Our experiments demonstrate that the local massaging and the local preferential sampling classify future data with reasonable accuracy and maintain low discrimination.

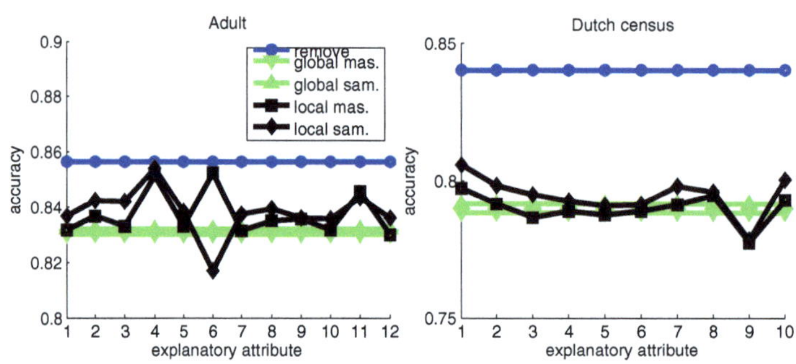

Fig. 8.10 Accuracy with *the local* techniques

8.5 Conclusion

In this chapter we discussed the issue of conditional non-discrimination in classifier design, where different treatment of sensitive groups can be explainable by other attributes and hence tolerable. We presented a methodology for quantifying the explainable differences in treatment and the illegal discrimination in data.

We argued that the techniques, that do not take into account the explainable part of the discrimination, may tend to overshoot and thus introduce a reverse discrimination, which is undesirable as well. We explained how to measure discrimination in data or decisions output by a classifier by explicitly considering explainable and illegal discrimination.

Finally, we presented the *local* techniques that remove exactly the illegal discrimination, allowing the differences in decisions to be present as long as they are explainable. These techniques preprocess the training data in such a way that it no longer contains illegal discrimination. After preprocessing classifiers that are trained using this data are expected not to capture the illegal discrimination. Our computational experiments demonstrated the effectiveness of the local preprocessing techniques.

References

Ahearn, T.: Discrimination lawsuit shows importance of employer policy on the use of criminal records during background checks (2010), `http://www.esrcheck.com/wordpress/2010/04/12/`

Asuncion, A., Newman, D.: UCI machine learning repository (2007), `http://archive.ics.uci.edu/ml/`

Bickel, P., Hammel, E., O'connell, J.: Sex bias in graduate admissions: Data from Berkeley. Science 187(4175) (1975)

Calders, T., Kamiran, F., Pechenizkiy, M.: Building classifiers with independency constraints. In: IEEE ICDM Workshop on Domain Driven Data Mining, pp. 13–18 (2009)

Calders, T., Verwer, S.: Three naive bayes approaches for discrimination-free classification. Data Mining and Knowledge Discovery 21, 277–292 (2010)

Dutch Central Bureau for Statistics, 'Volkstelling' (2001), `http://easy.dans.knaw.nl/dms`

Hart, M.: Subjective decisionmaking and unconscious discrimination. Alabama Law Review 56, 741 (2005); University of Colorado, Law Legal Studies Research Paper 06-26

Kamiran, F., Calders, T.: Classification with no discrimination by preferential sampling. In: Proceedings of the 19th Annual Machine Learning Conference of Belgium and The Netherlands (BENELEARN 2010), pp. 1–6 (2010)

Kamiran, F., Calders, T., Pechenizkiy, M.: Discrimination aware decision tree learning. In: Proceedings of IEEE ICDM International Conference on Data Mining (ICDM 2010), pp. 869–874 (2010)

Legislation: The us equal pay act (1963), `http://www.eeoc.gov/laws/statutes/epa.cfm`

Pedreschi, D., Ruggieri, S., Turini, F.: Discrimination-aware data mining. In: Proceedings of the 14th ACM SIGKDD International Conference on Knowledge Discovery and Data Mining (KDD 2008), pp. 560–568 (2008)

Pedreschi, D., Ruggieri, S., Turini, F.: Measuring discrimination in socially-sensitive decision records. In: Proceedings of the SIAM International Conference on Data Mining SDM 2009, pp. 581–592 (2009)

Ruggieri, S., Pedreschi, D., Turini, F.: DCUBE: discrimination discovery in databases. In: Proceedings of the International Conference on Management of Data (SIGMOD 2010), pp. 1127–1130 (2010)
Simpson, E.H.: The interpretation of interaction in contingency tables. Journal of the Royal Statistical Society (Series B) 13, 238–241 (1951)
Tootell, G.: Redlining in boston: Do mortgage lenders discriminate against neighborhoods? The Quarterly Journal of Economics 111, 1049–1079 (1996)
Zliobaite, I., Kamiran, F., Calders, T.: Handling conditional discrimination. In: Proceedings of IEEE ICDM International Conference on Data Mining, ICDM 2011 (2011) (in press)

Chapter 9
Knowledge-Based Policing: Augmenting Reality with Respect for Privacy

Jan-Kees Schakel, Rutger Rienks, and Reinier Ruissen

Abstract. Contemporary information-led policing (ILP) and its derivative, knowledge-based policing (KBP) fail to deliver value at the edge of action. In this chapter we will argue that by designing augmented realities, information may become as intertwined with action as it can ever get. To this end, however, the positivist epistemological foundation of the synthesized world (and ILP and KBP for that matter) has to be brought into line with the interpretive-constructivist epistemological perspective of every day policing. Using a real-world example of the Dutch National Police Services Agency (KLPD) we illustrate how augmented reality may be used to identify and intercept criminals red-handedly. Subsequently we discuss how we think that the required data processing can be brought into line with the legislative requirements of subsidiarity, proportionality, and the linkage between ends and means, followed by a discussion about the consequences for, among other things, privacy, discrimination, and legislation.

9.1 Introduction

The increasing digitization of services and goods, in combination with the expanding possibilities to interlink and provide them through networks such as the Internet, affects modern-day society in unprecedented ways (Castells 2000). From a policing perspective these developments result in new challenges to be met. For example, compared to criminals the police is slow in adapting to this new digital (or synthetic, or virtual) environment. An environment, which becomes more and more integrated with our real environment. At the same time, we observe that contemporary information-led policing (ILP), and knowledge-based policing (KBP) for that matter, is mostly restricted to strategic and tactical information, including overviews of hot crimes, hot times, hot spots, and hot shots. So far, ILP fails to deliver operational value in action. Designing augmented realities is one way to

Jan-Kees Schakel · Rutger Rienks · Reinier Ruissen
National Policing Services Agency, The Netherlands
e-mail: Jan.Kees.Schakel@klpd.politie.nl

B. Custers et al. (Eds.): Discrimination & Privacy in the Information Society, SAPERE 3, pp. 171–189.
springerlink.com

fill this void. Before we can discuss the application of augmented reality, however, we have to unify the underlying epistemological bases (philosophy concerning the nature of knowledge) underlying the 'real environment' and the 'synthetic environment' of policing. Within the policing profession the distinction between the two epistemological bases becomes clear by examining perspectives on 'old' and 'new' knowledge. In policing 'old' knowledge refers to knowledge related to traditional case-specific criminal investigations; 'New' knowledge refers to knowledge gained through digital data analysis, including the identification of trends, hotspots, 'hot moments', and other patterns (Ratcliffe 2008). Exploiting the potential of 'new' knowledge has become known as ILP, which in the Netherlands developed into a doctrine (Kop and Klerks 2009). As information within the ILP doctrine is treated as 'data given meaning and structure' (Ratcliffe 2008a: 4), its epistemological basis is clearly positivistic. Indeed, information is treated as an object that may be stored, enriched, and disseminated. In traditional case-specific criminal investigations, on the other hand, police officers are utterly aware of context, antecedents, and idiosyncratic perspectives. Hence, 'old' knowledge clearly has an interpretive-constructivist epistemological basis.

The emerging concept of KBP, which is an offspring of ILP (Brodeur and Dupont 2006), is thought to bring the worlds of 'old knowledge' and 'new knowledge' together (Ratcliffe 2008; Williamson 2008). To do so successfully, however, we argue that the epistemological bases have to be unified first. The positivist basis of ILP does not provide for the divers, dynamic, and complex nature of information which characterizes 'old' knowledge-based police work. Hence, we propose to follow Innes et al. (2005) and adopt an interpretive-constructivist perspective for both 'old' and 'new' forms of knowledge. We make this proposal specific by redefining the key-concepts (data, information, knowledge, and intelligence) accordingly. We then introduce the concept of boundary objects. This serves two purposes. First, boundary objects help to bridge the gap between information analysts, police officers, and other fields of discipline (Bechky 2003; Carlile 2003). Second, we use boundary objects as knowledge-structure to augment reality.

After our modest attempt to create a more suitable and holistic basis for KBP we turn our attention to augmented reality. Where people are limited to five highly sophisticated senses to observe their environment, virtuality provides opportunities to augment reality. Not necessarily by algorithms and routines that unearth hidden trends or other patterns in large databases, but in particular by processing data which is automatically obtained, filtered, and analyzed in real-time (through profiles) and integrated in policing practice, much like our human senses do. Where human senses, however, are intrinsically restricted to a given time and place, this is not necessarily so for artificial sensors. For example, a series of discrete observations of multiple geographically distributed sensors may be united into one composite observation. Such an endeavor requires the processing of large quantities of -often privacy related- data. Moreover, the observations include data related to people that legally may not be suspected of any legal offense. Hence, the application of augmented reality has to be brought in accord with the juridical framework of policing. We believe that KBP, as we are presenting it, provides a suitable basis for developing augmented realities that comply with the guiding juridical principles of proportionality, subsidiarity, and the linkage between ends and means.

In the following section we first provide a brief overview of ILP, followed by an epistemological reorientation of its key-concepts: data, information, knowledge, and intelligence (henceforth: key-concepts). In the section Knowledge-based policing we explain where current KBP theorizing falls short in uniting old and new forms of knowledge, followed by an explanation of what we see as needed to develop successful augmented realities. After creating the foundation for augmented reality we discuss what it takes to create augmenting realities and illustrate this using a case of the KLPD concerning the fight against drug-trafficking. This chapter ends with a discussion about, amongst things, privacy, discrimination, legal consequences, and a conclusion.

9.2 Intelligence-Led Policing

9.2.1 Origin and Epistemological Basis

ILP is a concept that originated in the 90s in England. A description of what Maguire called 'intelligence-led crime control' became the widely used definition of ILP (Lint 2006):

"a strategic, future-oriented and targeted approach to crime control, focusing upon the identification, analysis and 'management' of persisting and developing 'problems' or 'risks'" (Maguire 2000:316).

ILP is formally incorporated in the National Intelligence Model (NIM) of England, the core business model for policing, which is being described as an information-based deployment system aimed at 'identifying patterns of crime and enabling a more fundamental approach to problem solving in which resources can be tasked efficiently' (Centrex 2005:10). The British NIM functions as role-model for the creation of a NIM in the Netherlands, where ILP took flight in the beginning of this century (Abrio 2005, Hert et al. 2005). As the NIM is being used by the government (both in England and in the Netherlands) to implement standard ILP-practices, its influence on policing is substantial.

The resulting ILP-practices, at least in the Netherlands, are highly focused on the creation of information products (mostly in textual and numerical form) to direct police action. They are either focused on individual cases (reports), or driven (and limited) by statistics based on recorded data, such as criminal trends, hot spots, hot-moments, and social network analysis. Although these figures and facts may be useful to prioritize work (e.g. to select hot-spots that deserve additional attention), and as such may help to reduce crime (cf. Makkai et al. 2004), the information products offer little insight in the structure of criminal phenomena, the functioning of criminal networks, or the distinguishing signals that indicate (red-handed) criminal activity.

Before we illustrate how policing practice can benefit from products such as augmented reality that prescribe contextual actions (complementing the existing products that confine themselves to descriptions of the outer world), one should notice that the epistemological basis of the NIM and ILP, and thus, these products, is highly positivistic. From a positivist perspective, data, information, knowledge, and intelligence are treated as objects that can be obtained, recorded, enriched, analyzed, and disseminated. Moreover, the positivist line of reasoning is that (digital) data is the start of a chain from data to information to knowledge to actionable knowledge (intelligence) (Carter 2004; Kop and Klerks 2009; Ratcliffe 2008a; Williamson 2008). This data-driven logic can easily be reversed to a knowledge-driven logic. Indeed, knowledge must exist before information needs can be articulated, after which data can be collected, structured, and analyzed (Tuomi 1999). Both forms of logic, however, treat data, information, knowledge, and intelligence as concrete classes of objects, which after forming are little affected by context, personal interpretations, the social negotiation of meaning, or time. In the resulting mechanistic view attention is lop-sided to processing explicated, decontextualized, and encoded data. In practice this comes at the cost of losing the social richness of 'soft' knowledge (implicit and tacit forms of knowledge) (Innes et al. 2005). This negatively affects what information should be about: inform action. As we consider a positivist perspective on data, information, knowledge, and intelligence a too limitative foundation for creating augmented realities, we propose to reorient these key-concepts from an interpretive-constructivist epistemological perspective.

9.2.2 Reorienting Data, Information, Knowledge, and Intelligence

In this section we make an attempt to describe the ILP key-concepts from an interpretive-constructivist perspective. The descriptions should be viewed as ideal-types in a Weberian sense. Hence, a real-world instance will often represent a combination of these ideal types.

Data

Based on ones interests one may decide to encode signals by using e.g. text, sound recordings, photographs, films, or models. Although these artifacts may serve different purposes (e.g. art), within the context of KBP they represent data. Thus, data are intentionally created and structured representations (or abstractions) of reality as we observe it. As perspective, methods and means used in the process of creating the data are related to an idiosyncratic purpose, the abstractions may or may not be useful for other purposes, or within other contexts. For example, a police register of Police Reports may be used to study criminal patterns (or generalizations). The design of the data collection, however, would have been quite different if it had had that purpose from the start. Moreover, it should be noted that data is dynamic as it may be deleted or become illegible, as keys, formats, or carriers get lost, obsolete, or damaged.

Information

Taking an interpretive-constructivist perspective (cf. Orlikowski 2002; Tsoukas 2000), information is viewed primarily from the perspective of informing. The process of informing may be intentional or unintentional, successful or not successful. For example, accidentally overhearing a conversation of your neighbors may inform you about an event (unintentional); a report designed to inform a management team about the state of affairs may inform some, but leave others in confusion (unsuccessful); observing your colleague may inform you about his mood today, but may as well leave you wondering (unintentional and unsuccessful), et cetera. In all cases the information process involves the transmission of signals (or data) from an information source (e.g. person, text, images, or sounds), and the registration and interpretation of these signals by a receiver (Shannon 1948). While these signals or data may be informational to some, they may be non-informational or even confusing to others, or interpreted in another fashion than intended by the sender. Reasons may include differences in idiosyncratic perspectives on reality, a lack of interest or attention, and the lack of knowledge or trained senses to notice the difference between the various signals or data. Moreover, the meaning, intent, or validity of data sent more then once might shift (Shannon 1948). Thus, information is highly idiosyncratic, context specific, and dynamic. It is based in differences that make a difference (Bateson 1979) for a given person or group in a given context at a given time. Moreover, it represents that selection of the data that modifies our state of knowing (Boisot and Li 2005) and informs action (Orlikowsi 2002).

Knowledge

Following an interpretive-constructivist perspective, knowledge is viewed from the perspective of knowing to emphasize the role of know-how (Dean et al. 2008; Orlikowski 2002; Tsoukas 2000) in addition to know-what and know-why in acquiring and using knowledge through practice and experience (Dean et al. 2008; Gottschalk et al. 2009). In this perspective knowing how and practice are mutually constitutive, thus dynamically co-evolve through time (Nissen 2006; Orlikowski 2002). For example, knowing how to recognize a criminal in action may include the identification of a number of indicators that may be used to augment reality. As many officers will confirm, however, as soon as criminals learn about these indicators, they will try to conceal them or change their modus operandi all together. As a result, the recognition of criminals in action is to be regarded as a highly dynamic and knowledge-intensive game of 'cat and mouse'. As experience-records are idiosyncratic, so is the knowledge gained (and forgotten) in the process. It can only in part be encoded in generalized knowledge-rules. Moreover, some forms of knowledge (also called ecological knowledge (Walsh and Ungson 1991)) can only be remembered through direct interaction with the environment (one may recognize the case of a forgotten pin-code which may be remembered by visualizing the keyboard and remembering the pattern while typing it). Thus, knowledge is highly idiosyncratic, context specific, and dynamic.

Intelligence
Intelligence, defined by most police organizations as actionable knowledge, will typically proof to be actionable within a given context and time frame in the creative process of for instance the construction of a tactical plan to gather evidence or execute an arrest. As a consequence, intelligence can only be identified as intelligence in case-specific processes of planning, while it may lose its status of intelligence as the opportunity passes. Thus, like information and knowledge, what constitutes intelligence is idiosyncratic (related to a person or professional role), contextual, and dynamic.

9.3 Knowledge-Based Policing

9.3.1 The Need for a New Foundation

Like NIM and ILP literature, current KBP-literature has a highly positivist inclination, reflected in the positioning of knowledge as just another processing level in the data-information-knowledge-intelligence chain (cf. Williamson 2008). The consequence is that 'old' knowledge (of case-specific criminal investigations) became detached from 'new' knowledge (digital forms of pattern analysis). The division of work effectuates this. Where police investigation officers are dealing with 'old' knowledge, (sworn) desk officers deal with 'new knowledge' (Ratcliffe 2008). This division is not only effectuated in role, but also in organizational structure (Gottschalk 2008; Kop and Klerks 2009). The Dutch police consists of a large 'operating core' of police officers and criminal investigators that are organized in a hierarchical and geographical manner (neighborhood, district, region, (inter-) national). Following the NIM-structure (Centrex 2005) each level has a selection of facilities dealing with information and intelligence, based on the economical principles of specialization. Although this division may be justifiable from a specialization perspective, it comes at the cost of integration (Galbraith 1973). We hypothesize that the dominant positivist epistemological stance hinders the alignment and integration of old and new knowledge, which may explain why many police officers regard ILP as inadequate (KLPD 2011).

The solution as we see it does not rest in collecting more data, defining better information products, or improving the chain from producer to consumer (also known as sequential collaboration (Puonti 2007)), as may be expected from an positivist perspective. Instead, we advocate to adopt an interpretive-constructivist perspective and start approaching knowing and practice as mutually constituent (Orlikowski 2002). This means that forms of cooperation among desk-officers and executive police officers need to be stimulated to start the process of mutual informing, learning, and acting, i.e., to learn to work as a team (also known as parallel collaboration (Puonti 2007)). Key in this process is the identification and utilization of boundary objects, which we will discuss next.

9.3.2 *The Role of Boundary Objects in Augmented Reality*

Sharing the same epistemological basis is not sufficient for successful collaboration across different fields of discipline, such as information officers organized in information-units, executive police officers organized in squads, and technicians configuring augmented realities. To increase understanding across different fields of discipline people deploy boundary objects (Bechky 2003; Carlile 2002; Star and Griesemer 1989). A boundary object is an artifact that has meaning across practices and as such has the potential to improve coordination and synthesis across heterogeneous disciplines. Building on Star (1989), Carlile (2002) distinguishes three types of boundary objects, i.e. repositories; standardized forms and methods; and objects, models, and maps. Although all three play a role in augmenting reality, for the purpose of this chapter we elaborate on the last category, and more specifically, on models. This category of boundary objects is fundamental in structuring augmented realities.

We define a model as a generalized abstraction of a real world phenomenon, such as burglary, cargo theft, or drug trafficking. A phenomenon can be described in terms of e.g. 'business' processes and supply chains, social networks, favorite locations, and modus operandi. Such knowledge is typically distributed across different fields of discipline. Building a shared model aids participants to contribute what they know about the phenomenon through experience, observations, experiments, or desk studies. The model can then be used to (jointly) devise tactics, methods and means to approach the phenomenon and to find ways in which reality may be augmented. One such means is the application of profiles. Following Marx and Reichman (1984:4) we define profiling as a method 'to correlate a number of distinct data items in order to assess how close a person or event comes to a predetermined characterization or model of infraction'. Thus, each model may be translated into a number of (contextualized) profiles. If these profiles are used to augment reality, they have the potential to selectively make police officers 'in the field' aware of ongoing criminal activity and direct their attention accordingly.

To prevent boundary objects from becoming static and detached from practice, both models and profiles need to be subject of constant debate, stimulating the exchange of lessons learned, the creation of new intervention strategies and tactics, and the formulation of actionable hypotheses that can be tested in policing practice.

9.3.3 *Realizing the Augmented Reality Potential*

For the purpose of this chapter we define reality as the real world as one conceives it through ones natural senses, while virtuality (or artificial reality (Kruger 1991)) represents a fully (re)constructed, or synthesized world. Following Migram et al. (1994:283) we define augmented reality as 'augmenting natural feedback to the operator with simulated cues'. This is achieved by techniques that overlay reality with a stream of computer-synthesized data (virtual reality) (Fritzmaurice 1993). Although many characterizations (or indicators) of criminal phenomena can only be uncovered through interaction, some indicators may be detectable by means of technological sensors. These are potential candidates for augmenting reality. Examples include sensors for measuring weight, heat, speed, direction or route, or sensors which can be used to recognize texts on license plates, voices or faces. As policing takes place in public space, sensors used to augment reality need to be either integrated in the personal gear or equipment of police officers, or made available through a web of geographically distributed and interconnected sensors. In an ideal situation these sensors are seamlessly integrated in our natural environment, thus becoming transparent to natural persons (also known as ubiquitous computing (Weiser 1993)).

To facilitate the contextual presentation and interpretation of the real-time data-streams being produced by this sensor-network, knowledge-based systems are used (based on models) that can be customized to personal roles and contexts (using profiles) (cf. Feiner et al. 1993). The latter is of utmost importance to manage the volume of data being processed. For effectively augmenting reality, only those data have to be processed that are related to a particular officer who is working at a given location, at a given time, in a given context. Rather than storing all 'observations' of all sensors in perennial databases for offline analysis, for augmented reality only a selection of the ephemeral data-streams have to be analyzed in real-time. Based on explicated profiles these data-streams are used to assess how close a person or event comes to a predetermined characterization or model of infraction (Marx and Reichman 1984:4), thus rigorously endorsing the principle of ‘select before you collect’ (Jacobs 2005). Fitting a profile oftentimes does not provide sufficient legal grounds for treating someone as a criminal suspect. Like natural observations, however, synthesized observations just aid officers ‘to select cases worth inspection’ (Holgersson and Gottschalk 2008).

Augmenting reality in such real-time real-life environment is complex. This complexity is caused by several factors, including technical, financial, organizational, cultural, and not in the least of juridical and ethical factors. To name a few, networks of distributed sensors have to be integrated in the (fortified) ICT-network of the police, while the location and configuration of the sensors need to be attuned to the (fluid) criminal phenomenon under study; Police officers have to learn how to mentally integrate the augmented part of reality with their own observations and common sense, while the organization has to learn how to contextualize signals from their sensor-network and organize a response. Last but not least, the application of profiles has to be incorporated in the legal framework of policing. This includes the organization of mandates to act upon synthetic observations, defining the legal status of the data being processed, and the justification for breaching privacy due to the processing of person-related data. These particularities and the potential value of augmented reality in policing practice are illustrated in the grey box below.

Large numbers of drug-seeking tourists have been causing major problems in the public domain in the Dutch city of Maastricht for years. Traveling up and down the nearby borders of Germany and Belgium they flooded the city, visiting Dutch coffee shops, looking for drugs. These large numbers of customers attract criminal groups like drugs-traffickers and hard-drug retailers. As coffee shops, selling small amounts of soft drugs, are legal under Dutch law, local police have limited means to stop this criminal expansion. Retailing grams of cannabis and other drugs to thousands of tourists, however, requires a supply-chain of larger quantities.

As the problem had a clear community-transcending character, meetings were organized between local and national police officers to share knowledge about the phenomenon and discuss intervention strategies. Participants included criminal investigators, neighborhood police officers, highway patrol officers, and information and intelligence specialists. Through these meetings it became clear that Rotterdam was an important distribution center for heroine and other drugs. Moreover, it appeared that certain groups had specialized in import or retail, while trafficking the drugs from hidden stashes to retailers was the domain of other groups. While the import of larger quantities of drugs is very irregular and well hidden, and because the ultimate retail of small amounts of soft drugs is legal under Dutch law, it was reasoned that if the police would be able to discriminate between normal traffic and drug trafficking the criminal chain could potentially be most vulnerable during transport. Given the limited stock and high turnover of coffee shops, trafficking would be routine. Moreover, if the network of drug-traffickers could be made visible, it could provide clues about the location of drug-stashes, middlemen, and routes.

Follow-up sessions resulted in the construction of a 'drug-trafficker model' and profiles consisting of lists of indicators that officers could apply in specific contexts. Aided with these profiles several control actions on highways were organized, stopping and checking hundreds of vehicles. The results were very poor. Just grams of heroine were found. Clearly, human senses were not well suited to discriminate drug-trafficking behavior in large traffic flows, while most indicators were only assessable after a vehicle was stopped. As a consequence, the next question was how the spatial behavior of drug-traffickers could be discriminated from other vehicles 'in the flow' (with a density of circa 4000 per hour). What data was to be assessed and analyzed and how could this contribute to the ability of police officers to discriminate red-handed drug-traffickers from other travellers?

The solution was found in using a real-time complex event processing system, fed by live data-streams of automatically read license plates, assessed at four strategically chosen points along the route. One of the constructed profiles was aimed at detecting vehicles that travelled to and from Rotterdam and Maastricht within short periods of time, a pattern that investigation officers knew to be typical for drug-traffickers. The profile was further strengthened by combining this information with a list of license plates of vehicles that frequented coffee shops in Maastricht. This list could also have been generated automatically, if sufficient sensors would have been available. For the above profile data needed to be kept in memory for a short period only (number of hours). To reduce the impact of privacy invasion, reads of license plates that within this period did not score on the profile were automatically removed. Moreover, as the profile was based on time-spatial behavior only, the profile was discrimination-free.

At the beginning of the operation more seasoned policemen were sceptical about the idea that technology could complement their sentience. Initially, thousands of cars passed our sensors, with no results. After about an hour the first alert was generated. It appeared to be an ambulance of the Maastricht University Hospital, which had just delivered a patient at the Rotterdam Hospital. The driver was not amused, nor was the motor-policeman who carried out the selection, while "knowing perfectly well that this is not what he was looking for". At the end of that day however, out of as few as ten vehicles that were stopped as many as six proved to be serious drugs-traffickers. Caught red-handed, each carrying over a kilogram of soft and hard-drugs. These results convinced even our most sceptical colleagues. As one of them stated, "It is almost like Christmas day, the presents are delivered and we only have to unwrap them".

One week later, at a similar occasion, a taxi was selected based on information generated by the profile. Under normal circumstances taxis are never stopped. This time, too, the intuitive response was to let it pass. But it met all criteria of the profile. Additional information from a cop with local knowledge indicated that despite second thoughts the taxi could be worth inspection. It resulted in a find of over 1.7 kilograms of hard-drugs.

After using this profile for a sustained period, results started to decline. The drug-traffickers started to deviate from their usual routes, thus avoiding the temporal sensor-network of the police. The most logical route, however, had been compromised, forcing them to take deviant (and a bit awkward) routes. If the police would be able to deploy sensors on these routes as well, their detection would be easy.

The operations did fuel a healthy public debate about the application of sensor-technology by the police, the mandate of the police to stop a vehicle based on automated knowledge-rules, and the violation of privacy regulations due to the processing and alleged storage of large quantities of privacy-related data. Although the profile-based approach was approved in court, it was concluded that current law did not provide sufficient clarity for using augmented reality applications in police operations.

9.4 Discussion

Policing science distinguishes various knowledge- and intelligence-disciplines that are specialized in dealing with various abstraction levels, focal areas, and analysis techniques (Gottschalk 2008; Holgersson and Gottschalk 2008; Innes et al. 2005; Ratcliffe 2008). Discussing KBP, profiling, and augmented reality in relation to these disciplines is beyond the point and scope of this chapter. Instead, we limit the discussion to the implications of our contribution to the emerging concept of KBP with respect to the handling of large data-collections in relation to the real-time discrimination of criminal phenomena, including the impact on privacy.

9.4.1 Databesity: The Ever Present Hunger for Larger Databases

The positivist epistemological foundation of ILP and KBP falls short in acknowledging and dealing with the idiosyncratic, contextual, and dynamic nature of their key-concepts: data, information, knowledge, and intelligence. The positivist epistemological perspective is reflected in statements such as: 'information is data given meaning and structure' (Ratcliffe 2008a: 4). Although such definitions are straightforward and deeply ingrained in policing practice, in day-to-day business we observe that they tend to lead to discussions about format rather than informing, to unilateral processes of requesting and receiving information products (rather than dialogue), and to data-driven rather than knowledge-driven explorations. This data-driven focus often leads to something we call 'databesity', which we define as an urge to collect data for the sake of assembling (large) data collections in which –potentially– informative patterns may be found (cf. Innes et al. 2005). Finding and not finding these patterns urges to extend the data collection even more, either in depth or in breadth. Where the hunger of people with obesity, however, cannot be satisfied by eating (as the true cause is of another nature), so the data-hunger of organizations with databesity cannot be cured by collecting more data. Instead, KBP from an interpretive-constructivist perspective urges 'old' knowledge-type of questions to be raised when dealing with 'new' knowledge, such as: what is the problem to be solved? How can the criminal phenomenon be recognized? How are the criminals organized? What would be a good strategy to solve the problem? Who and what (including what data) are needed to execute the strategy? Approaching problems from this perspective leads to more focused data processing which is not limited to encoded data or computerized analysis techniques, but involves on-going sense making and natural world feedback based on behavioral and other cues. Indeed, to achieve augmented reality, ICT-affordances have to be integrated seamlessly into the construed reality of natural persons (Feiner et al. 1993).

9.4.2 Augmented Reality: Real-Time Processing of Data-Streams

Augmenting reality is a way for police organizations to bring ILP as close to practice as it can possibly get. To this end, however, focus has to be shifted from (persistent) databases to ephemeral data streams. The aim of creating and assessing data-streams is not to collect evidence, but to augment reality with a data-layer that complements the sentience and awareness of police officers, support sense making and decision making processes, and thus, inform action. Modelling augmented reality (i.e. configuring sensor-networks and data-layers) for such purpose is a continuous effort, taking into account the characteristics of the police officers, the phenomenon involved, its context, and its development through time.

Like shown in the grey box the creation of an augmented reality for fighting drug-trafficking was achieved by explicating and sharing knowledge of police workers, jointly developing hypotheses about the modus operandi of drug-traffickers, instructing officers to look out for related (mostly behavioral) signs, creating access to appropriate data-streams to identify generalized time-spatial behavior, and construct automated profiles to analyzing these live data-streams in order to create real-time feed-back for the police officers conducting the control. Little to none of this knowledge can be traced back to an existing database. But even if such data-collection would have been present, its predictive value would have been short-lived, as caught criminals learn fast. Short after the police successfully started to act on the hypothesized pattern, criminals adapted their modus operandi, creating new patterns that were slow to appear through the analysis of Police Report collections, if at all. Instead, criminal investigations, interrogation of caught drug-traffickers, qualitative analysis of Police Reports (rather than statistical), and common sense (read: experience) led to new understandings of criminal modus operandi. The result of the augmented reality experiment was that six from the ten selected vehicles deemed 'worth inspection' contained more then 1kg drugs – an unprecedented success for the Dutch police.

This being said, we do not claim that quantitative analysis of large data-collections is ineffective or of minor importance. To the contrary. We assert, however, that it is far from sufficient. At most it is complementary. For example, in the drug-trafficking case combining data sources and conducting social network analyses shed light on parts of the organizational structure that would have been overlooked otherwise (e.g. shared phone numbers, addresses, bank accounts). This effort contributed to the success of the operations because the analysts collaborated with the investigations team in a parallel fashion (Puonti 2007).

9.4.3 Developing an Ubiquitous Sensor-Network

As we have illustrated, augmented reality may already function with as few as 3 to 4 networked sensors. This augmented reality, however, can be extended substantially if more sensors can be used. This is not a matter of buying and deploying more sensors per se, as many (if not most) locations are already equipped with suitable sensors. Most of these sensors are government-owned, but few are interconnected. As a consequence one may observe that many locations are equipped with multiple systems, one for each governmental authority. We expect it to be a matter of time and economical sense before these sensors become interconnected.

Rather than fighting the rear-guard, we suggest to start thinking about formulating access and use regulations. As food for thought we suggest one initial measure. In contrast with current practice, in our view it would make sense to distinguish between the sensor, the data it produces, and the governmental authorities that are allowed to use the data (preferably in a real-time fashion, as we did). Where the format of the data and the physical location are determined by the sensor, the data-stream may be managed in terms of activation period and retention-time, while the use of the data by governmental authorities is determined by their legal mandate.

The governing principles of proportionality, subsidiarity, and linkage between ends and means (hereafter: the ruling juridical principles), rule out unrestricted access to and use of the network. Indeed, the ruling juridical principles signify that the means (i.e. mandates, resources, action) deployed by the police have to be proportional in relation to the offense, that there are no other means available with less impact, and that the means are deployed to reach a specified goal (e.g. restore order, catch the criminal). These ruling juridical principles imply that the use of means is always context-specific, never generic. Thus, when deploying a sensor-network to augment reality to combat (a specific form of) crime, all sensors that are activated through profiles need to contribute to this end. This is also in accord with the principle of 'select before you collect' (Jacobs 2005). Moreover, where citizens' related data is being assessed, the impact of large-scale privacy invasion has to be weighted against the ruling juridical principles, which is ultimately judged upon in court. Continuously collecting all data of all available sensors is obviously out of proportion.

9.4.4 Dealing with Privacy Invasion

In addition to the ruling juridical principles, all-inclusive access to the sensor-network is undesirable as it may lead to 'a feeling of omnipresent control' (Bentham 1843). Such feeling causes people to adjust their behavior, also known as 'chilling effect'. This effect is at odds with the right to be let alone, by some equated with privacy (Skousen 2002).

The invasion of privacy is absolute: it is invaded or it is not. Whether it is legal is case specific, involving the juridical principles of proportionality, subsidiarity, and the linkage between ends and means. Notwithstanding this harsh formulation it is possible to minimize the impact of privacy invasion. Within the context of augmented reality we distinguish three factors that together determine the impact of privacy invasion. The impact is influenced by the amount, detail and person-relatedness of the data being collected (we call this intrusion); the number of parties that may have access to this data (we call this spread); and the period that the data is being kept (we call this persistence). In the Dutch Data Protection Act, spread and persistence would be categorized under protection measures, which are aimed at minimizing the chance of unauthorized access and use of the data.

One of the reasons we propose to use ephemeral data-streams rather than persistent databases is that it allows for minimizing the impact of privacy invasion by minimizing spread and persistence, thus contributing to proportionality. Minimizing persistence contributes to 'the right to be forgotten' (Reding 2011), as well as to the prevention of 'function creep'. The latter is a concept of privacy scholars to denote that data obtained for one function tends to be used for other functions as well. By focusing on ephemeral data-streams, potential function-creep is limited to forward creep only (as data is removed from memory after real-time analysis). Moreover, with respect to spread, 'data controllers' (i.e. the police) 'must prove that they need the data rather than individuals having to prove that controlling their data is not necessary' (Reding 2011). In the Netherlands such proof is being approved by the Public Prosecution Officer and, in case of a trial, judged upon in court. Moreover, an important effect of this approval process is the self-correcting inclination to work with high integrity data and hypotheses.

As augmented reality, based on an interpretive-constructivist perspective, minimizes on intrusion, persistence and spread, with respect for privacy we present our version of KBP as a minimalist approach. Intrusion is being limited by the application of profiles, which means that the needed data is carefully selected before it is collected. Spread is being limited as the data is being tied to (immediate and localized) action. And persistence is being limited to the time required to complete the observation.

9.4.5 Dealing with Discrimination

Discrimination has a strong association with generalizations related to identity-related characteristics, such as race, ethnicity, religion, gender, social class, political affiliation, and so on, upon which action is illegal. Discrimination, however, can also be approached from a less-loaded mathematical angle, i.e. being able to differentiate.

Because in our case the synthesized part of augmented reality produces the leads to direct police attention, the impact of personal bias has to be eliminated during model and profile construction. Discriminating tendencies that are nonetheless encoded in the algorithms can further be neutralized by using corrective techniques to create discrimination-free classifiers (chapter 14). Most contributing to discrimination-free selection, however, is the fact that profiles are geared to detecting (time-spatial) behavior, rather than personal or social-economical characteristics (Alpert et al. 2005). Our rational is that being a drug-trafficker is not an offense: only the act of drug trafficking is. In the endeavor of identifying criminals red-handedly, reference to a single hotspot, hot moment, or hotshot observation may be used to strengthen a profile. The profile grows stronger, however, both in terms of effectiveness and in reduced bias, if multiple observations are used to determine behavioral pattern. Notwithstanding these efforts, behavioral profiling cannot completely prevent discrimination. For example, drug-trafficking related behavior, such as cruising a particular route, may also be characteristic to other (non-criminal) groups (Warren et al. 2006). In such cases profiles may produce too many false positives. This renders the profile less economical and, thus, mounts pressure to adjust the profile (which is true, of course, for all profiles).

9.4.6 Dealing with Group-Think

Group-think is a single-minded self-confirming pattern of thinking, not receptive for conflicting signals of the outside world (Cannon-Bowers et al. 1993; Janis 1972). The risk of group-think when working with augmented realities increases when feedback on profiles is not organized and subsequently used to update the profiles, or when a few dominant participants in model construction leave little room for others to discuss alternative explanations. Group-think shields police officers from identifying criminal behavior that does not fit their pattern of thinking. It reduces their creativity, their adaptivity, and, thus, their effectiveness. Measures to avoid group-think include diversifying the team (also in time), prevent dominant leadership, and building in randomness in the selection process (Cannon-Bowers et al. 1993). Moreover, police officers involved in creating layers of augmented reality as well as police officers making use of it during their operations need to nurture a critical attitude towards illegal discriminating bias that may have crept into their augmented reality. Just like they do in their not-augmented reality. As described by Thatcher (2005), failing to do so is bound to lead to 'trust-decay' in police operations.

9.4.7 Sustaining Trust

The way the police handles the above issues of data-collections, the sensor-network, privacy, discrimination, and group-think, all add up to trust: trust of police officers working within an augmented reality; and trust of citizen in police operations.

To start with the first, where in their daily practices police officers are utterly aware of the subtleties of language, context, and antecedents, this sensitivity seems to be absent when dealing with digital data (Innes et al. 2005). Artifacts, however, often lack context and history, unless such awareness functions are explicitly incorporated in its code. Ultimately an artifact acts as a 'finite state machine' (Arbib 1969): it does what it is programmed to do. If virtual reality is to be integrated into the reality of policing practices, these limitations of artifacts have to be engraved in the psyche of police officers whose reality is being augmented. During decision making each virtual representation of reality has to be validated and valuated against the context of that moment. Not complying with this baseline may severely impact the safety and security of police officers and the public, as decision making and action may be compromised by 'objective' yet erroneous signals. Thus, as (non-critical) artifacts are situated in practice, their agency has to be controlled by officers that are aware of the abilities and limitations of the artifacts used.

In democratic nations, citizens' trust in police operations is of utmost importance for the legitimacy of police organizations (Tyler and Wakslak 2004). At the moment most civilians (in the Netherlands) seem to experience the control measures of the government as soothing rather than undermining their sense of privacy (Boutellier 2007). This trust, however, has to be earned on a daily basis. By discussing the above issues we have tried to demonstrate that the development of augmented reality in police operations is a delicate enterprise. In our opinion, in contemporary society, in which the physical and the virtual world are becoming more and more interwoven, the question is not if the police should engage in augmented reality, but how it can do so in a responsible manner. To ensure that the benefits (catching more criminals through less control actions) outweigh the risks (culminating in public trust decay), this on-going development should be embedded in public debate. Moreover, efforts should be undertaken to formally regulate and supervise its use, obvious candidates being the Public Prosecutor and the Dutch Data Protection Authority (Cbp). To this end we discuss some legal issues related to acting within an augmented reality, followed by two considerations for future legislation.

9.4.8 Consequences for Legislation

Using augmented reality within the context of actions along roads is legally complicated. We will explain this based on the case described (see grey box).

In the Netherlands, the inspection of vehicles on roads is organized in the Road Traffic Law (Wegenverkeerswet). Possession of drugs, on the other hand, is organized in the Law on Opium (Opiumwet). If someone is suspected of carrying drugs, (s)he may be stopped for a check based on the Law on Opium. If there is no such suspicion, however, an officer is not allowed to evoke the Road Traffic Law for the sole purpose of conducting a drug-control. Such action would lead to *détournement de pouvoir* (misuse of powers), which often results in related evidence being excluded from the lawsuit. Thus, the augmented reality-adagio of 'aiding officers to select vehicles worth inspection' does not always hold. It only does if there are sufficient legal grounds for suspicion. If an officers observes a number of signs that (s)he relates to drug-trafficking, the officer has the right to approach that person as a formal suspect and use the mandates that follow from that suspicion. The question is whether clues detected by artificial sensors can lead to the same mandates. Of course, the value of a profile is as good as the intelligence used to make it. The drug-trafficking case, in which 6 of the 10 vehicles selected for inspection resulted in the detection of major drug offenses, proves that human intelligence of some experienced officers can successfully be used to augment the reality of many fellow officers.

At the moment of writing, the Ministry of Security and Justice (Ministerie van Veiligheid en Justitie) is drafting a law in which police officers are mandated to register license plate information using technology. To further the public debate on the incorporation of technology in policing we forward two thoughts for consideration. First, although in the drug trafficking case license plates have been used to recognize criminal behavior it should be considered to draft the law in more abstract terms. This would expand the range from the public road to other public domains such as railways, waterways, or cyberspace, Characteristics other than license plates (e.g. RFID) could be more but also less privacy intrusive, providing the police with alternatives to bring their approach in line with the juridical principles of subsidiarity, proportionality, and the linkage between ends and means. And second, neither current laws nor the law-in-preparation do provide for the forming of legal grounds for suspicion based on 'technological' assessments. The question is, of course, how purely technical these observations are. Indeed, in an augmented reality, based on an interpretive-constructivist epistemology, the division between the physical and the virtual world is diminishing. As a consequence, the question should not be whether an observation is physical or synthetic, but up to what degree the observation can be verified and trusted.

9.5 Conclusion

Like Ratcliffe, we believe that KBP 'requires police leaders to learn and embrace a new way of thinking about knowledge' (2007:2). In this chapter we made this plea specific by proposing to adopt an interpretive-constructivist epistemological perspective, reformulating the key-concepts of ILP and KBP accordingly, and forward boundary objects as means to 1) help bridging the gap between different fields of discipline, and 2) serve as knowledge structure for the creation of augmented realities. Moreover, what counts for police leaders is just as true for society at large: augmenting reality in police operations has consequences that should be considered with care. In this chapter we provided the first sketches of the consequences of augmented reality for large-scale data-processing, the protection of privacy, group-think, discrimination, and trust. It is our belief that the principles forwarded in this chapter represent a minimalist approach with respect for privacy.

Acknowledgments. The valuable comments and suggestions of the two anonymous peer-reviewers and Jaap-Henk Hoepman on earlier drafts have greatly contributed to the quality of this text. All remaining errors are the authors' responsibility.

Abbreviations

ANPR	Automatic Number Plate Recognition
CBP	College Bescherming Persoonsgegevens
ILP	Information-led policing
KBP	Knowledge-based policing
KLPD	Dutch National Policing Services Agency
NIM	National Intelligence Model
OCR	Optical Character Recognition
RFID	Radio Frequency Identifier

References

Abrio. Informatiegestuurdepolitie: sturen op resultaat. Abrio, Houten (2005)

Alpert, G.P., MacDonald, J.K., Dunham, R.G.: Police suspicion and discretionary decision making during citizen stops. Criminology 43(2), 407–434 (2005)

Arbib, M.A.: Memory limitations of stimulus-response models. Psychological Review 76, 507–510 (1969)

Bateson, G.: Mind and nature: a necessary unity, advances in systems theory, complexity, and the human sciences. Hampton Press (1979)

Bechky, B.A.: Sharing meaning across occupational communities: the transformation of understanding on a production floor. Organization Science 14(3), 312–330 (2003)

Bentham, J.: The panopticon writings. In: Bozovic, M. (ed.) Republished, Verso, London, pp. 29–95 (1843-1995)

Boisot, M., Li, L.: Codification, abstraction, and firm differences: a cognitive information-based perspective. Journal of Bioeconomics 7, 309–334 (2005)

Boutellier, H.: Nodaleorde: veiligheid en burgerschap in eennetwerkmaatschappij. InaugureleRede 19 September 2007. VrijeUniversiteit Amsterdam (2007)

Brodeur, J.P., Dupont, B.: Knowledge workers or "knowledge" workers? Policing and Society 16(1), 7–26 (2006)

Cannon-Bowers, J.A., Salas, E., Converse, S.: Shared mental models in expert team decision making. In: Castellan Jr., N.J. (ed.) Individual and Group Decision Making, pp. 221–246. Lawrence Erlbaum Associates, Hillsdale (1993)

Carlile, P.R.: A pragmatic view of knowledge and boundaries: boundary objects in new product development. Organization Science 13(4), 442–455 (2002)

Carter, D.L.: Law enforcement intelligence: a guide for state, local, and tribal law enforcement agencies. Washington, DC: Office of Community Oriented Policing Services, U.S. Department of Justice (2004)

Castells, M.: The information age, economy, society and culture: the rise of the network society, Second revised edn. Blackwell Publishers (2000)

Centrex: Guidance on the National Intelligence Model. ACPO (2005)
Dean, G., Fahsing, I.A., Glomseth, R., Gottschalk, P.: Capturing knowledge of police investigations: towards a research agenda. Police Practice and Research 9(4), 341–355 (2008)
Feiner, S., MacIntyre, B., Seligmann, I.: Knowledge-based augmented reality. Communications of the ACM 36(7), 53–62 (1993)
Fritzmaurice, G.W.: Situated information spaces and spatiality aware Palmtop computers. Communications of the ACM 36(7), 39–49 (1993)
Galbraith, J.R.: Designing complex organizations. Addison-Wesley Publishing Company, Massachusetts (1973)
Gottschalk, P.: Organizational structure as predictor of intelligence strategy implementation in policing. International Journal of Law, Crime and Justice 36, 184–195 (2008)
Gottschalk, P.: Knowledge management in policing: enforcing law on criminal business enterprises. Hindawi Publishing Corporation, New York (2009)
Hert, P., de, H.W., Vis, T.: Intelligence led policing ontleed. Tijdschriftvoor Criminologie 47(4), 365–375 (2005)
Holgersson, S., Gottschalk, P.: Police officers' professional knowledge. Police Practice and Research 9(5), 365–377 (2008)
Innes, M., Fielding, N., Cope, N.: The appliance of science? The theory and practice of crime intelligence analysis. British Journal of Criminology 45, 39–57 (2005)
Jacobs, B.: Select before you collect. Ars Aequi 54 (2005)
Janis, I.L.: Victims of groupthink: a psychological study of foreign-policy decisions and fiascoes. Houghton Mifflin College, Boston (1972)
KLPD: Tussenraportage II: Flexibele teams. Internal report KLPD (2011)
Kop, N., Klerks, P.: Doctrine intelligencegestuurdpolitiewerk. Politieacademie. LectoraatRecherchekunde, Apeldoorn (2009)
Kruger, M.: Artificial Reality H. Addison-Wesley, Reading (1991)
Lint, W.: Intelligence in policing and security: reflections on scholarship. Policing and Society 16(1), 1–6 (2006)
Maguire, M.: Policing by risks and targets: Some dimensions and implications of intelligence-led crime control'. Policing and Society 9, 315–336 (2000)
Makkai, T., Ratcliffe, J.H., Veraar, K., Collins, L.: ACT recidivist offenders. Research and Public Policy Series 54, 83 (2004)
Marx, G.T., Reichman, N.: Routinizing the discovery of secrets: computers as informants. American Behavioral Scientist 27(4), 423–452 (1984)
Migram, P., Takemura, H., Utsumi, A., Kishino, F.: Augmented reality: a class of displays on the reality-virtuality continuum. Telemanipulator and Telepresence Technologies 2351, 282–292 (1994)
Nissen, M.E.: Harnessing knowledge dynamics: principled organizational knowing and learning. IRM Press (2006)
Orlikowski, W.J.: Knowing in practice: enacting a collective capability in distributed organizing. Organization science 13(3), 249–273 (2002)
Puonti, A.: Foreword. In: Gottschalk, P. (ed.) Knowledge Management Systems in Law Enforcement, Idea Group Publishing, Hershey (2007)
Ratcliffe, J.H.: Integrated intelligence and crime analysis: enhanced information management for law enforcement leaders. In: COPS, Police Foundation, Washington DC (2007)

Ratcliffe, J.H.: Intelligence-led policing. In: Wortley, R., Mazerolle, L., Rombouts, S. (eds.) Environmental Criminology and Crime Analysis, Willan Publishing, Cullompton (2008a)

Ratcliffe, J.H.: Knowledge management challenges in the development of information-led policing. In: Williamson, T. (ed.) The Handbook of Knowledge-Based Policing: Current Conceptions and Future Directions, John and Wiley and Sons, Chihester (2008)

Reding, V.: Privacy platform: the review of the EU data protection framework. Press Releases, Speech/11/183, March 16 (2011)

Shannon, C.E.: A mathematical theory of communication. Reprinted with corrections from The Bell System Technical Journal 27, 379–423, 623–656 (1948)

Skousen, M.: The right to be left alone. Ideas on Liberty, pp. 4–5 (May 2002), http://www.fee.org/pdf/the-freeman/skousen0502.pdf (accessed August 30, 2011)

Star, S.L.: The structure of ill-structured solutions: boundary objects and heterogeneous distributed problem solving. In: Huhns, M., Gasser, L. (eds.) Readings in Distributed Artificial Intelligence, Morgan Kaufman, Menlo Park (1989)

Star, S.L., Griesemer, J.R.: Institutional ecology, "translations" and boundary objects: amateurs and professionals in Berkeley's Museum of Vertebrate Zoology, 1907-39. Social Studies and Science 19, 387–420 (1989)

Thatcher, D.: The local role of homeland security. Law and Society Review 39, 635–676 (2005)

Tsoukas, H.: Knowledge as action, organization as theory: reflections on organizational knowledge. Emergence 2(4), 104–112 (2000)

Tuomi, I.: Corporate knowledge: theory and practice of intelligence organization, Helsinki, Metaxis (1999)

Tyler, T.R., Wakslak, C.J.: Profiling and police legitimacy: procedural justice, attributions of motive, and acceptance of police authority. Criminology 42(2), 253–281 (2004)

Walsh, J.P., Ungson, G.R.: Organizational memory. The Academy of Management Review 16(1), 57–91 (1991)

Warren, P., Tomaskovic-Devey, D., Smith, W., Zingraff, M., Mason, M.: Driving while black: bias processes and racial disparity in police stops. Criminology 44(3), 709–738 (2006)

Wbp, Wet beschermingpersoonsgegevens. Article 12, http://wetten.overheid.nl/BWBR0011468/geldigheidsdatum_17-11-2011 (accessed November 17, 2011)

Weiser, M.: Some computer science issues in ubiquitous computing. Communications of the ACM 36(7), 75–84 (1993)

Williamson, T. (ed.): The handbook of knowledge-based policing, current conceptions and future directions (2008)

Chapter 10
Combining and Analyzing Judicial Databases

Susan van den Braak, Sunil Choenni, and Sicco Verwer

Abstract. To monitor crime and law enforcement, databases of several organizations, covering different parts of the criminal justice system, have to be integrated. Combined data from different organizations may then be analyzed, for instance, to investigate how specific groups of suspects move through the system. Such insight is useful for several reasons, for example, to define an effective and coherent safety policy. To integrate or relate judicial data two approaches are currently employed: a data warehouse and a dataspace approach. The former is useful for applications that require combined data on an individual level. The latter is suitable for data with a higher level of aggregation. However, developing applications that exploit combined judicial data is not without risk. One important issue while handling such data is the protection of the privacy of individuals. Therefore, several precautions have to be taken in the data integration process: use aggregate data, follow the Dutch Personal Data Protection Act, and filter out privacy-sensitive results. Another issue is that judicial data is essentially different from data in exact or technical sciences. Therefore, data mining should be used with caution, in particular to avoid incorrect conclusions and to prevent discrimination and stigmatization of certain groups of individuals.

10.1 Introduction

In the Netherlands, many organizations work together to ensure the enforcement of law and public safety of people. Each of these organizations covers a specific area in the field of crime and law enforcement. For instance, the police focus on reported crime and hand over suspects to the prosecution service. The Public Prosecution Service then decides whether to prosecute or drop a case. The court can either convict or acquit a suspect, and may impose sanctions such as imprisonment. When a sentence is pronounced by court, the execution of sanctions

Susan van den Braak · Sunil Choenni · Sicco Verwer
Research and Documentation Centre (WODC) of the Ministry of Security and Justice, The Netherlands
e-mail: {s.w.van.den.braak,r.choenni}@minjus.nl, siccoverwer@gmail.com

B. Custers et al. (Eds.): Discrimination & Privacy in the Information Society, SAPERE 3, pp. 191–206.
springerlink.com

follows. Together, these steps from reporting a crime to the execution of sanctions are referred to as the *criminal law chain*. This chain thus consists of four phases: investigation, prosecution, trial, and execution. The police, prosecution service, courts, and the organizations that execute sanctions collaborate in this chain. Each organization registers relevant data, for instance, about the case and the suspect, in its own data source.

To define an effective and coherent safety policy, policymakers have a practical need for statistical insights into the registered data.[1] Such insights can only be gained by relating and integrating the data in a coherent manner. For instance, when data from the different co-operating organizations are integrated and compared, it can be investigated how specific groups of suspects or criminal proceedings move through the chain. Also, by monitoring flows within or between organizations in the chain, policymakers are able to observe whether there are potential problems in a certain part of the chain.

In the Netherlands, combined crime data have already been distributed offline (in book form) for several years.[2] Although the statistical yearbook is very useful in its current form, there is a growing demand for online data from different groups of users. Therefore, several attempts have been made to develop tools or information systems that collect and process safety-related data from relevant sources and present them in an integrated and uniform way to the users.[3] Such tools obviously have potential, but should be developed with care, as they may also provoke undesired effects. One of the core issues here is the protection of the privacy of individuals. Data should be processed, collected, and combined in a way that respects privacy laws and regulations. In general, privacy has a subjective nature and is open to different interpretations depending on its context. In the context of public safety, privacy is primarily focused on the non-disclosure of the identity of individuals. A related issue is the discrimination of groups of individuals, that is, the prejudiced treatment of individuals because they belong to a certain group. To minimize the risk of discrimination or stigmatization, combined crime data should be presented and analyzed with caution.

In this chapter, it will be described how judicial data can be collected, combined, and analyzed such that the privacy of individuals in society is not violated. It is explained that although IT offers great potentials to automate the collection and combination of data, still a significant manual effort is required to ensure data quality and to avoid undesired effects. A dataspace approach is presented that allows one to efficiently relate and exploit data from different sources. It is demonstrated how the information needs of judicial policymakers can be fulfilled using this approach. To analyze data, besides traditional statistical techniques, contemporary techniques such as data mining can be employed. However, it is argued that the straightforward application of such data analysis techniques on judicial

[1] Choenni, S., van Dijk, J. & Leeuw, F. (2010), Choenni, S. & Leertouwer, E. (2010), Kalidien, S., Choenni, S. & Meijer, R. (2009), Kalidien, S., Choenni, S. & Meijer, R. (2010).

[2] De Heer-de Lange, N.E.& Kalidien, S. (2010), Kalidien, S. & De Heer-de Lange, N.E. (2011).

[3] Choenni, S. & Leertouwer, E. (2010), Choenni, S., Kalidien, S., Ariel, A. & Moolenaar, D. (2001).

data is not without risk. The main reason for this is that the nature of these data is essentially different from the nature of data in exact or technical sciences.

The remainder of this chapter is organized as follows. Section 10.2 is devoted to a brief description of the major databases in the Dutch criminal justice system. In Section 10.3, it is described how data from these databases are currently collected and combined. In this section two approaches to combining judicial data are presented: a data warehouse approach and a dataspace approach. Section 10.4 elaborates upon the problems that may occur in the data integration process due to the nature of crime data. Subsequently, in Section 10.5, potential privacy-related risks of integrating and presenting crime data are described and methods that enforce privacy laws and regulations are listed. Section 10.6 explains how combined crime data may be analyzed and which risks are entailed by applying data analysis techniques to them. Finally, Section 10.7 concludes this chapter.

10.2 Databases in the Dutch Criminal Justice System

The Dutch criminal law chain consists of various organizations, each of which operates relatively autonomously and independently. This means that each organization registers data in its own way and in its own operational system. The most important databases of these organizations are described below.

The national database of the Dutch police is called the Identification Service System (*Herkenningsdienstsysteem*, HKS). HKS contains information about crime reports and suspects. Additional information is provided by Statistics Netherlands (*Centraal Bureau voor de Statistiek* (CBS), a national institute that provides statistical information). CBS Police Statistics also contains information about crime reports and suspects.

Information about judicial cases is stored in the registration system of the Public Prosecution Service (*Openbaar Ministerie* (OM); the information system is called OM-data) and in CBS Court Statistics. Note that these databases register information on a case level, while the police databases register crime reports. As more than one crime report may be handled in a single case, numbers obtained from these different sources should be combined and compared with care.

Sanctions are registered by the different organizations involved in the execution of sanctions. Among these organizations are the Custodial Institutions Agency (*Dienst Justitiële Inrichtingen*, DJI), the Child Care and Protection Board (*Raad voor de Kinderbescherming*, RvdK), after-care and resettlement organizations, and the Central Fine Collection Agency (*Centraal Justitieel Incassobureau*, CJIB). All of them have their own information system. For instance, DJI uses the Execution of Sanctions Program (*Tenuitvoerleggingprogramma*, TULP) to register the duration of detentions, while the Dutch Probation and After-care Organization (*Reclassering Nederland*, RN) uses a Client Follow System (*Cliënt Volg Systeem*, CVS). A schematic overview of the databases maintained in the Dutch criminal law chain is given in Figure 10.1.

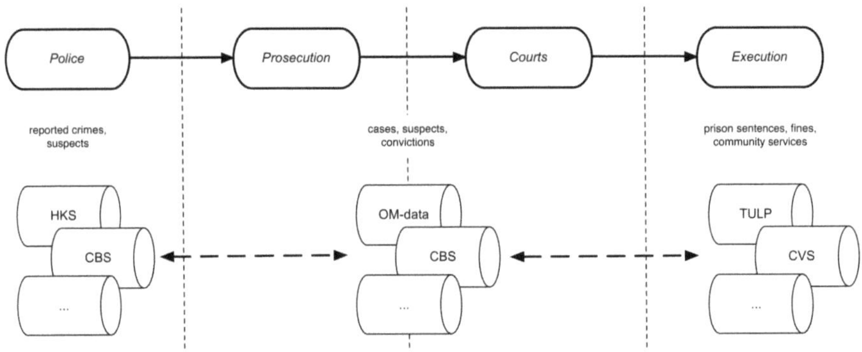

Fig. 10.1 Databases in the Dutch criminal law chain

10.3 Collecting and Combining Judicial Data

The database systems described above have in common that they contain data about individuals and their actions. Each of these individuals came into contact with the police or the criminal justice system. Each organization involved registers privacy-sensitive attributes such as name, address, and identifying numbers, but also other data regarding a person. The different databases thus store the same, or similar, information and, therefore, they are partially redundant. Consider, for instance, the database systems of the police and the prosecution service that both contain data about people who are suspected of a crime and about the crimes they supposedly committed. If someone is suspected of a murder, both the database of the police and the database of the prosecution service will contain information regarding the date and place where the body is found and (if known) the date and place where the murder is committed. Other information, however, is registered in only one of the databases. For example, the police database contains detailed information about the suspect (such as whether he is first offender or not), while the database of the prosecution service contains detailed information about the case (such as the sections of the law that were violated). This is due to the fact that the police and the organizations involved in the execution of sanctions are individual-oriented, while the prosecution service and the courts are case-oriented.

To perform their tasks in an effective and efficient manner, the police and justice organizations not only require access to their own data; they also have a great demand for a combination of relevant data from other organizations in the criminal law chain. Organizations with operational tasks (such as the police and the prosecution service) require combined data at an individual level, while organizations with strategic or knowledge transfer tasks (such as policymakers and criminologists) require data at a higher aggregation level.

As an example of the former, assume that a Public Prosecutor wants to prosecute a suspect for his actions, then all relevant data (from different sources) that pertain to this suspect should be collected and combined. In this way, the prosecutor can build the strongest case possible, because all information about the suspect is gathered; including evidence for the fact that he is a criminal. Thus, integrating data on an individual level involves data reconciliation, that is, the identification of data in different sources that refer to the same entity. In Subsection 10.3.1 it is shown how this can be established using a data warehouse approach.

Alternatively, policymakers need combined data at an aggregate level. They want to gain insight into the criminal law system as a whole, for instance, to answer the question of which kinds of suspects are brought to court and which kinds of cases are settled out of court. Such insight may be relevant to them in order to be able to define an effective policy. To provide them with this information, the different databases also have to be combined, but not on an individual level, in this case a higher level view is more useful as will be shown in Subsection 10.3.2. In this subsection, a dataspace approach will be presented in which aggregate data are related.

10.3.1 A Data Warehouse Approach to Combining Judicial Data

A data warehouse is a central repository of data collected from different sources.[4] These data are stored and structured in such a way that querying and reporting are facilitated. It provides a uniform data model for all data regardless of their source. Generally, a data warehouse consists of three layers that provide storage of the original data sources, integration, and access (see Figure 10.2). First, the raw data from different databases are extracted. Subsequently, these data are cleaned, transformed, and loaded into the data warehouse. The data warehouse then contains data from different databases that are combined and ordered. In addition, information about the data in the data warehouse is stored in a metadatabase. This database contains information about the sources and history of the data. Finally, as a last step, data from the data warehouse are provided to end-users through data marts. The key step in developing a data warehouse is data integration; therefore, data reconciliation is of crucial importance.[5]

The main problem with combining and integrating crime data is that only a few organizations with an operational task are allowed (by law) to combine data based on unique identifiers or a set of privacy-sensitive attributes. For this reason, before making crime data available for research purposes, privacy-sensitive attributes are stripped from the databases. Hence, for data reconciliation other overlapping information in the to-be-combined databases has to be exploited. This can either be information about the database schemata or information that is extracted from the database content. Furthermore, in order to be able to utilize this information, domain knowledge from experts is needed.

In practice, to establish whether two records from different database system denote the same object, the following general rule of thumb can be applied:[6] the larger the number of common attributes with the same values for two records from two different systems, the higher the chance that the records relate to the same

[4] Kimball, R. & Ross, M. (2002).

[5] Choenni, S., van Dijk, J. & Leeuw, F. (2010).

[6] Choenni, S., van Dijk, J. & Leeuw, F. (2010), Choenni, S. & Meijer, R. (2011).

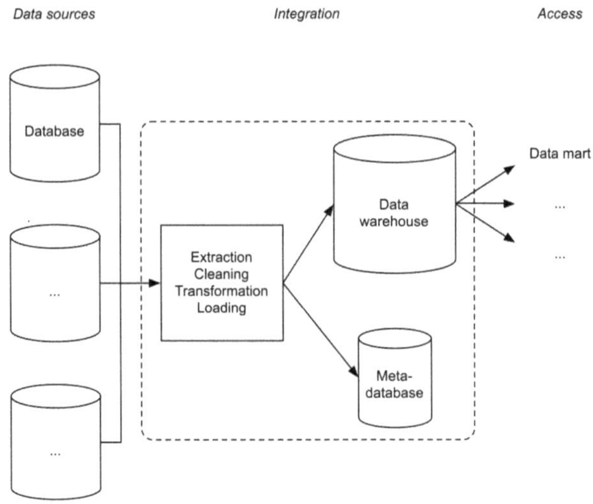

Fig. 10.2 An overview of the data warehouse approach

object in reality. Note that this rule of thumb requires that the selectivity factors of the common attributes are small.[7]

Example: An offender-oriented data warehouse
An example of a data warehouse in the Dutch criminal law chain is the offender-oriented data warehouse.[8] *In this data warehouse, data from different judicial databases (HKS and OM-data) were integrated by applying the rule of thumb explained above. Additionally, the data was structured and combined in such a way that all data relate to individuals.*

In the data warehouse the 'intersection' of the to-be-combined databases was exploited. HKS stores information on three entities: suspects, the official reports about them, and the offences of which they are suspected. OM-data also records information about suspects and offences. Additionally, it registers case-related information. Thus, the databases were integrated based on the attributes concerning the two common entities, that is, suspects and offences. To do so, the databases were compared to each other and the probability that two records relate to the same person based on common attributes was determined. While doing so, domain knowledge was considered, for instance, the fact that an offence is usually reported to the police on the same day as it is committed. The date of an official report in HKS was, therefore, considered to be the same as the date of an offence in OM-data.

As an example, assume that HKS contains a record relating to a person who resides in Amsterdam and in respect to whom an official report has been filed on September 1, 2010. Additionally, assume that OM-data contains a record of a

[7] Choenni, S., Blanken, H. & Chang, T. (1993).
[8] Choenni, S., van Dijk, J. & Leeuw, F. (2010), Choenni, S. & Meijer, R. (2011).

person residing in Amsterdam who committed an offence on September 1, 2010. Then, it is likely that both records concern the same person. Alternatively, if HKS would show that the date of the official report is unknown because it is not entered correctly, the probability would be considerably lower. Note that in the example the residence of the suspect is not very selective and that it is surely possible that multiple residents of Amsterdam commit an offence on the same day. If this is the case, additional or different attributes are needed to ensure that the records are combined properly. After all, if more attributes overlap, the probability that the two records denote the same person increases.

The data in a data warehouse can be made available through data marts. An example that is based on the offender-oriented data warehouse is the Drug Crime Data Mart which consists of a selection of the data concerning drug-related crime.[9] *This data mart can be used for analysis and reporting purposes, such as National Drug Monitor publications.*

10.3.2 A Dataspace Approach to Combining Judicial Data

In a dataspace approach,[10] also three layers are distinguished (see Figure 10.3): a dataspace layer, a space manager layer, and an interface layer. The dataspace layer contains a set of (cleaned) databases that are complement to each other and may be related. Although these databases are related there is no need for data reconciliation. Alternatively, the relations that exist between the databases are stored in a relationship manager in the space manager layer. This layer maintains data quality (the plausibility and consistency of the data) by providing rules to which the data must adhere. For this purpose the relationship manager contains different types of rules:

1. Rules to handle similar data coming from different sources.
2. Rules to deal with missing data.
3. Rules to allow for incomplete or tentative data.
4. Rules to record semantic changes in attributes.
5. Rules to filter out results that should not be shown to the user.
6. Rules to determine whether large deviations exist between past and future data or between values from the same or different databases.

All in all, a combined set of rules in the relationship manager serves to complete incomplete data sets, determine whether they are acceptable, and warn users when they are less reliable. The relationship manager also serves to minimize the chances of misinterpreting data. To do so, this layer maintains the relations between attributes in the different databases and keeps track of changes in the meaning of these attributes. Based on this history of changes, the space manager may decide to reorganize or convert a database, in particular if the semantics of major attributes changed considerably over the years.

---

[9] Choenni, S. & Meijer, R. (2011), Meijer, R., van Dijk, J., Leertouwer, E. & Choenni, S. (2008).

[10] Franklin, M., Halevy, A. & Maier, D. (2005).

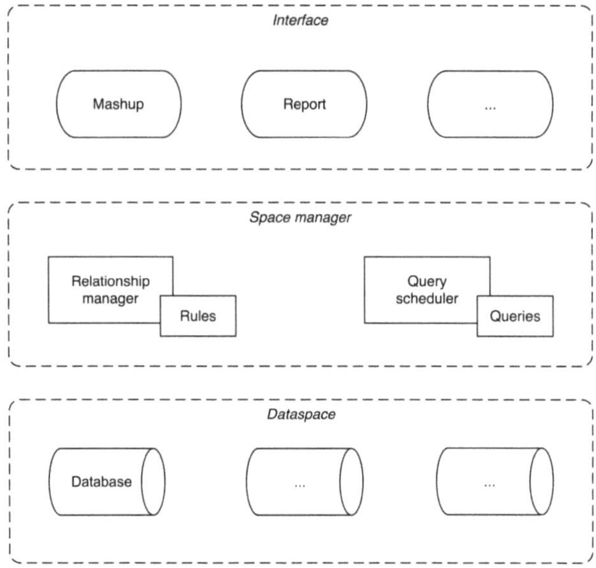

Fig. 10.3 An overview of the dataspace approach

Another task of the space manager (besides providing a relationship manager) is to serve as a communicator between the database and the user interface. As users define questions at the interface level, the query scheduler of the space manager decides which databases to query in order to answer each question. Once the answer is retrieved from the databases, it is displayed to the user through the interface. Before presenting the output, rules (of the fifth type) may be applied to check whether it can be shown to the user. This can, for instance, be used to preserve the privacy of individuals as will be explained in Section 10.5.

The interface layer not only contains mashups of crime data, but also provides features that are more tailored to the needs of specific users. An example is a publishing on demand module which provides users with the possibility to generate and print reports. Such a module should insert automatically updated tables and graphs in a preformatted report. A feature of this kind is particularly useful for standardized research reports (see, for instance, Meijer et al., 2008).

Example: A public safety monitor

An example of the dataspace approach at work in the Dutch justice system is the public safety monitor.[11] *This monitor shows the development of the input and output of cases in the different organizations in the criminal law chain. In addition, it provides a comparison of the actual data with forecasts. The data in the monitor's dataspace is extracted and aggregated from the various databases in the field of crime and law enforcement described in Section 10.2.*

[11] Choenni, S., Kalidien, S., Ariel, A. & Moolenaar, D. (2001).

With respect to the relationship manager of the monitor, it should be clear that the data in the monitor are closely related. The output of one organization in the criminal law chain often serves as the possible input of another organization. For instance, the input of cases into the prosecutorial level largely depends on the number of suspects handed over by the police. Therefore, a plausible rule in the manager would be: number of suspects handled by the police ≥ number of case handled by the prosecution service; meaning that the police usually do not send all cases to the prosecution service. Using such rules the plausibility and consistency of the data is maintained. Similar rules can be formulated in order to handle variables coming from different sources. Take, for instance, the number of community services imposed by the Public Prosecutor. This information comes from two organizations in the chain: the prosecution service and the organization responsible for executing sentences. As a rule, the number of community services registered by the executing organization is lower than the number of community services registered by the prosecution service, as suspects may die or 'disappear' before the sentence is executed. Such rules are typically based on historical data and domain knowledge and can be extended with error values to allow for incomplete or tentative data (see Choenni et al., 2001).

The primary goal of the monitor is to alert users when there are large differences between input and output, or between the actual input or output and the forecasted input or output. In this way, policymakers are able to indentify potential capacity problems at an early stage. Therefore, rules are added to the relationship manager that detect large deviations. Based on these rules, three types of alerts are provided to the users:

1. *large deviations in the proportion of organization X's output to its own input;*
2. *large deviations in the proportion of organization X's input to organization Y's output;*
3. *large forecasting errors.*

The user interface presents the user with an overview of the input and output data and the corresponding alerts in either table or graph format, depending on the user's preference. In this way a quick overview of the irregularities in the data is provided. In these views, the user is able to zoom in on specific parts of the criminal law chain by selecting a subset of data categories. More specifically, the user can subdivide the data into various categories including the age and gender of the suspect, the region in which the crime was committed, and the type of crime committed by the suspect. Thus, the user can, for instance, choose to only show the input and output of male suspects who are older than 18 years or the input and output in a specific region. Additionally, the monitor periodically produces written reports through a printing on demand module as described above.

In this section it was shown how data from various judicial databases may be combined and integrated using two different approaches: a data warehouse and a dataspace. In the first approach, data is linked explicitly on an individual level. In the second approach, more dynamic relations or rules are established to link data and maintain data quality. Thus, a dataspace differs from a data warehouse in the sense that a common data model is not required and that there is no need to link data based on unique identifiers. As a result, in a dataspace approach not only microdata but also aggregate data may be used. This does not alter the fact that a dataspace layer may contain a data warehouse as a data source.

The worked-out examples from the Dutch criminal justice chain illustrate that data integration can be executed in a variety of ways. For instance, depending on the needs of the users or the availability of the data, parts in this process may have to be altered. In the next section it is shown how potential problems associated with linking (crime) data affect the data integration process and the choices made in it.

10.4 Challenges in Combining Judicial Data

The main problem with data integration in the field of justice is that, although it can be automated for a large part, a significant amount of manual effort is still required. The main reason for this is the nature of crime data: redundancy, inconsistencies, dependencies, and semantic changes are not uncommon. In the remainder of this section, these potential problems and their consequences for the data integration process are described in detail.

Taking care of quantitative and qualitative dependencies
One of the problems with reconciling judicial data is the fact that quantitative dependencies between different data sources exist. For example, the date on which a crime is reported is usually the same as the date on which the crime is committed or the output of the police is usually greater than the input into the prosecutorial level. Though some of this knowledge may be exploited for data reconciliation (to compare records from different sources), it requires manual effort and the participation of domain experts.

Qualitative dependencies also exist within databases. For instance, it is generally assumed that the value of a certain attribute does not change dramatically in a few years. Therefore, it is recommended to compare the value of an attribute in a certain year to its value in preceding years in order to detect large deviations.

Thus, when data from different sources are combined, both quantitative and qualitative dependencies have to be managed in order to avoid unreliable data. In a data warehouse this has to be done manually by domain experts. In a dataspace approach it can be automated fully using dynamic rules that check the reliability of the data and detect deviations.

Managing semantic dependencies
Besides quantitative and qualitative dependencies, also semantic dependencies exist in and between judicial databases. These arise because different organizations in the criminal law chain store data about the same events, but often label or classify these data differently. For example, in case of a robbery a victim may classify it as a violent crime, while the police may classify it as a crime against property. Additionally, for a single case in court that contains several offences, the severest

case is taken as the classification criterion. As a result, less severe offences 'disappear' in the data reported by the court.

It is important that existing semantic dependencies between attributes (if any) are preserved while integrating data. Therefore, in a data warehouse domain experts need to keep track of semantic dependencies. In a dataspace these may be captured in rules.

Resolving inconsistencies

The different judicial databases have overlapping or redundant attributes. Redundancy may introduce inconsistencies that have to be detected and solved manually based on domain expertise. Take for example the nationality of a suspect that is recorded by different organizations. It is known that, in practice, foreigners tend to provide a wrong nationality when they are not able to show identification papers. As a result, inconsistencies may arise between different databases of different organizations. This can be resolved by utilizing the domain knowledge.

Prior to loading data into a data warehouse, inconsistencies have to be indentified and resolved. This means that all values of overlapping or redundant attributes have to be in agreement with each other. In a dataspace approach inconsistencies can be detected automatically and on the fly using rules that check attributes coming from different sources.

Handling semantic changes

Data evolve over time as rules and regulations are changing. Therefore, certain values on certain attributes may have gotten a different meaning over time. For instance, due to municipal reorganizations in the Netherlands, names of municipalities and cities have changed, while the old registered names were not always updated. Over time, the meaning of the old names may become unknown. Moreover, in case cities are expanded, their names mean something different before the reorganization than after. If these changes are not recorded, data may be combined improperly or wrong conclusions may be drawn based on them. To keep track of the 'history' of the attributes, semantic changes have to be recorded. In a dataspace this can be done in the relationship manager.

Concluding example

In general, a dataspace approach may be considered to be more efficient and practical than a data warehouse approach, because in the former it is easier to combine data and add new sources, as there is no need for data reconciliation. Additionally, using a dataspace approach dependencies, inconsistencies, and changes can be managed more effectively.

As an illustration, assume that one wants to know how many of the suspects questioned by the police are handed over to the prosecution and how many of them are actually prosecuted. To answer this question, the databases of the police (HKS) and the prosecution (OM-data) have to be integrated. However, OM-data only contains data of cases that are handled by the prosecution. This means that not all individuals in HKS are present in OM-data and, therefore, combining on an individual level, which is needed in a data warehouse approach, is impossible for these individuals. In a dataspace approach, however, aggregate data can be used, so the database may contain the total number of suspects questioned by the police (aggregated from HKS) and the total number of suspects (cases) handled by the prosecution (from OM-data). Then, a comparison can be made between the two totals, and the difference between output and input can be calculated. This task can be performed easily by the public safety monitor (described in Subsection 10.3.2). Thus, for this type of questions, a dataspace is more efficient as the heavy computational and troublesome task of uniquely linking individuals does not have to be performed.

10.5 Protecting Privacy When Combining Judicial Data

Tools or information systems that collect, relate, and present safety-related data, pose a serious privacy threat as the identity of individuals or groups of individuals may be exposed. For instance, assume that in the public safety monitor (see Subsection 10.3.2) the number of sex offenders is presented, and that it is possible to categorize them by age, gender, and city. If there is only one female sex offender in a certain city, then the age of this female is exposed. Depending on the additional information that is shown about her, or the information that can be gathered from alternative sources, it is likely that her full identity is exposed. If this is indeed the case, privacy laws are violated.

In the data integration process several precautions can be taken to respect the privacy of individuals and to minimize the risk of exposing someone's identity. First, a data source that contains crime data should only record attributes that are in line with the Dutch Personal Data Protection Act (PDPA). This act defines a set of sensitive attributes that should be handled with care, namely data on someone's religion or life conviction, ethnic origin, political opinions, health, sexual orientation, and memberships of (trade) unions.[12] Such sensitive attributes should not be stored. Second, aggregate data has a clear advantage over microdata as data on a higher aggregation level does not provide personal information. Therefore, for privacy reasons, it is recommended to use aggregate data instead of microdata when possible. Finally, whenever there is a risk of exposing the identity of an individual to a user of a tool, the result of the user's question or selection should not (or only in part) be shown. For example, if a user wants to view the number of sexual offenders per region, and if there are just two offenders in a certain region, this number should not be presented to the user. After all, in this case there is a reasonable chance that with additional information, the identity of the offenders concerned can be deduced. When all three precautions are followed, the risk of disclosing personal data and thereby violating the privacy of individuals is minimized.

The preceding sections focused on ways to combine and integrate data from various judicial databases. Combined crime data may help in gaining insight into the criminal law chain and in developing new policies. An even deeper understanding of crime and delinquency may be acquired by applying data analysis

[12] Sauerwein, L.B. & Linnemann, J.J. (2001).

techniques to such data. In this way, profiles of criminals or offenders may be constructed. In the next section, potentials and challenges of analyzing combined crime data will be described.

10.6 Risks of Analyzing Judicial Data

Statistics may be considered as a standard tool for the analysis of police and justice data. However, as in many organizations, the amount of data collected and stored by the judicial organizations has grown exponentially. In many fields, especially technically oriented fields, data mining has been proven to have an added value over statistics in analyzing large amounts of data.[13] See Choenni et al. (2005) for a summary of the differences between statistics and data mining. Data mining is the process of searching for statistical relations, or patterns, in large data sets. It is often used to gain a different perspective on the data and to extract useful information from them. Commonly used methods include rule learning (searching for relationships in the data), clustering (discovering groups in the data that are similar), and classification (generalizing known structures to new data). Thus, data mining is able to reveal useful knowledge that is hidden in a large amount of data. Therefore, there is a growing interest in applying data mining techniques to crime data.

However, the straightforward application of statistical techniques, and data mining in particular, may be risky. As has been pointed out in the literature (Hand, 1998), data mining results need to be evaluated by experts to determine whether they hold in the real world. The main reason for this is that data mining is based on induction and, therefore, the results may be true given the data, but not in the real world. For example, assume that all swans in a given databases are white, then it may be induced from the database that all swans are white. However, it is very well possible that only features of white swans are stored in the databases and that the very small group of black swans is neglected. As a result, the induced knowledge with regard to swans does not hold in the real world. Therefore, it is of vital importance to evaluate the truthfulness of data mining results.

For police and justice data, evaluation is even more important and that because of the following reasons. Opposed to findings in exact or technical sciences, findings in social sciences may be subject to change in the course of time. For instance, Newton's laws of motion were true decades ago and do still hold today, while the age-crime distribution in crime science is changing over time. For instance, in 2000 minors were responsible for roughly 17% of the committed crimes (that is, of all interrogated suspects, 17% was between 12 and 17 years old); while in 2007 they were responsible for around 19% of the committed crimes.[14]

Another reason to be cautious with data mining results in social sciences is the fact that, since data collection is a time-consuming and difficult process, often legacy databases are used for data mining. Such databases contain large amounts of

[13] Choenni, S., Bakker, R., Blok, H. & De Laat, R. (2005), Hand, D.J. (1998), Tan, P., Steinbach, M. & Kumar, V. (2005).

[14] De Heer-de Lange, N.E. & Kalidien, S. (2010).

data that were collected and stored in the past; sometimes decades ago. As a result, these databases mostly reflect the situation in the past, so mining these databases results in knowledge about the past. Evaluation of such data mining results is important for three reasons:

1. It has to be determined whether this knowledge corresponds with the real world of the past.
2. It has to be determined whether the knowledge still holds in the real world of today.
3. It has to be determined whether it is useful to apply the obtained knowledge (for instance, in developing new policies).

As an example, assume that data mining is applied to a database containing data about juveniles and nuisance offences from 1975 to 2005. By doing so, profiles of youngsters who cause annoyance may be found. A hypothetical result may be that young men born in a particular country have a higher probability to cause nuisance. However, it may be the case that this was true in the seventies, but not today, as since then they may have adapted their behavior to the Dutch society and norms. Thus, although the result corresponds to the real world of the past, it does not correspond to the real world of today. It is surely possible that nowadays young men from other countries show nuisance behavior. In this case, the fact that young men in general cause nuisance does hold in today's world, and can be usefully applied, but using the country of origin of these men is dangerous.

Contrary to data mining, the chances that such issues are encountered when applying statistics on crime data are small. Statistics requires carefully formulating hypotheses that are tested on newly collected data. Thus, the data used for standard statistical analyses always reflect the real world as it is today and do not involve the issues relating to legacy data.

Another important issue is that, since data mining tools are developed to find patterns based on any correlation in data, they can find patterns that use personal characteristics of groups of individuals. This may lead to discrimination and stigmatization of these groups. For instance, assume that data mining algorithms are employed on a database of sex offenders that is enriched with demographical and economical data. A likely data mining result may be that unemployed white men are responsible for 80% of the sex offences. There are two problems with such a statement. First, it could lead to stigmatization as the relation to the total population of unemployed white men is not made clear. Second, using it to discover new (unknown) sex offenders leads to discrimination because suddenly all unemployed white men are suspects, while only a few of them are actual sex offenders.

In sum, in this section it was shown that although applying data mining techniques to crime data seems promising, there are some issues regarding the applicability and generalizability of the obtained results. Additionally, data mining may lead to discrimination and stigmatization. Therefore, data mining methods should be used with caution.

10.7 Concluding Remarks

In this chapter it was illustrated how data from judicial databases in the Netherlands are currently processed, combined, and analyzed. It was explained how precautions in the data integration process should be taken to better respect privacy laws and regulations. When such measures are taken, the risks of exposing the identity of individuals are minimized. Subsequently, it was shown that applying data analysis methods to judicial data is not straightforward and that data mining results should be considered with caution. When these reservations are taken into account and the precautions mentioned are taken, applications that exploit combined crime data and provide statistical overviews are valuable tools for judicial policymakers in developing new and effective policies. An example is the recently developed public safety monitor. This monitor fulfills the information needs of policymakers and advisors and allows them to timely identify potential capacity problems.

References

Choenni, S., Bakker, R., Blok, H., de Laat, R.: Supporting technologies for knowledge management. In: Baets, W. (ed.) Knowledge Management and Management Learning: Extending the Horizons of Knowledge-Based Management-Part 2. Integrated Series in Information Systems, vol. 9, pp. 89–112 (2005), doi:10.1007/0-387-25846-9_6

Choenni, S., Blanken, H., Chang, T.: Index selection in relational databases. In: Computing and Information, Proceedings of ICCI 1993, Fifth International Conference on Computing and Information, pp. 491–496 (1993), doi:10.1109/ICCI.1993.315323

Choenni, S., van Dijk, J., Leeuw, F.: Preserving privacy whilst integrating data: Applied to criminal justice. Information Polity 15(1,2), 125–138 (2010), doi:10.3233/IP-2010-0202

Choenni, S., Leertouwer, E.: Public Safety Mashups to Support Policy Makers. In: Andersen, K.N., Francesconi, E., Grönlund, Å., van Engers, T.M. (eds.) EGOVIS 2010. LNCS, vol. 6267, pp. 234–248. Springer, Heidelberg (2010), doi:10.1007/978-3-642-15172-9_22.

Choenni, S., Kalidien, S., Ariel, A., Moolenaar, D.: A framework to monitor public safety based on a data space approach. In: Janssen, M., Macintosh, A., Scholl, H.J., Tambouris, E., Wimmer, M.A., de Bruijn, H., Tan, Y.-H. (eds.) Electronic Government and Electronic Participation. Joint Proceedings of Ongoing Research and Projects of IFIP EGOV and ePart 2011, Schriftenreihe Informatik, vol. 37, pp. 196–202. Trauner, Linz (2001)

Choenni, S., Meijer, R.: From police and judicial databases to an offender-oriented data warehouse. In: Proceedings of the IADIS International Conference on e-Society, pp. 98–105 (2011)

Franklin, M., Halevy, A., Maier, D.: From databases to dataspaces: a new abstraction for information management. SIGMOD Record 34(4), 27–33 (2005)

Hand, D.J.: Data mining: statistics and more? The American Statistician 52(2), 112–118 (1998)

De Heer-de Lange, N.E., Kalidien, S.: Criminaliteit en rechtshandhaving 2009: ontwikkelingen en samenhangen [crime and law enforcement 2009] Onderzoek en Beleid. 279. Boom Juridische uitgevers, The Hague (2010)

Kalidien, S., Choenni, S., Meijer, R.: Towards a tool for monitoring crime and law enforcement. In: Proceedings of ECIME 2009, the 3rd European Conference on Information Management and Evaluation, pp. 239–247 (2009)

Kalidien, S., Choenni, S., Meijer, R.: Crime statistics online: potentials and challenges. In: Proceedings of the 11th Annual International Digital Government Research Conference on Public Administration Online: Challenges and Opportunities (dg.o 2010), pp. 131–137 (2010)

Kalidien, S., de Heer-de Lange, N.E.: Criminaliteit en rechtshandhaving 2010: ontwikkelingen en samenhangen [Crime and law enforcement 2010] Onderzoek en Beleid. 298. Boom Juridische uitgevers, The Hague (2011)

Kimball, R., Ross, M.: The data warehouse toolkit: the complete guide to dimensional modeling, 2nd edn. John Wiley & Sons, Inc., New York (2002)

Meijer, R., van Dijk, J., Leertouwer, E., Choenni, S.: A drug crime data mart to support publication on demand. In: Proceedings of the 2nd European Conference on Information Management and Evaluation, pp. 277–286 (2008)

Sauerwein, L.B., Linnemann, J.J.: Guidelines for personal data processors: personal data protection act. Ministry of Justice, The Hague (2001)

Tan, P., Steinbach, M., Kumar, V.: Introduction to data mining. Addison Wesley, London (2005)

Part IV
Solutions in Code

Chapter 11
Privacy-Preserving Data Mining Techniques: Survey and Challenges

Stan Matwin

Abstract. This chapter presents a brief summary and review of Privacy-preserving Data Mining (PPDM). The review of the existing approaches is structured along a tentative taxonomy of PPDM as a field. The main axes of this taxonomy specify what kind of data is being protected, and what is the ownership of the data (centralized or distributed). We comment on the relationship between PPDM and preventing discriminatory use of data mining techniques. We round up the chapter by discussing some of the new, arising challenges before PPDM as a field.

11.1 Introduction

Exponential growth of information and communication technologies in the last thirty years, and their deep penetration of every segment of the society, raised no fundamental opposition or critique. There is, however, social sensitivity related to one aspect of those technologies: it is their potentially negative influence on personal privacy. Data that – in themselves – are not jeopardizing individual privacy, can be instantaneously and freely combined with other data and used in sophisticated inference. New information produced in that manner becomes available to parties completely unknown to the original "owner" of the data. This has been aptly described by James Moor as the "greased data" phenomenon (Moor 2004), and it is only possible thanks to information technology. It is therefore reasonable to look for technological solutions to potential privacy breaches that are enabled by modern advances in information and data transmission. In the eyes of some, there is a "moral imperative" for Computer Science as a field: at least some researchers should work on finding solutions to problems that the field itself may have exacerbated. Research focusing on privacy aspects of data mining is known as Privacy-Preserving Data Mining (PPDM). This article provides a high-level review of the accomplishments of this research, as well as a brief discussion of the

Stan Matwin
University of Ottawa, Canada
e-mail: stan@site.uottawa.ca

B. Custers et al. (Eds.): Discrimination & Privacy in the Information Society, SAPERE 3, pp. 209–221.
springerlink.com

questions awaiting solutions and the forthcoming challenges. For a more technical and a more complete presentation, the reader may consult (Vaidya, Zhu et al. 2006), or more recent, in-depth technical tutorials (Fung, Wang et al. 2010), (Chen, Kifer et al. 2009).

Data privacy is often seen as an aspect of, or appendix to, data security. This is not a correct view, as the goals of the two fields are divergent. On the one hand, security protects the data against unauthorized access, e.g. reading the data while it is transmitted across a network. But once the data reaches an authorized recipient, security does not impose additional constraints having to do with revealing personal information of an individual. This is, on the other hand, the goal of data privacy. Such divergence of goals is well illustrated by public key cryptography that protect the data encrypted using a person's private key, but also make the data tightly linked to an individual whose public key is used to decipher it, thereby identifying that individual. It is therefore correct to describe the relationship between data security and data privacy as the former being a prerequisite of the latter. Data must be protected in storage and transmission by data security methods (e.g. with cryptographic techniques), but if data privacy is a goal, then additional steps, some of them described below, must be taken to protect privacy of the individuals represented in the data.

Before reviewing current work in PPDM, we need to establish dimensions that will structure this review. In order to identify those dimensions, we need to ground the discussion in the process that PPDM addresses, mainly sharing data and results of a data mining operation between users $u_1,\ldots u_m$, $m\geq 2$. Furthermore, it is useful to view the data as a database of n records, each consisting of l fields, where each record represents an individual i_i, and describes i_i in terms of its fields. The usual simplified representation is a table T, in which rows represent individuals $i_1,\ldots i_n$, and columns – referred to as attributes – represent the fields $a_1,\ldots a_l$. This assumes a fixed representation, i.e. each individual is represented by a vector of values of $a_1,\ldots a_l$.

For a holistic view of PPDM, the first useful dimension is to consider privacy in terms of what is being protected, or conversely – what does an attacker want to obtain from T. The second useful dimension is the ownership structure of the data – does it belong to one entity and has to be shared with another entity ($m = 2$) or is it built from parts owned by different entities? We therefore propose to consider the following dimensions:

- What is being protected:
  - the data: an attacker, given T,
    - will not be able to link any row in T to a specific i [**identity disclosure**]
    - will not be able to obtain a value a_{ij} of a sensitive attribute a_j of i_i [**attribute disclosure**]
  - the inferred data mining result: an attacker, not knowing T but given the results of the data mining operation, e.g. an association rule learned from T, will be able to identify some attributes of a specific i_i [**model-based identity disclosure**]

- Is the data
  - centralized: T is owned by one party u_i, and is to be shared with another party (or parties) u_k , e.g. so that u_k can perform a data mining operation on T ?
  - or distributed: each u_i knows only certain rows (or columns) of T, but all of u_i's need a result of a data mining operation performed on the whole T?

In remainder of this chapter, we will follow these taxonomical dimensions in our review of the existing PPDM research.

We need to introduce some further definitions useful in the presentation of the PPDM concepts. In particular, an *explicit identifier* is an attribute that allows direct linking of an instance (a row in T) to a person i, e.g. a knowing a cellular phone number or a driver's license number will unambiguously link the row in T in which this explicit identifier occurs to a person i. A *quasi-identifier* is a set of attributes which individually are not explicit identifiers, but which jointly may link a row in T to a specific person. For instance, (Sweeney 2002) shows that in the United States the quasi-identifier triplet <date of birth, 5 digit postal code, gender> uniquely identifies 87% of the population of the country. As a convincing application of this observation, using quasi-identifiers and combining a public healthcare information dataset with a publicly available voters' list, Sweeney was able to obtain health records of the governor of Massachusetts from a published dataset of health records of all state employees in which only explicit identifiers have been removed.

For the sake of completeness, it has to be mentioned that there can also be a so called "membership" privacy attack: given a table T and an individual i, is i in T? We can observe that this is a form of an identity disclosure attack, in terms of the PPDM dimensions proposed above.

11.2 Identity Disclosure

In general, the main PPDM identity protection methods draw on simple ideas known to humans throughout history and amply presented in literature and film. These paradigms can be described as "hiding in the crowd" and "camouflage".

One "hiding in the crowd" approach to data privacy is k-anonymity. The k-anonymity method (Sweeney 2001) (Ciriani, Capitani di Vimercati et al. 2007) modifies the original data T to obtain T' such that for any quasi-identifier q that can be built from attributes of T there are at least k instances in T' such that q matches these instances. Datasets need to be generalized to satisfy k-anonymity. See Fig. 1 for an example of k-anonymized data. Conceptually, such data generalizations correspond to clustering of datasets, and to using clusters instead of the original elements. These clusters can also be viewed as equivalence classes of the attribute generalization. Clearly, generalizations cause deterioration of the quality of the data as the original values of at least some attributes are lost. k-anonymization can be therefore seen as a task of minimal data generalization of

PID	Zip Code	Date of Birth	Nationality	Disease
1	120**	1967	*	Heart Disease
2	120**	1967	*	Bronchitis
3	120**	1967	*	Viral Infection
4	120**	1970	*	Viral Infection
8	120**	1970	*	Cancer
9	120**	1970	*	Cancer
5	118**	1964	*	Cancer
6	118**	1964	*	Heart Disease
7	118**	1964	*	flu
10	118**	1964	*	Diabetes

Fig. 11.1 Example of a *k*-anonymized data table *T'*, *k*=3. Attributes Zipcode and Nationality have been generalized to ensure *3*-anonymity. From (Fung, Wang et al. 2010).

the data in *T* that will satisfy *k*-anonymity for a given *k*. It has been shown (Bonizzoni, Vedova et al. 2009) that such task is *NP*-complete, and therefore the existing, practical *k*-anonymization methods (Sweeney 1998) (El Emam, Dankar et al. 2009) are not necessarily optimal in the above sense.

It needs to be observed that *k*-anonymity does not fully resolve data privacy problems. With additional domain knowledge, which the attacker will often possess, successful attacks, albeit of different type, are still possible. For instance, if all the records in an equivalence class in a *k*-anonymized *T'* have the same value of a sensitive attribute (e.g. the medical diagnosis), then mapping an instance *i* to that equivalence class will also inevitably give away the value of this attribute for *i*. This would then become a successful attribute disclosure attack. In order to avoid this kind of privacy attack, *k*-anonynymity is often extended to require *l*-diversity: every equivalence class in *T'* must have at least *l* values of the sensitive attributes. *l*-diversity, however, is also prone to attacks: consider a two-class problem assigning a sensitive medical diagnosis to people. Being put in the positive class may be stigmatizing an individual and may lead to discrimination. But if the cluster contains only negative individuals, there is no need for diversity: nobody will mind being in this cluster as no negative inference can be associated with this membership. On the other hand, knowing that one is in a cluster with 49 positive and one negative individual makes is highly likely (98%) that one has the condition, while knowing that one is in a cluster with 49 negative and 1 positive individual is completely different. Both clusters, however, have the same *2*-diversity. (Li and Li 2007) have therefore proposed yet another privacy model, known at *t*-closeness, attempting to fix these shortcomings of *l*-diversity. A cluster (a result of data generalization) satisfies *t*-closeness if the distance between the distribution of a sensitive attribute in this cluster and the distribution of this attribute in the whole table *T* is no more than a threshold *t*. In that manner *t*-closeness may, in principle, prevent discrimination by making it impossible to assert negative inferences about the sensitive attribute based on a cluster membership, such that these inferences would be stronger than the ones for the entire table (the whole population). It is clear, however, that requiring *t*-closeness imposes a very strong constraint on the generalization process, resulting in a potentially very significant distortion of data, thereby decreasing the quality of the data (and any model obtained from it) unacceptably.

It is worth observing that the attack model behind data *k*-anonymity is somewhat unrealistic. It assumes that the attacker has a total knowledge of all values of the attributes for a given instance, which will normally not be the case. Starting with this observation, more realistic models have been proposed. For instance, in (Mohammed, Fung et al. 2009) the attack model assumes that the attacker's knowledge is limited to *L* quasi-identifiers, and the *k*-anonymization is limited to those identifiers.

k-anonymization is often the method of choice in data publishing, particularly for medical data. The reason is that, unlike other perturbative methods discussed in the next section, the approach does not distort the data: even the generalized data is "true", i.e. it represents true (even though possibly imprecise) statements about the original data.

A completely different identity disclosure attack is possible when the model build using data mining techniques such as classification or association rules is so granular (on a specific data set) that it identifies a specific individual. Publishing such model alone, even without access to data from which it has been obtained, would then disclose data values that the model represents for that specific individual. Rule-hiding is an approach attempting to solve this problem. For instance, (Verykios, Elmagarmid et al. 2004) present strategies prevening association rules with a sensitive attribute in the consequent from being produced by the association rule mining algorithms. Another approach to rule hiding is described in (Oliveira, Zaïane et al. 2004). These strategies are based on reducing the support and confidence of rules with such attributes in the consequent. (Atzori, Bonchi et al. 2008) show how such disclosure can be avoided by elegantly generalizing to models the concept of *k*-anonymity discussed above for the data.

11.3 Attribute Disclosure

A different set of methods protecting against disclosure of a value of sensitive attribute are the *perturbative* methods. They implement the "camouflage" paradigm. The seminal work in this area is due to (Agrawal and Srikant 2000). The main idea is simple: an attribute (say, a *j*-th column in *T*) is systematically changed by adding to each a_{ij}, $i=1 \ldots n$, a value obtained from a probability distribution. Individual attribute values therefore bear no similarity to the original values, e.g. salary values may become negative numbers. The distribution of such randomized attribute is of course totally different from the original distribution of a_j, but (Agrawal and Srikant 2000) show how its distributional properties can be reconstructed from the randomized distribution so that meaningful data mining operations (e.g. classification, or association rules) can be performed. This is illustrated in Fig. 2. The results of these data mining operations are close to the results obtained on the original data. This approach of (Agrawal and Srikant 2000) has been shown to be prone to an attack, using sophisticated control theory methods (Kargupta, Datta et al. 2003). Another perturbative approach introduces multiplicative rather than additive noise in the data (Kun, Kargupta et al. 2006), with privacy guarantees stronger than those given by additive noise.

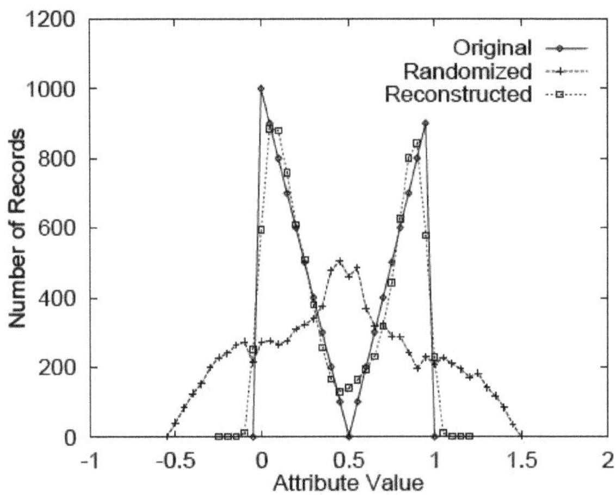

Fig. 11.2 The effects of the perturbation the distribution of the attribute being perturbed, to protect the disclosure of its values (from (Aggarwal and Yu 2008)). The curve labeled "randomized" has a distribution very different from the original one. Data mining is performed on the reconstructed distribution, obtained using the algorithm described in the paper. The reconstructed distribution close is to the original, and therefore the results of data mining are close to what they would have been on the original, confidential.

Perturbative methods have the disadvantage of modifying the original data, which can be difficult to accept in certain classes of applications, e.g. in working with medical data.

A different kind of perturbative approach is known as *rank swapping* (Nin, Herranz et al. 2008). The main idea is to swap values of a given attribute among records in a dataset. The swapping is controlled by the distance between the swapped values – values that are close are more likely to be swapped. The advantage of this approach is that, unlike with the noise-injecting approaches described above, the entire set of values of a given attribute and its distribution are preserved. The disadvantage is that potentially implicit relationships between values of attributes can be broken.

An ultimate method for protection against attribute disclosure is based on the idea that the original data is replaced, in its entirety, by a synthetic data set with the same statistical properties (e.g. mean, variance, etc.) as the ones of the original dataset. (Krishnamurty Muralidhar and Sarathy 2008) present a method which, besides preserving the mean vector and the covariance matrix, also guarantees similarity of the synthetic confidential values to the original confidential values. This somewhat radical approach may encounter some resistance in applications in which veracity of the data is important, e.g. in medical research. On the other hand, it may be acceptable in areas where use of the aggregated data is already a norm, e.g. in large-scale social sciences research.

A number of attribute disclosure attacks, and methods to protect against them, have been described in the literature. We can mention here (Loukides, Gkoulalas-Divanis et al. 2011), (Martin, Kifer et al. 2007), (Chen, LeFevre et al. 2007). The generality of these attacks is questionable and leads to high-granularity privacy protection approaches in which multiple transformations are applied to the data, resulting in potentially significant decrease in data quality while still leaving the resulting data vulnerable to privacy attacks of novel kind, which are not yet known or described in the literature. This is analogous to multi-layered anti-virus patches, which themselves may open vulnerabilities to novel, yet unknown viruses to come in the future.

11.4 Privacy of Decentralized Data

As described in sec. 1, we address here an important scenario in which the ownership structure of data in T is shared among multiple parties in order to obtain a meaningful data mining result of interest to all parties. This is a frequent phenomenon, as groups of users may be interested in performing data mining on the union of their data, but cannot share the data for legal or commercial (competitive) reasons. We are then talking about the data being *partitioned*. As shown in Fig. 3, the partitioning may be either vertical or horizontal. In the *vertical* partitioning, all the parties have data referring the same instances, but each party will have a different subset of attributes describing the instances. An example of such a situation is a scenario in which one wants to perform an extensive association rule mining on a dataset describing vehicles involved in certain types of accidents. Data (attributes) pertaining to performance of different subcomponents (tires, engine, brakes) will belong to different manufacturers who do not want to share it with others, but are interested in the results. In the *horizontal* scenario, different parties have different subsets of instances, but they all have the same attributes. An example of such situation is a medical study performed jointly by a number of hospitals. Each of the hospitals may have its own limited set of patients participating in the study, but results drawn from the much larger union of all the data from different hospitals will achieve a much higher level of credibility. Finally, mixed horizontal-vertical scenarios are also possible.

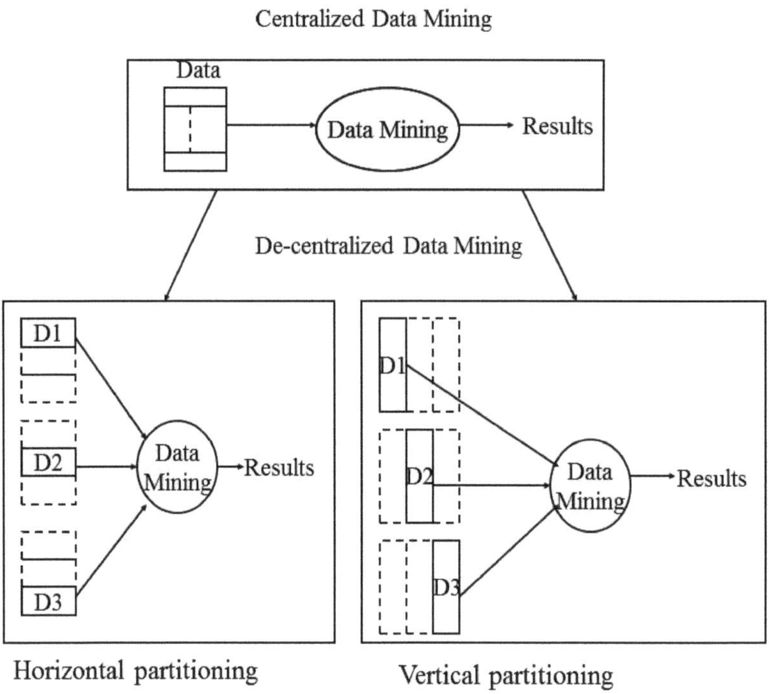

Fig. 11.3 De-centralized data mining with horizontal and vertical partitioning. In the former, all parties share the same attributes but have different instances. In the latter, parties have different attributes of the same instances.

Methods addressing these important scenarios are mainly based on the use of cryptographic techniques (see Yang et al. (2006)) for a discussion of the computational performance of these kinds of methods). The main idea is to encrypt each party's data, and share the encrypted data with other parties so that a dedicated algorithm working on encrypted data can produce a result that can then be made available to all the parties (or, in some cases, to just one selected party). The key idea in these approaches is the concept of homomorphic encryption of the data (Paillier 1999)[1]. It will be best explained with the use of a very simple example. Suppose that we have two parties, A and B, each having a data vector $a_1, \ldots a_n$ and $b_1, \ldots b_n$, respectively. They cannot show their vector to each other, but they want to

[1] Another approach is the use of the Secure Multiparty Computation (SMC) framework. SMC offers algorithms in which parties compute function results on arguments each of them owns, through the use of especially designed circuitry, without sharing the argument values with each other (see e.g. Lindell, Y. and B. Pinkas (2009). "Secure Multiparty Computation for Privacy-Preserving Data Mining." Journal of Privacy and Confidentiality **1**(1): 59-98. for a thorough presentation of privacy-oriented computation in a distributed environment we consider here). While theoretically elegant and secure, this approach has not yet produced computationally acceptable implementations.

compute the scalar product of their two vectors: $A \bullet B$. A may encrypt the data with homomorphic encryption h and compute $h(a_1), h(a_2), \cdots, h(a_n)$, and send those to B. B will then multiply each of these encryptions by his b_i, but only for those i for which $b_i \neq 0$, and due to the property of homomorphic encryption that $h(x) \times h(y) = h(x+y)$, B obtains $e(A_{j_1} B_{j_1} + \ldots + A_{j_m} B_{j_m})$. This expression is then passed back to A for decoding of the result, which is $A \bullet B$ (we have omitted some details here, having to do with the use of digital envelopes and with doing all the arithmetic operations modulo and agreed X). This is generalized from two to multiple parties, and from dealing with binary values to any symbolic or numerical values. Furthermore, complex protocols are designed to implement in the above manner many data mining operations: different classifiers, e.g. decision trees (Vaidya, Clifton et al. 2008), support vector machines (Zhan 2007), Bayesian classifiers (Yang and Wright 2006), *k*-nearest neighbour (Zhan, Chang et al. 2005), clustering (Vaidya and Clifton 2003), etc., in both horizontal and vertical partitioning settings.

A big advantage of these approaches is that data remains unchanged, and therefore there is no loss of data quality. A disadvantage is the significant overhead, due to the multiple encryption/decryption operations, and even more importantly to the need to communicate securely between the parties to exchange the keys for these encryptions. (Yang, Wright et al. 2006) presents an empirical study that analyzes the computational and communication overhead of the cryptographic approaches outlined above, and shows that the computational overhead is quite heavy: vector product for vector of size O(10**5) was exceeding 1 hour (although specialized optimization would decrease this to seconds). For these reasons that there are no reported large-scale implementations and applications of these approaches with real-life data.

A recent important cryptographic result by (Gentry 2010) generalizes the concept of cryptographic encryption to the functionality of computation of any type on encrypted data, and then decrypting the result. While there is ongoing research to build an efficient (or even feasible) implementation of these results, it is clear that even if only partially successful, the approach will go a long way towards ensuring data security in a cloud environment. It will also support data privacy in the distributed context that we are discussing here.

11.5 New Challenges for Data Privacy

As connectivity becomes ubiquitous and computing technology permeates human life, more and more data is produced as people go about their daily life. Mining of this data can result in novel and unexpected threats to privacy.

Data from mobile devices may represent one such threat (Wang, Pedreschi et al. 2011). Collecting this kind of data may result in highly useful services. For instance, collecting data from vehicles' or persons' trajectories in large cities may provide traffic and urban planners with patterns that can be used to build new roads, manage traffic, introduce corrective highway tariffs etc. At the same time, analysis of this kind of data can result in identifying person's movements without the person being aware of this identification. In extreme cases, combining trajectories of an individual with external knowledge (e.g. the address of their home and workplace) may identify a person uniquely from a large set of trajectories, obtained from a huge population. Early work in this area has been a topic of a focused research project (Giannotti, Pedreschi et al. 2009). Some of the techniques proposed aggregate trajectories into groups before releasing them for data mining, in the spirit of k-anonymization applied on "static" data and described above.

Another enormous challenge is the growing universal use of social networks. Clearly, there is a basic contradiction between privacy and the goal of social networks, which is to present information about a person, their opinions and their activities. From a Computer Science perspective, social networks are often described and analyzed as graphs. There exists a body of recent literature exploiting social network mining as graph data mining, and proposing techniques that make unique identification of a person from purely *structural* information hard. Numerous papers follow this approach, see e.g. (Liu and Terzi 2008) and (Hay, Miklau et al. 2008). Real social networks, however, supply a wealth of non-structural information – names, photos, email addresses etc. – which can be used as explicit identifiers. Therefore the practical value of graph-based social network privacy protection research remains to be proven.

As no technical solutions for protecting privacy in social networks exist, it seems this is not a purely technical problem. Perhaps the main tool to mitigate potentially disastrous effects of social networks for privacy remains education. Users, especially the teenage population, need to be explained the basic facts, e.g. that posting anonymous photos of people for the world to see may cause automatic tagging and identification of people in the photos. Oftentimes, many privacy breaches could be prevented if the users of social networks were taking advantage of setting privacy of their personal information using the existing privacy settings, provided by social networks. For instance, users may allow only their direct friends to see their tagged photos. Most users, however, never learn about these privacy settings and never use them. In that realm, novel work by (Fang, Kim et al. 2010) seems very interesting. The gist of it is to give users tools based on machine learning and recommender systems, and that make it relatively painless to set the existing privacy settings in social networks such as Facebook. It is something most users are not doing, and it would protect against many privacy breaches by limiting access to information the users provide.

Finally, cloud computing is a major challenge for data security, and hence data privacy. In a cloud the data owner lose control over their data. The existing legal safeguards are jurisdictional, and the cloud makes it hard, if possible at all, to determine where the data resides and where is it processed, and therefore which legal constraints – if any – on collecting, storing, and using the data apply. It has to be observed, however, that if the research initiated by the paper (Gentry 2010) succeeds, it could provide a comprehensive solution for the privacy issues in a cloud setting.

11.6 Conclusion

In this chapter, we review the existing Privacy-preserving Data Mining methods. General knowledge and understanding of these methods and techniques is highly relevant for preventing discriminatory effects of modern data mining techniques. When appropriate, we have underscored the usability of specific techniques to prevent discriminatory use of data mining. Some of the techniques presented in this chapter generalize the data, so that any stigmatized group would not be more targeted in the generalized data than it is in the general population. All data generalizations, however, incur a cost in data quality. The cryptographic approaches, on the other hand, preserve the data but impose a heavy computational overhead. The chapter is complete with a discussion of some of the challenges before PPDM as a field.

Acknowledgments. Research support by the Natural Sciences and Engineering Research Council of Canada and the Ontario Centres of Excellence is kindly acknowledged by the author.

References

Aggarwal, C.C., Yu, P.S.: A framework for condensation-based anonymization of string data. Data Mining and Knowledge Discovery 16, 251–275 (2008)

Agrawal, R., Srikant, R.: Privacy-preserving data mining. ACM SIGMOD Record 29, 439–450 (2000)

Atzori, M., Bonchi, F., et al.: Anonymity preserving pattern discovery. VLDB Journal 17(4), 703–727 (2008)

Bonizzoni, P., Della Vedova, G., Dondi, R.: The *k*-Anonymity Problem is Hard. In: Kutyłowski, M., Charatonik, W., Gębala, M. (eds.) FCT 2009. LNCS, vol. 5699, pp. 26–37. Springer, Heidelberg (2009)

Chen, B.-C., Kifer, D., et al.: Privacy-Preserving Data Publishing. Found Trends Databases 2(1-2), 1–167 (2009)

Chen, B.-C., LeFevre, K., et al.: Privacy skyline: privacy with multidimensional adversarial knowledge. In: Proceedings of the 33rd International Conference on Very Large Data Bases, Vienna, Austria. VLDB Endowment (2007)

Ciriani, V., Capitani di Vimercati, S., et al.: k-Anonymity. Secure Data Management in Decentralized Systems 33, 323–353 (2007)

El Emam, K., Dankar, F.K., et al.: A Globally Optimal k-Anonymity Method for the De-Identification of Health Data. Journal of the American Medical Informatics Association 16(5), 670–682 (2009)

Fang, L., Kim, H., et al.: A privacy recommendation wizard for users of social networking sites. In: Proceedings of the 17th ACM Conference on Computer and Communications Security, Chicago, Illinois, USA. ACM (2010)

Fung, B.C.M., Wang, K., et al.: Privacy-preserving data publishing: A survey of recent developments. ACM Comput. Surv. 42(4), 1–53 (2010)

Gentry, C.: Computing arbitrary functions of encrypted data. Commun. ACM 53(3), 97–105 (2010)

Giannotti, F., Pedreschi, D., Turini, F.: Mobility, Data Mining and Privacy the Experience of the GeoPKDD Project. In: Bonchi, F., Ferrari, E., Jiang, W., Malin, B. (eds.) PinKDD 2008. LNCS, vol. 5456, pp. 25–32. Springer, Heidelberg (2009)

Hay, M., Miklau, G., et al.: Resisting structural re-identification in anonymized social networks. Proc. VLDB Endow. 1(1), 102–114 (2008)

Kargupta, H., Datta, S., et al.: On the privacy preserving properties of random data perturbation techniques. In: Third IEEE International Conference on Data Mining, ICDM 2003, pp. 99–106 (2003)

Muralidhar, K., Sarathy, R.: Transactions on Data Privacy 1(1), 17–33 (2008)

Kun, L., Kargupta, H., et al.: Random projection-based multiplicative data perturbation for privacy preserving distributed data mining. IEEE Transactions on Knowledge and Data Engineering 18(1), 92–106 (2006)

Li, N., Li, T.: t-Closeness: Privacy Beyond k-Anonymity and ℓ-Diversity. In: Proceedings of IEEE International Conference on Data Engineering (2007)

Lindell, Y., Pinkas, B.: Secure Multiparty Computation for Privacy-Preserving Data Mining. Journal of Privacy and Confidentiality 1(1), 59–98 (2009)

Liu, K., Terzi, E.: Towards identity anonymization on graphs. In: Proceedings of the 2008 ACM SIGMOD International Conference on Management of Data, Vancouver. ACM (2008)

Loukides, G., Gkoulalas-Divanis, A., Shao, J.: Anonymizing Transaction Data to Eliminate Sensitive Inferences. In: Bringas, P.G., Hameurlain, A., Quirchmayr, G. (eds.) DEXA 2010. LNCS, vol. 6261, pp. 400–415. Springer, Heidelberg (2010)

Martin, D.J., Kifer, D., et al.: Worst-Case Background Knowledge for Privacy-Preserving Data Publishing. In: IEEE 23rd International Conference on Data Engineering, ICDE 2007 (2007)

Mohammed, N., Fung, B.C.M., et al.: Anonymizing healthcare data: a case study on the blood transfusion service. In: Proceedings of the 15th ACM SIGKDD International Conference on Knowledge Discovery and Data Mining, Paris, France. ACM (2009)

Moor, J.: Towards a theory of privacy in the information age. In: Bynum, T., Rodgerson, S. (eds.) Computer Ethics and Professional Responsibility. Blackwell Publishing (2004)

Nin, J., Herranz, J., et al.: Rethinking rank swapping to decrease disclosure risk. Data Knowl. Eng. 64(1), 346–364 (2008)

Oliveira, S.R.M., Zaïane, O.R., Saygın, Y.: Secure Association Rule Sharing. In: Dai, H., Srikant, R., Zhang, C. (eds.) PAKDD 2004. LNCS (LNAI), vol. 3056, pp. 74–85. Springer, Heidelberg (2004)

Paillier, P.: The 26th International Conference on Privacy and Personal Data Protection. In: Advances in Cryptography - EUROCRYPT 1999, pp. 23–38 (1999)

Sweeney, L.: Datafly: A System for Providing Anonymity in Medical Data. In: Proceedings of the IFIP TC11 WG11.3 Eleventh International Conference on Database Securty XI: Status and Prospects, pp. 356–381 (1998)

Sweeney, L.: Computational Disclosure Control: A Primer on Data Privacy Protection, Ph.D. thesis. Massachusetts Institute of Technology (2001)

Sweeney, L.: k-anonymity: a model for protecting privacy. Int. J. Uncertain. Fuzziness Knowl.-Based Syst. 10(5), 557–570 (2002)

Vaidya, J., Clifton, C.: Privacy-preserving k-means clustering over vertically partitioned data. In: Proceedings of the Ninth ACM SIGKDD International Conference on Knowledge Discovery and Data Mining, Washington, D.C., ACM (2003)

Vaidya, J., Clifton, C., et al.: Privacy-preserving decision trees over vertically partitioned data. ACM Trans. Knowl. Discov. Data 2(3), 1–27 (2008)

Vaidya, J., Zhu, Y.M., et al.: Privacy Preserving Data Mining. Springer (2006)
Verykios, V.S., Elmagarmid, A.K., et al.: Association Rule Hiding. IEEE Trans. on Knowl. and Data Eng. 16(4), 434–447 (2004)
Wang, D., Pedreschi, D., et al.: Human mobility, social ties, and link prediction. In: Proceedings of the 17th ACM SIGKDD International Conference on Knowledge Discovery and Data Mining, San Diego, California, USA, pp. 1100–1108. ACM (2011)
Yang, Z., Wright, R.N.: Privacy-Preserving Computation of Bayesian Networks on Vertically Partitioned Data. IEEE Trans. on Knowl. and Data Eng. 18(9), 1253–1264 (2006)
Yang, Z., Wright, R.N., et al.: Experimental analysis of a privacy-preserving scalar product protocol. Comput. Syst. Sci. Eng. 21(1) (2006)
Zhan, J., Chang, L., et al.: Privacy preserving k-nearest neighbor classification. International Journal of Network Security (1), 46–51 (2005)
Zhan, J., Matwin, S.: Privacy-preserving support vector machine classification. International Journal of Intelligent Information and Database Systems 1(3-4), 365–385 (2007)

Chapter 12
Techniques for Discrimination-Free Predictive Models

Faisal Kamiran, Toon Calders, and Mykola Pechenizkiy

Abstract. In this chapter, we give an overview of the techniques developed ourselves for constructing discrimination-free classifiers. In discrimination-free classification the goal is to learn a predictive model that classifies future data objects as accurately as possible, yet the predicted labels should be uncorrelated to a given sensitive attribute. For example, the task could be to learn a gender-neutral model that predicts whether a potential client of a bank has a high income or not. The techniques we developed for discrimination-aware classification can be divided into three categories: (1) removing the discrimination directly from the historical dataset before an off-the-shelf classification technique is applied; (2) changing the learning procedures themselves by restricting the search space to non-discriminatory models; and (3) adjusting the discriminatory models, learnt by off-the-shelf classifiers on discriminatory historical data, in a post-processing phase. Experiments show that even with such a strong constraint as discrimination-freeness, still very accurate models can be learnt. In particular, we study a case of income prediction, where the available historical data exhibits a wage gap between the genders. Due to legal restrictions, however, our predictions should be gender-neutral. The discrimination-aware techniques succeed in significantly reducing gender discrimination without impairing too much the accuracy.

Faisal Kamiran
Lahore Leads University, Pakistan
e-mail: faisal.kamiran@gmail.com

Toon Calders · Mykola Pechenizkiy
Eindhoven University of Technology, The Netherlands
e-mail: {t.calders,m.pechenizkiy}@tue.nl

B. Custers et al. (Eds.): Discrimination & Privacy in the Information Society, SAPERE 3, pp. 223–239.
springerlink.com

12.1 Introduction

Classifier construction is one of the most popular data mining and machine learning techniques (see also Chapter 2 of this book). We assume that a training set in which labels are assigned to the instances is given. The labels indicate the *class* the training examples belong to, and will hence often be called the *class labels*. The training examples are represented by tuples over a set of attributes; that is, every example will be described by values for the same set of attributes. The attribute containing the label will be called the *class attribute*. The label of an example is hence its value for the class attribute. In Table 12.1 an example training set is given. Every example corresponds to a person and is described by the attributes *gender*, *ethnicity*, *highest degree*, *job type*, and the *class* attribute determining whether or not this person belongs to the class of people with a high income (label '+'), or a low income (label '–'). A classifier construction algorithm learns a predictive model for labeling new, unlabeled data. For the given example, a classifier construction algorithm would learn a model for predicting if a person has a high income or not, based upon this person's gender, ethnicity, degree, and job type. Many algorithms for learning various classes of classification models have been proposed during the last decades. The quality of a classifier is measured by its predictive accuracy when classifying previously unseen examples. To assess the accuracy of a classifier, usually a labeled test-set is used; test samples from which the label is removed are classified by the model and the predicted label is compared to the true label.

For the vast majority of these classification techniques maximizing *accuracy* is the only objective; i.e, when the classifier is applied on new data, the percentage of correctly labeled instances should be as high as possible. As explained in detail in Chapter 3 of this book, however, blindly optimizing for high accuracy may lead to undesirable side-effects such as discriminatory classifiers. In this chapter we study the following fictitious case: a bank wants to attract new, preferably rich customers.For this purpose, the dataset of Table 12.1 of its current clients is gathered and labeled according to their income. On the basis of this dataset, a classifier is learnt and applied on the profiles of some prospective clients. If the classifier predicts that the candidate has a high income, a special promotion will be offered to him or her. Such promotional schemes targeting particularly profitable groups are not uncommon in commercial settings. In the dataset of Table 12.1, however, we can clearly observe that the positive label is strongly correlated to males and to the native people. As a result, the promotional scheme will mainly benefit the group of native males, potentially leading to ethical and legal issues. We will use this scenario as a running example.

In this chapter, we concentrate on the very specific case in which the input data for training a classifier can be discriminatory; for instance due to historical discrimination in decision making. And, it is either forbidden by law, or ethically unacceptable, that a classifier learns and applies this discrimination on new instances. We assume that the class label that needs to be predicted can take two values: $+$ and $-$. Furthermore, there is only one sensitive attribute S that can take two values; one for the deprived community (f for "female"), and one for the favored community (m for "male"). This setting represents the simplest possible of all situations and marks the starting point of the recent discrimination-aware research. For a discussion on more elaborated settings which builds upon this base case, but involves a more complex ecology of attributes, see Chapter 8 of this book.

First we motivate the problem of discrimination-free classification by relating it to existing anti-discrimination laws that prohibit discrimination in housing, employment, financing, insurance, and wages on the basis of race, color, national origin, religion, sex, familial status, and disability (Section 12.2.1). For a more in-depth discussion on anti-discrimination and privacy legislation, we refer the interested reader to Chapter 4 of this book. we give a measure for discrimination on which the problem of *classification without discrimination* will be based (Section 12.2.2). Then, we show how to learn accurate classifiers on discriminatory training data that do not discriminate in their future predictions (Section 12.3). Particularly, we discuss three types of techniques that lead to discrimination-free classifiers. The three classes of techniques and where in the classifier learning process they take place is illustrated in Figure 12.1.

Input Training data	→	**Learning** Induce classifier	→	**Output** Predictive Model
(Section 3.1) - Instance relabeling (Massaging) - Reweighing & Resampling *(Chapter 13)* *- Rule hiding*		(Section 3.2) - DA-Decision trees *(Chapter 14)* *- EM for Bayesian nets*		(Section 3.3) - Leaf Relabeling in decision trees *(Chapter 14)* *- Adjusting thresholds in Naïve Bayes*

Fig. 12.1 Graphical illustration of the three classes of discrimination-free techniques for classification

The first class of techniques removes the discrimination from the input data, either by selectively relabeling some of the instances (we call this *massaging*); for instance, in the example above, some of the unsuccessful females could be labeled as successful and some of the successful males as unsuccessful, or by resampling the input data; that is, some of the successful males are removed from the input data, and some of the successful females' records get duplicated, or by reweighing, that is assigning higher weights for unsuccessful females and lower weight for successful males(Calders, Kamiran, & Pechenizkiy,2009; Kamiran & Calders, 2009a). Another approach that belongs to this class is described in Chapter 13 of this book; based on a collection of discriminative rules detected by discrimination discovery techniques as described in Chapter 5 of this book, rule hiding techniques from privacy preserving data mining (Chapter 11 of this book) are used to suppress the discriminative rules in the input data.

Table 12.1 Sample relation for the income class example

Sex	Ethnicity	Highest Degree	Job Type	Class
m	native	university	board	+
m	native	high school	board	+
m	native	university	education	+
m	non-native	university	healthcare	+
m	non-native	none	healthcare	-
f	non-native	high school	board	-
f	native	university	education	-
f	native	none	healthcare	+
f	non-native	high school	education	-
f	native	university	board	+

The second class of techniques is based upon the modification of the classifier learning procedure itself. We show how a decision tree learning algorithm can be adapted for inducing discrimination-free predictive models. Technical details of this approach can be found in (Kamiran et al., 2010a). Another approach that belongs to this class, a non-discriminating Bayesian classifier, can be found in Chapter 14 of this book.

The third class of techniques is based upon the post-processing of the learnt models. We explain one decision tree leaves relabeling approach that allows to make an already induced decision tree, with an off-the-shelf approach like C4.5 on biased historical data, discrimination-free (Kamiran et al., 2010b). Another technique in this class, but for Bayesian models is presented in Chapter 14 of this book.

We illustrate the behavior of these different types of techniques in Section 12.4 using the well-known *Adult* dataset (Frank & Asuncion, 2010). The goal associated with this dataset is to predict, for promotional purposes, whether a person falls into the high or the low income class. The dataset, however, exhibits a significant gender-gap with respect to income; there are substantially less females with a high income than males. Nevertheless, as sketched in the example above, we want to learn a classifier which is gender-neutral. The sensitive attribute is thus gender, and the deprived community are the females, the favored community – the males. For the discussed techniques, we show that they clearly outperform the traditional classification approaches for this task; without trading in too much accuracy, the discrimination in the learnt classifier's predictions is reduced to an acceptable level.

12.2 Problem Statement: Discrimination-Aware Classification

The input to our problem consists of a dataset in tabular format, such as the one in Table 12.1. Every row in the table represents one instance, and there is a special column *Class*, indicating the class label that we need to learn to predict for new instances. Based upon the dataset it is expected that a model is learnt that can predict the class based upon the other attributes of a previously unseen instance. Furthermore, in the discrimination-aware paradigm, we assume that a *sensitive attribute*, here "sex" and a *sensitive attribute value*, in this case "female" are set to indicate a subset of the instances which should not be discriminated against. The goal is now to learn a predictive model that will classify future instances as accurately as possible into the high or low income class, under the constraint that the predictions should not be discriminative with respect to the sensitive attribute sex.

In the example dataset of Table 12.1, we can see that 4 out of 5 males have the positive class label, whereas for the females, only 2 out of 5 have the positive class label. Nevertheless, our classifier should learn a predictive model which will, overall, assign to the same proportion of males and females the positive class. Notice that in the problem statement we do not consider the potential existence of other attributes that can explain (part of) the discrimination. For a discussion on explanatory attributes and how they influence the problem we refer to Chapter 8 of this book. In this chapter we concentrate only on the case in which none of the other attributes can be used to justify the discrimination.

Before the formal definition of the discrimination-free classification we give a discussion of anti-discrimination legislation followed by an explanation of how the discrimination should be measured.

12.2.1 Motivation: Links to Legislation

There are many anti-discrimination laws that prohibit discrimination in housing, employment, financing, insurance, wages, etc. on the basis of race, color, national origin, religion, sex, familial status, and disability etc. For instance, the Australian Sex Discrimination Act 1984 (Australian Law, 1984) prohibits discrimination in work, education, services, accommodation, land, clubs on the grounds of marital status, pregnancy or potential pregnancy, and family responsibilities. The US Equal Credit Opportunity Act 1974 (US Legislation, 1968) declares unlawful for any creditor to discriminate against any applicant, with respect to any aspect of a credit transaction, on the basis of race, color, religion, national origin, sex or marital status, or age. Similarly there are many other laws which prohibit discriminatory practices. Our discrimination-aware classification paradigm clearly applies to these situations. If we are interested to apply classification techniques and our available *historical* data contains discrimination, it will be illegal to use traditional classifiers without taking the discrimination aspect into account.

The problem of classification with non-discrimination constraints is not a trivial one. The straightforward solution of removing the sensitive attribute from the training-set does in most cases not solve this problem at all. Consider, for example, the German Credit Dataset available in the UCI ML-repository (Frank & Asuncion, 2010). This dataset contains demographic information of people applying for loans and the outcome of the scoring procedure. The rating in this dataset correlates with the age of the applicant. Removing the *age* attribute from the data, however, does not remove the age-discrimination, as many other attributes such as, *own_house*, indicating if the applicant is a home-owner, turn out to be good predictors for *age*. A parallel can be drawn with the practice of *redlining*: denying inhabitants of particular racially determined areas from services such as loans. It describes the now abolished practice of marking a red line on a map to delineate the area where banks would not invest; later the term was used for indirect discrimination against a particular group of people (usually by race or sex) no matter the geography[1].

12.2.2 Measuring Discrimination

There are many different ways in which discrimination could be quantified, and each of them has its own advantages and disadvantages. Here, in this chapter, and in our earlier works (Calders et al., 2009; Kamiran & Calders, 2010; Kamiran et al., 2010b; Kamiran & Calders, 2009a,b; Kamiran et al., 2010a; Calders & Verwer, 2010), we define the level of discrimination in a dataset as the difference between the probability that someone from the favored group gets a positive class and the probability that someone from the deprived community gets a positive class. For alternative measures of discrimination, see Chapters 5 and 6 of this book.

For the running example of Table 12.1, the discrimination with respect to the deprived community *Sex=female* is 4/5 - 2/5 = 40%. Formally, for a *sensitive attribute* S, *deprived community (sensitive attribute value)* f, *favored community* m, the discrimination in D with respect to the group $S = f$, denoted $disc_{S=f}(D)$, is defined as:

$$disc_{S=f}(D) := \frac{|\{X \in D \mid X(S) = m, X(Class) = +\}|}{|\{X \in D \mid X(S) = m\}|} - \frac{|\{X \in D \mid X(S) = f, X(Class) = +\}|}{|\{X \in D \mid X(S) = f\}|} .$$

When measuring the discrimination of a classifier, we want to assess how the classifier will act on new, previously unseen examples. We assume a setting in which one example comes at a time, and the classifier needs to assign a label to them immediately. In order to assess the level of discrimination of the classifier when it would be applied to unseen examples, we use a test-set; that is, following standard machine learning practice, before learning a classifier, we split the dataset in two parts; one for learning the classifier, and one for measuring its quality. The examples of the test-set (with their labels removed) are passed one by one to the classifier and its decisions are recorded. After that, the discrimination of the classifier can be assessed as follows. The discrimination of the classifier C with respect to the group $S = f$ on a test dataset D_{test}, denoted $disc_{S=f}(C, D_{test})$, is defined as:

[1] Source: `http://en.wikipedia.org/wiki/redlining`, November 17th, 2011.

$$disc_{S=f}(C, D_{test}) := \frac{|\{X \in D_{test} \mid X(S) = m, C(X) = +\}|}{|\{X \in D_{test} \mid X(S) = m\}|} - \frac{|\{X \in D_{test} \mid X(S) = f, C(X) = +\}|}{|\{X \in D_{test} \mid X(S) = f\}|} .$$

12.3 Techniques for Discrimination-Free Classification

In this section we discuss different techniques for discrimination-aware classification. First, we discuss data pre-processing techniques to make the training data unbiased before learning a classifier. Second, we discuss the adaptation of a classifier learning procedure itself to make it discrimination-free. Third, we discuss the modification of the post-processing phase of a learnt classifier to make it unbiased.

12.3.1 Pre-processing Techniques

The first kind of solutions are based on removing the discrimination from the training dataset. If we can remove discrimination directly from the source data, a classifier can be learnt on a cleaned, discrimination-free dataset. Our rationale for this approach is that, since the classifier is trained on discrimination-free data, it is likely that its predictions will be (more) discrimination-free as well, as the classifier will no longer generalize the discrimination. The first approach we discuss here is called *massaging* the data (Kamiran & Calders, 2009a). It is based on changing the class labels in order to remove the discrimination from the training data. The second approach is less intrusive as it does not change the class labels in the training data. Instead, weights are assigned to the data objects to make the dataset discrimination-free. This approach is called *reweighing* (Calders et al., 2009). Since reweighing requires the learner to be able to work with weighted tuples, we propose another variant, in which we re-sample the dataset in such a way that the discrimination is removed. We refer to this approach as *Sampling* (Kamiran & Calders, 2010).

12.3.1.1 Massaging

In *massaging* we change the class labels in the training set; some objects of the deprived community change from class − to +, and the same number of objects of the favored community change from + to −. In this way the discrimination decreases, yet the overall class distribution is maintained; the same number of people has the positive class as before. This strategy reduces the discrimination to the desirable level with the least number of changes to the dataset while keeping the overall class distribution fixed. Notice that we do not randomly pick the objects to relabel. Instead, first we learn a regular, possibly discriminative (i.e. not discrimination-free) classifier. This classifier, although not acceptable as a final result, still provides useful information. Based on this classifier we can see, for the deprived and favored communities separately, which instances are closest to the *decision boundary*. Many classifiers assign a probability of being in the positive class to the instances, and if

Table 12.2 Sample relation for the income class example with positive class probability

Sex	Ethnicity	Highest Degree	Job Type	Class	Prob
m	native	university	board	+	.99
m	native	high school	board	+	.90
m	native	university	education	+	.92
m	**non-native**	**university**	**healthcare**	**+**	**.76**
m	non-native	none	healthcare	-	.44
f	non-native	high school	board	-	.09
f	**native**	**university**	**education**	**-**	**.66**
f	native	none	healthcare	+	.66
f	non-native	high school	education	-	.02
f	native	university	board	+	.92

this probability exceeds 0.5, the object is assigned to the positive class. The objects close to the decision boundary are those with a probability close to 0.5. We select these objects first to relabel.

Example 1. *Consider again the dataset D given in Table 12.1. We want to learn a classifier to predict the class of objects for which the predictions are non-discriminatory towards Sex = f. In this example we rank the objects by their positive class probability given by a* Naive Bayes *classification model. In Table 12.2 the positive class probabilities as given by this ranker are added to the table for reference (calculated using the "NBS" classifier of Weka (Hall et al., 2009)).*

In the second step, we arrange the data separately for female *applicants with class − in descending order and for* male *applicants with class + in ascending order with respect to their positive class probability. Relabeling one promotion candidate and one demotion candidate makes the data discrimination-free. Hence, we relabel the top promotion candidate; that is, the highest scoring female with a negative class label, and the top demotion candidate; that is, the lowest scoring male with a positive class label (the bold examples in Table 12.2). After the labels for these instances are changed, the discrimination decreases from* 40% *to* 0%*. The resulting dataset is used as a training set for classifier induction.*

12.3.1.2 Reweighing and Resampling

The *massaging* approach is rather intrusive as it changes the class labels of the objects. Our second approach does not have this disadvantage. Instead of relabeling the objects, different weights are attached to them. For example, the deprived community objects with $X(Class) = +$ get higher weights than the deprived community objects with $X(Class) = -$ and the favored community objects with $X(Class) = +$ get lower weights than the favored community objects with $X(Class) = -$. We refer to this method as *massaging*. Again we assume that we want to reduce the discrimination to 0 while maintaining the overall positive class probability. We now discuss the idea behind the weight calculation.

If the dataset D would have been unbiased; that is, S and $Class$ were statistically independent, the expected probability of being non-native and having the positive class $P_{exp}(f \wedge +)$ would be:

$$P_{exp}(f \wedge +) := \frac{|X(S) = f|}{|D|} \times \frac{|X(Class) = +|}{|D|} \ .$$

For instance in the example dataset of Table 12.1, 50% of people are female, and 60% of people have a positive class. Therefore, if the dataset was non-discriminatory, one would expect also 60% of females to have the positive class, which gives in total $50\% \times 60\% = 30\%$ of people being female and having the positive class. In reality, however, the observed probability in D,

$$P_{obs}(f \wedge +) := \frac{|X(S) = f \wedge X(Class) = +|}{|D|}$$

might be different. If the expected probability is higher than the observed probability value, it shows the bias towards class '$-$' for those objects X with $X(S) = f$. Continuing the example, in the dataset of Table 12.1, we observe that only 2 people in the dataset are female and have a positive class label, so the observed probability of female and positive is 20%, which is considerably lower than the expected 30%, thus indicating discrimination.

To compensate for the bias, we assign weights to objects. If a particular group is under-represented, we give members of this group a higher weight, making them more important in the classifier training process. The weight we assign to an object is exactly the expected probability divided by the observed probability. In the example this would mean that we assign a weight of 30% divided by 20% = 1.5 to females with a positive class label. In this way we assign a weight to every object according to its S- and $Class$-values. We call the dataset D with the added weights, D_W. It can be proven that the resulting dataset D_W is unbiased; that is, if we multiply the frequency of every object by its weight, the discrimination is 0. On this balanced dataset the discrimination-free classifier is learnt.

Since not every classification algorithm can directly work with weights, we may also use the weights when resampling the dataset; that is, we randomly select objects from our training set to form a new dataset. When forming the new dataset, some objects may be omitted and some may be duplicated. In the sampling procedure, the weight of an object represents its relative chance of being chosen from the dataset; that is, an object with a weight of 2.4 in every selection step has a 4 times higher probability of being chosen than an object with a weight of 0.6. This variant is called *resampling*.

Example 2. *Consider again the dataset in Table 12.1. The weight for each data object is computed according to its S- and $Class$-value, e.g. for instances with values $X(Sex) = f$ and $X(Class) = +$:*

$$W(X) = \frac{0.5 \times 0.6}{0.2} = 1.5 \ .$$

Similarly the weights of all other combinations is as follows:

$$W(X) := \begin{cases} 1.5 & \textit{if } X(Sex) = f \textit{ and } X(Class) = + \\ 0.67 & \textit{if } X(Sex) = f \textit{ and } X(Class) = - \\ 0.75 & \textit{if } X(Sex) = m \textit{ and } X(Class) = + \\ 2 & \textit{if } X(Sex) = m \textit{ and } X(Class) = - \end{cases} .$$

12.3.1.3 Related Approaches

The authors of (Luong et al., 2011) propose a variant of k-NN classification for the discovery of discriminated objects. They consider a data object as discriminated if there exists a significant difference of treatment among its neighbors belonging to the deprived community and its neighbors not belonging to it (that is, the favored community). They also propose a discrimination prevention method by changing the class labels of these discriminated objects. This discrimination prevention method is very close to our massaging technique (Kamiran & Calders, 2009a), especially when the ranker being used is based upon a nearest neighbor classifier. There is, however, one big difference: whereas in massaging only the minimal number of objects is changed to remove all discrimination from the dataset, the authors of (Luong et al., 2011) propose to continue relabeling until all labels are consistent. From a legal point of view, the cleaned dataset obtained by (Luong et al., 2011) is probably more desirable as it contains less "illegal inconsistencies." For the task of discrimination-aware classification, however, it is unclear if the obtained dataset is suitable for learning a discrimination-free classifier.

The authors of (Hajian, Domingo-Ferrer, & Martinez-Balleste, 2011; Hajian, Domingo-Ferrer, & Martínez-Ballesté, 2011) also propose methods similar to massaging to preprocess the training data in such a way that only potentially non-discriminatory rules can be extracted. For this purpose they modify all the items in a given dataset that lead to the discriminatory classification rules by applying rule hiding techniques on either given, or discovered discriminative rules. For an extensive description of this technique, see Chapter 13 of this book.

12.3.2 Changing the Learning Algorithms

In this section, we discuss the discrimination-aware techniques in which we modify the classification model learning process itself to produce discrimination-free classifiers. For this purpose, we discuss the discrimination-aware decision trees construction in which we modify the decision tree construction procedure to make them discrimination-free.

12.3.2.1 Discrimination-Aware Decision Tree Induction

Traditionally, when constructing a decision tree (Quinlan, 1993), we iteratively refine a tree by splitting its leaves until a desired objective is achieved. Consider the dataset given in Table 12.1. Suppose we want to learn a tree over this dataset in

Table 12.3 Gini Index for different possible splits of the data from Table 12.2

Condition	left branch		right branch		Gini Index
	# pos	# neg	# pos	# neg	
sex=m	4	1	3	2	0.4
ethnicity=native	5	1	1	3	0.32
diploma=none	1	1	5	3	0.48
...	...	...	...	...	...

order to predict the *Class*. Initially, we start with a tree consisting of only one node, predicting the majority class '+'. Then, iteratively, we refine the tree by considering all possible splitting criteria, and evaluating which split is the best. Selecting the best split is done by observing how the split condition separates the positive class from the negative class. A split that is better at separating the classes will score higher on the quality measure. For the dataset of Table 12.1, the different splits are as follows: The split $sex = m$ would divide the dataset into those instances that satisfy the condition (the left branch), including 4 positive and 1 negative instance, and those instances that do not satisfy the condition (the right branch), having 3 negative and 2 positive examples. Based on these figures, a degree of impurity can be computed, in this case, based upon the Gini index (Lerman & Yitzhaki, 1984): to compute the Gini-index of a split, we first separate the dataset according to the split criterion. For each partition, the relative frequencies of the positive and negative class, f_+ and f_- respectively, are counted. The Gini-index is then the weighted average of the Gini-score $1-(f_+^2+f_-^2)$. If a partition is pure, this implies that either $f_+ = 1$ and $f_-^2 = 0$, or $f_+ = 0$ and $f_-^2 = 1$. In both cases, the partition contributes $1-(f_+^2+f_-^2) = 0$ to the gini-score of the split. The contribution of a partition is the highest if it is maximally impure; i.e., $f_+ = f_-^2 = 0.5$. For the example split $sex = m$, the partition containing the males contributes $1-((1/5)^2+(4/5)^2) = 8/25$, while the partition with the females contributes $1-((2/5)^2+(3/5)^2) = 12/25$. The Gini-index for the split is now the weighted average over the two partitions, being: $0.5(8/25)+0.5(12/25) = 10/25 = 0.4$.

The better the split separates positive from negative, the lower the impurity. From all splits the one with the lowest impurity is selected. The dataset is split in two parts, according to the splitting criterion and the procedure continues on both parts until a stopping condition is met. In (Kamiran et al., 2010b, 2010a) we show how the splitting criterion can be changed in such a way that not only the impurity with respect to the class label can be incorporated, but also the level of discrimination introduced by the split. In particular, we do not only compute how good the split predicts the class label, but also how good it predicts the sensitive attribute, using the same gini-index, but now with the relative frequencies of the deprived and favored communities in the partitions of the split. The good split will then be the one that achieves a high purity with respect to the class label, but a low purity with respect to the sensitive attribute. In the running example this means that we want splits that are good for distinguishing high income from low income people, without separating

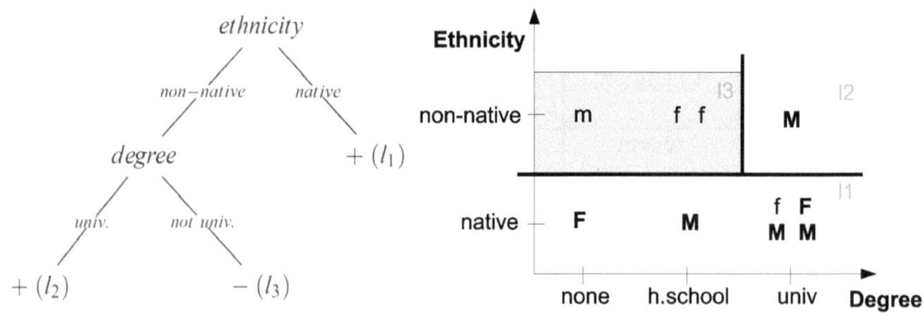

Fig. 12.2 Decision tree with the partitioning induced by it. The bold capital letters in the partitioning denote the positive examples, the lowercase letters the negative examples. m/M denotes a male, f/F denotes a female. The grey background denotes regions where the majority class is −. The discrimination of the tree is 20%.

too much the males from the females. In that way we can guide the iterative tree refinement procedure, disallowing steps that would increase discrimination in the predictions or explicitly adding a penalty term for increasing discrimination into the quality scores of the splits.

12.3.2.2 Related Approaches

Also for other learning algorithms a similar approach could be applied by embedding the anti-discrimination constraints deeply into the learning algorithm. Another example of such an approach is described in Chapter 14 of this book, where a Naïve Bayes model is learnt which explicitly models the effect of the discrimination. By learning the most probable model that leads to the observed data, under the assumption that discrimination took place, one can reverse-engineer the effect of the discrimination and hence filter it out when making predictions.

12.3.3 Post-Processing the Induced Models

Our third and last type of discrimination-aware techniques is based upon the modification of the post-processing phase of the learnt model. We discuss the decision tree leaf relabeling approach of (Kamiran et al., 2010b) where we assume that a tree is already given and the goal is to reduce the discrimination of the tree by changing the class labels of some of the leaves.

12.3.3.1 Decision Tree Leaf Relabeling

The rationale behind this approach is as follows. A decision tree partitions the space of instances into non-overlapping regions. See, for example, Figure 12.2. In this figure (left) a fictitious decision tree with 3 leaves is given, labeled l_1 to l_3. The right part of the figure shows the partitioning induced by the decision tree. For example, the third leaf in the tree corresponds to all non-native people without a university diploma. The leaves can hence be seen as non-overlapping "profiles" dividing up the space of all instances. Every example fits exactly one profile, and with every profile exactly one class is associated. When a new example needs to be classified by a decision tree, it is given the majority class label of the region/profile it falls into. If some of the profiles are very homogeneous with respect to the sensitive attribute; for instance, containing only members of the deprived community, then this may lead to discriminative predictions. In l_3, for instance, two thirds of the instances are from the deprived community. The relabeling technique now consists of changing the labels for those regions where this results in the highest reduction in discrimination while trading in as little accuracy as possible. Conceptually this method corresponds to merging neighboring regions to form larger, less discriminative profiles. The process of relabeling continues until the discrimination is removed.

Example 3. *Consider the example decision tree given in Figure 12.2. The discrimination of the decision tree is 20%. Suppose we want to reduce the discrimination to 5%. For each of the leaves it is given how much the discrimination changes ($\Delta disc$) when relabeling the node, and how much the accuracy decreases (Δacc). The node for which the tradeoff between discrimination reduction versus lowered accuracy is most beneficial, is selected first for relabeling.*

Node	Δacc	$\Delta disc$	$\frac{\Delta disc}{\Delta acc}$
l_1	-40%	0%	0
l_2	-10%	10%	1
l_3	-30%	10%	$1/3$

In this particular case, the reduction algorithm hence pick l_2 to relabel; that is, the split on degree is removed and leaves l_2 and l_3 are merged.

12.3.3.2 Related Approaches

The idea of model correction has been explored in different settings, particularly in cost-sensitive learning, learning from imbalanced data, and context sensitive or context-aware learning. Concrete examples of model correction include Naive Bayes prior correction (also in Chapter 14 of this book) and posterior probabilities correction based on a confusion matrix (Morris & Misra, 2002); nearest neighbor based classification or identification correction based on current context, e.g. in driver-route identification (Mazhelis, Zliobaite, & Pechenizkiy, 2011) or in context-sensitive correction of phone recognition output (Levit, Alshawi, Gorin, & Nöth, 2003). The tree node relabeling ideas have been used in recognizing textual entailments (Heilman & Smith, 2010) and probabilistic context-free grammar parsing (Johnson, 1998). But these are not related to the idea of decision tree learning. However, we are not aware of other approaches directly related to the discussed idea of leaf relabeling in decision trees applicable to our settings.

12.4 Experiments

The different techniques discussed in this chapter have been experimented with extensively. We refer the interested reader for the detailed discussion of the experimental studies and results to (Kamiran et al., 2010b,a; Kamiran & Calders, 2012; Kamiran, 2011). In this section we give an overview of the most important empirical results for the *Adult* dataset. This dataset has 48 842 instances and contains demographic information of people. The associated prediction task is to determine whether a person makes over 50K per year or not; that is, income class *High* or *Low* has to be predicted. The other attributes in the dataset include: age, type of work, education, years of education, marital status, occupation, type of relationship (husband, wife, not in family), sex, race, native country, capital gain, capital loss and weekly working hours. We consider *Sex* as sensitive attribute. In our sample of the dataset, 16 192 citizens have $Sex = f$ and 32 650 have $Sex = m$. The discrimination with respect to $Sex = m$ in the historical data is 19.45%: $P(X(Class) = + \mid X(Sex) = m) - P(X(Class) = + \mid X(Sex) = f) = 19.45\%$. The goal is to learn a classifier that has minimal discrimination and maintains high accuracy.

Figure 12.3 shows the result of experiments when we learn decision trees after applying our proposed discrimination-aware preprocessing techniques on the training data (label 'Preprocessing'), with discrimination-aware splitting criteria (label 'Learner-adaptation'), with leaf relabeling (label 'Postprocessing'), a Naïve Bayes model of Chapter 14 of this book (label '3-NaiveBayes') and learnt without any discrimination-aware technique (label 'Zero-treatment'). We observe in Figure 12.3 that the discrimination-aware techniques discussed in this chapter reduce the discrimination significantly while maintaining a high accuracy as compared to the ordinary methods. For instance, a traditional decision tree without using any discrimination removal method classifies the future data objects with 16.65%

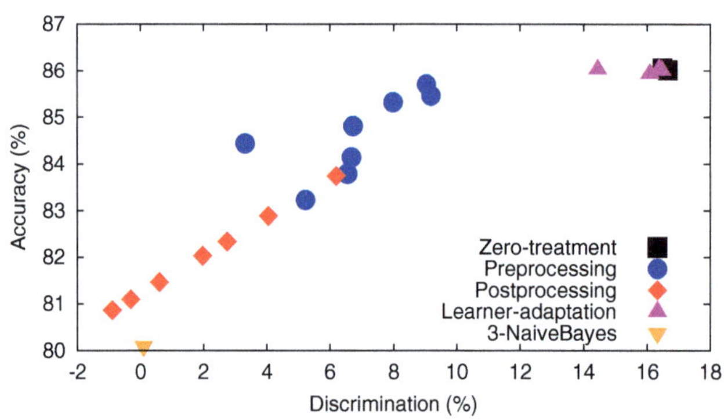

Fig. 12.3 Comparison of techniques discussed in Section 12.3.1 (label Preprocessing), Section 12.3.2 (label Learner-adaptation), Section 12.3.3 (label Postprocessing), Naïve Bayes model of Chapter 14 (label 3-NaiveBayes), and ordinary methods (label Zero-treatment) over the Adult dataset.

discrimination and 86.01% accuracy even though the sensitive attribute was not used at the prediction time. We observe in our experiments that learning a decision tree with modified splitting criterion, that is, using the second type of discrimination-aware classification alone does not significantly reduce the discrimination. However, when the decision trees are learnt on cleaner data obtained with discrimination-aware pre-processing techniques, the discrimination is reduced to 3.32% while keeping the accuracy at 84.44%. The decision trees with leaf relabeling were able in our experiment to reduce the discrimination to 0% while keeping a reasonably high accuracy. Figure 12.3 also shows that our proposed methods outperform the discrimination-aware Naïve Bayes model of Chapter 14 of this book with respect to the accuracy-discrimination trade-off.

12.5 Discussion and Conclusion

In this chapter we discussed the idea of discrimination-aware classification and introduced a procedural way to calculate the discrimination in a given dataset and in the predictions of a classifier. We also discussed three types of techniques to learn the discrimination-free classifiers which include data preprocessing techniques, an adapted classifier learning procedure and an approach for postprocessing of a learnt decision tree by changing the labels of some of its leaves to make the final predictive model discrimination-free. Finally, we presented empirical validation results showing that the discrimination-aware classification methods predict labels for the previously unseen data objects with no or significantly lower discrimination and with the minimal loss of accuracy.

Depending on the situation one of the proposed techniques may be better than another. First of all, if none of the other attributes is correlated to the sensitive attribute, clearly it suffices to just remove this attribute. Unfortunately this is seldomly the case, and even if it is the case, no guarantees can be given that no such correlations exist. The presented preprocessing techniques have the advantage that they make input data discrimination-free which can then be used by any classification algorithm, yet have the disadvantage of giving no guarantee about the degree of discrimination in the final classifier. The model post-processing techniques do not have this disadvantage; in principle the postprocessing is continued until a discrimination-free classifier (on a validation set) is obtained. The model post-processing techniques as well as the learner adaptation techniques on their turn, however, have the disadvantage of being model and even algorithm specific; for every classifier new algorithms will have to be invented. In the experiments it was further shown that the learner adaptation approach did not work as expected, unless it was combined with the post-processing techniques. This surprising failure calls for more research to better understand the reasons for it.

Despite of showing some promising results on discrimination-free classifier construction, our study is far from complete. For instance, often there is a much more complex ecology of attributes than what is assumed in the chapter. In the chapter we assume there is just one sensitive attribute, dividing the objects into one disadvantaged and one advantaged group. Often, however, there may be more than two groups, each of which are advantaged/disadvantaged to a different level. Consider, e.g., different ethnic minorities being treated in different ways. Furthermore, there may be multiple of such sensitive attributes; e.g., gender, age, and ethnicity. Removing gender-discrimination by the preprocessing techniques may introduce an age-discrimination. Furthermore, it could be the case that even if discrimination does not manifest itself at the general level, in some specialized niches or contexts, there might be discrimination present. Chapter 5 of this book deals with the detection of such subtle contexts for discrimination. Also, as discussed in Chapter 8 of this book, not all difference in acceptance rates between an advantaged and a disadvantaged group is due to discrimination. If people in the disadvantaged group are more likely to be lowly educated, as a result their salaries will be lower on average, without this difference necessarily indicating a discrimination. As a conclusion, the area of discrimination-aware classification remains a rich source of inspiration and application area for novel techniques in the data mining area, and we hope to see significant contributions in future to this ethically and societally important research area, leading towards providing companies and practitioners with the necessary toolkit for data-driven discrimination-free decision making.

References

Australian Law. Australian Sex Discimination Act 1984. Australian sex discimination act 1984. via: (1984) http://www.comlaw.gov.au/Details/C2010C00056

Calders, T., Kamiran, F., Pechenizkiy, M.: Building Classifiers with Independency ConstraintsBuilding classifiers with independency constraints. In: Saygin, Y., et al. (eds.) ICDM Workshops 2009, IEEE International Conference on Data Mining Workshops, Miami, Florida, USA, December 6, pp. 13–18 (2009)

Calders, T., Verwer, S.: Three naive Bayes approaches for discrimination-free classificationThree naive bayes approaches for discrimination-free classification. Data Mining and Knowledge Discovery 21, 277–292 (2010)

Frank, A., Asuncion, A.: UCI Machine Learning Repository. UCI machine learning repository (2010), http://archive.ics.uci.edu/ml

Hajian, S., Domingo-Ferrer, J., Martinez-Balleste, A.: Discrimination prevention in data mining for intrusion and crime detectionDiscrimination prevention in data mining for intrusion and crime detection. In: IEEE Symposium on Computational Intelligence in Cyber Security (CICS)IEEE Symposium on Computational Intelligence in Cyber Security (CICS), pp. 47–54 (2011)

Hajian, S., Domingo-Ferrer, J., Martínez-Ballesté, A.: Rule protection for indirect discrimination prevention in data miningRule protection for indirect discrimination prevention in data mining. In: Modeling Decision for Artificial Intelligence, pp. 211–222 (2011)

Hall, M., Frank, E., Holmes, G., Pfahringer, B., Reutemann, P., Witten, I.: The WEKA data mining software: An updateThe WEKA data mining software: An update. ACM SIGKDD Explorations Newsletter 11, 110–118 (2009)

Heilman, M., Smith, N.A.: Tree edit models for recognizing textual entailments, paraphrases, and answers to questionsTree edit models for recognizing textual entailments, paraphrases, and answers to questions. In: Human Language Technologies: The 2010 Annual Conference of the North American Chapter of the Association for Computational Linguistics, pp. 1011–1019. USAAssociation for Computational Linguistics, Stroudsburg (2010)

Johnson, M.: PCFG models of linguistic tree representationsPCFG models of linguistic tree representations. Comput. Linguist. 24, 613–632 (1998)

Kamiran, F.: Discrimination-aware ClassificationDiscrimination-aware classification. Doctoral dissertation, Eindhoven University of Technology, The Netherlands (2011)

Kamiran, F., Calders, T.: Classifying without discriminatingClassifying without discriminating. In: 2nd IEEE International Conference on Computer, Control and Communication (IC4), pp. 1–6 (2009a)

Kamiran, F., Calders, T.: Discrimination-Aware ClassificationDiscrimination-aware classification. In: 21st Benelux Conference on Artificial Intelligence (BNAIC), pp. 333–334 (2009b)

Kamiran, F., Calders, T.: Classification with No Discrimination by Preferential Sampling-Classification with no discrimination by preferential sampling. In: Proceedings Machine Learning Conference of Belgium and The Netherlands, BENELEARN (2010)

Kamiran, F., Calders, T.: Data Preprocessing Techniques for Classification without DiscriminationData preprocessing techniques for classification without discrimination. Knowledge and Information Systems (2012) (to Appear)

Kamiran, F., Calders, T., Pechenizkiy, M.: Discrimination Aware Decision Tree LearningDiscrimination aware decision tree learning Tech. Rep. No. CS 10-13. Eindhoven University of Technolgy. 16 Faisal Kamiran, Toon Calders, and Mykola Pechenizkiy (2010a)

Kamiran, F., Calders, T., Pechenizkiy, M.: Discrimination Aware Decision Tree LearningDiscrimination aware decision tree learning. In: IEEE International Conference on Data Mining IEEE International Conference on Data Mining, pp. 869–874 (2010b)

Lerman, R., Yitzhaki, S.: A Note on the Calculation and Interpretation of the Gini IndexA note on the calculation and interpretation of the gini index. Economics Letters 15(3-4), 363–368 (1984)

Levit, M., Alshawi, H., Gorin, A.L., Nöth, E.: Context-sensitive evaluation and correction of phone recognition outputContext-sensitive evaluation and correction of phone recognition output. In: Proc. of the 8th European Conference on Speech Communication and Technology, EUROSPEECH 2003, ISCA (2003)

Luong, B., Ruggieri, S., Turini, F.: k-NN as an Implementation of Situation Testing for Discrimination Discovery and Preventionk-NN as an Implementation of Situation Testing for Discrimination Discovery and Prevention. In: Proc. ACM SIGKDD International Conference on Knowledge Discovery and Data Mining, pp. 502–510 (2011)

Mazhelis, O., Žliobaitė, I.e., Pechenizkiy, M.: Context-Aware Personal Route Recognition. In: Elomaa, T., Hollmén, J., Mannila, H. (eds.) DS 2011. LNCS, vol. 6926, pp. 221–235. Springer, Heidelberg (2011)

Morris, A., Misra, H.: Confusion matrix based posterior probabilities correctionConfusion matrix based posterior probabilities correction Idiap-RR No. Idiap-RR-53-2002. IDIAP (2002)

Quinlan, J.: C4. 5: programs for machine learning. Morgan Kaufmann (1993)

US Law. The US Equal Credit Opportunity Act.The US equal credit opportunity act. via: (1968), http://www.fdic.gov/regulations/laws/rules/6500-1200.html

Chapter 13
Direct and Indirect Discrimination Prevention Methods

Sara Hajian and Josep Domingo-Ferrer

Abstract. Along with privacy, discrimination is a very important issue when considering the legal and ethical aspects of data mining. It is more than obvious that most people do not want to be discriminated because of their gender, religion, nationality, age and so on, especially when those attributes are used for making decisions about them like giving them a job, loan, insurance, etc. Discovering such potential biases and eliminating them from the training data without harming their decision-making utility is therefore highly desirable. For this reason, anti-discrimination techniques including discrimination discovery and prevention have been introduced in data mining. Discrimination prevention consists of inducing patterns that do not lead to discriminatory decisions even if the original training datasets are inherently biased. In this chapter, by focusing on the discrimination prevention, we present a taxonomy for classifying and examining discrimination prevention methods. Then, we introduce a group of pre-processing discrimination prevention methods and specify the different features of each approach and how these approaches deal with direct or indirect discrimination. A presentation of metrics used to evaluate the performance of those approaches is also given. Finally, we conclude our study by enumerating interesting future directions in this research body.

13.1 Introduction

Unfairly treating people on the basis of their belonging to a specific group, namely race, ideology, gender, etc., is known as discrimination. In law, economics and social sciences, discrimination has been studied over the last decades and anti-discrimination laws have been adopted by many democratic governments. Some examples are the US Employment Non-Discrimination Act (United States

---

Sara Hajian · Josep Domingo-Ferrer
University Rovira i Virgili, Tarragona, Catalonia, Spain
e-mail: {sara.hajian,josep.domingo}@urv.cat

B. Custers et al. (Eds.): Discrimination & Privacy in the Information Society, SAPERE 3, pp. 241–254.
springerlink.com

Congress 1994), the UK Sex Discrimination Act (Parliament of the United Kingdom 1975) and the UK Race Relations Act (Parliament of the United Kingdom 1976). There are several decision-making tasks which lend themselves to discrimination, *e.g.* loan granting, education, health insurances and staff selection. In many scenarios, decision-making tasks are supported by information systems. Given a set of information items on a potential customer, an automated system decides whether the customer is to be recommended for a credit or a certain type of life insurance. Automating such decisions reduces the workload of the staff of banks and insurance companies, among other organizations. The use of information systems based on data mining technology for decision making has attracted the attention of many researchers in the field of computer science. In consequence, automated data collection and a plethora of data mining techniques such as association/classification rule mining have been designed and are currently widely used for making automated decisions.

At first sight, automating decisions may give a sense of fairness: classification rules (decision rules) do not guide themselves by personal preferences. However, at a closer look, one realizes that classification rules are actually learned by the system based on training data. If the training data are inherently biased for or against a particular community (for example, foreigners), the learned model may show a discriminatory prejudiced behavior. For example, in a certain loan granting organization, foreign people might systematically have been denied access to loans throughout the years. If this biased historical dataset is used as training data to learn classification rules for an automated loan granting system, the learned rules will also show biased behavior toward foreign people. In other words, the system may infer that just being foreign is a legitimate reason for loan denial. A more detailed analysis of this fact is provided in Chapter 3.

Figure 13.1 illustrates the process of discriminatory and non-discriminatory decision rule extraction. If the original biased dataset *DB* is used for data analysis without any anti-discrimination process (*i.e.* discrimination discovery and prevention), the discriminatory rules extracted could lead to automated unfair decisions. On the contrary, *DB* can go through an anti-discrimination process so that the learned rules are free of discrimination, given a list of discriminatory attributes (*e.g.* gender, race, age, etc.). As a result, fair and legitimate automated decisions are enabled.

Despite the wide deployment of information systems based on data mining technology in decision making, the issue of anti-discrimination in data mining did not receive much attention until 2008 (Pedreschi *et al.* 2008). After that, some proposals have addressed the discovery and measure of discrimination. Others deal with the prevention of discrimination. The discovery of discriminatory decisions was first proposed by Pedreschi *et al.* (2008) and Ruggieri *et al.* (2010). The approach is based on mining classification rules (the inductive part) and reasoning on them (the deductive part) on the basis of quantitative measures of discrimination that formalize legal definitions of discrimination. For instance, the U.S. Equal

Pay Act (United States Congress 1963) states that: "a selection rate for any race, sex, or ethnic group which is less than four-fifths of the rate for the group with the highest rate will generally be regarded as evidence of adverse impact".

Discrimination can be either direct or indirect (also called systematic, see Pedreschi *et al.* (2008)). Direct discriminatory rules indicate biased rules that are directly inferred from discriminatory items (*e.g.* Foreign worker = Yes). Indirect discriminatory rules (redlining rules) indicate biased rules that are indirectly inferred from non-discriminatory items (*e.g.* Zip = 10451) because of their correlation with discriminatory ones. Indirect discrimination could happen because of the availability of some background knowledge (rules), for example, indicating that a certain zipcode corresponds to a deteriorating area or an area with a mostly black population. The background knowledge might be accessible from publicly available data (*e.g.* census data) or might be obtained from the original dataset itself because of the existence of non-discriminatory attributes that are highly correlated with the sensitive ones in the original dataset.

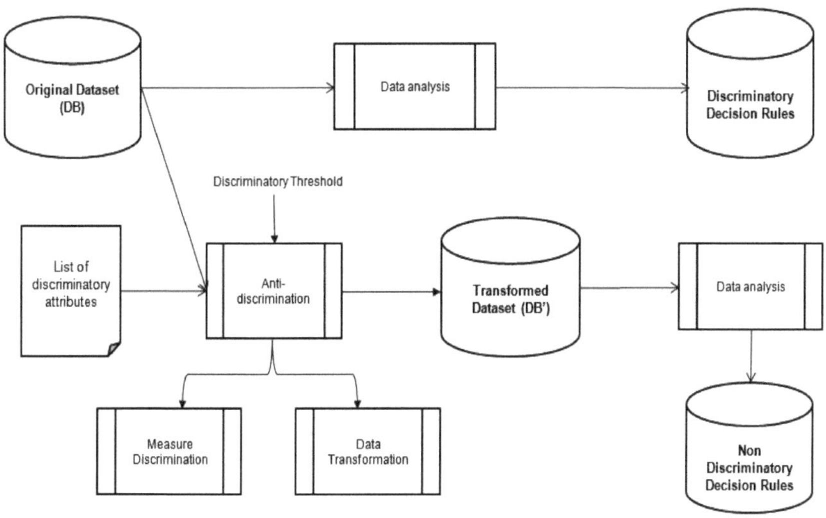

Fig. 13.1 The process of extracting biased and unbiased decision rules

One might conceive that, for direct discrimination prevention, removing discriminatory attributes from the dataset and, for indirect discrimination prevention, removing non-discriminatory attributes that are highly correlated with the sensitive ones could be a basic way to handle discrimination. However, in practice this is not advisable because in this process much useful information would be lost and the quality/utility of the resulting training datasets and data mining models would substantially decrease.

The rest of this chapter is as follows. Section 13.2 contains notation and background on direct and indirect discriminatory rules. Section 13.3 gives a taxonomy of discrimination prevention methods. Section 13.4 describes several pre-processing discrimination prevention methods we have proposed in recent papers.

Metrics to measure the success at removing discriminatory rules are given in Section 13.5. Data quality metrics are listed in Section 13.6. Section 13.7 contains experimental results for the direct discrimination prevention methods proposed. Conclusions and suggestions for future work are summarized in Section 13.8.

13.2 Preliminaries

In this section we briefly recall some basic concepts which are useful to better understand the study presented in this chapter.

13.2.1 Basic Notions

- A *dataset* is a collection of data objects (records) and their attributes. Let *DB* be the original dataset.
- An *item* is an attribute along with its value, *e.g.* {Race=black}.
- An *itemset*, *i.e.* *X*, is a collection of one or more items, *e.g.* {Foreign work er=Yes, City=NYC}.
- A *classification rule* is an expression $X \rightarrow C$, where *C* is a class item (a yes/no decision), and *X* is an itemset containing no class item, e.g. {Foreign worker=Yes, City=NYC} $\rightarrow$ {hire=no}. *X* is called the premise of the rule.
- The *support* of an itemset, *supp(X)*, is the fraction of records that contain the itemset *X*. We say that a rule $X \rightarrow C$ is *completely supported* by a record if both *X* and *C* appear in the record.
- The *confidence* of a classification rule, *conf*($X \rightarrow C$), measures how often the class item *C* appears in records that contain *X*. Hence, if *supp(X)*> 0

$$conf(X \rightarrow C) = \frac{supp(X,C)}{supp(X)}$$

1. Support and confidence range over *[0,1]*.

- A *frequent classification rule* is a classification rule with a support or confidence greater than a specified lower bound. Let *FR* be the database of frequent classification rules extracted from *DB*.
- *Discriminatory attributes and itemsets (protected by law):* Attributes are classified as discriminatory according to the applicable anti-discrimination acts (laws). For instance, U.S. federal laws prohibit discrimination on the basis of the following attributes: race, color, religion, nationality, sex, marital status, age and pregnancy (Pedreschi *et al.* 2008). Hence these attributes are regarded as discriminatory and the itemsets corresponding to them are called discriminatory itemsets. {Gender=Female, Race=Black} is just an example of a discriminatory itemset. Let DA_s be the set of predetermined discriminatory attributes in *DB* and DI_s be the set of predetermined discriminatory itemsets in *DB*.
- *Non-discriminatory attributes and itemsets*: If A_s is the set of all the attributes in *DB* and I_s the set of all the itemsets in *DB*, then nDA_s (*i.e.* set of

non-discriminatory attributes) is A_s - DA_s and nDI_s (*i.e.* set of *non-discriminatory itemsets*) is I_s - DI_s. An example of non-discriminatory itemset could be {Zip= 10451, City=NYC}.

- The *negated itemset, i.e.* ~*X* is an itemset with the same attributes as *X*, but such that the attributes in ~*X* take any value except those taken by attributes in *X*. In this chapter, we use the ~ notation for itemsets with binary or categorical attributes. For a binary attribute, *e.g.* {Foreign worker=Yes/No}, if *X* is {Foreign worker=Yes}, then ~*X* is {Foreign worker=No}. Then, if *X* is binary, it can be converted to ~*X* and vice versa. However, for a categorical (non-binary) attribute, *e.g.* {Race=Black/White/Indian}, if *X* is {Race=Black}, then ~*X* is {Race=White} or {Race=Indian}. In this case, ~*X* can be converted to *X* without ambiguity, but the conversion of *X* into ~*X* is not uniquely defined, which we denote by $\sim X \Rightarrow X$. In this chapter, we use only non-ambiguous negations.

13.2.2 Direct and Indirect Discriminatory Rules

As more precisely discussed in Chapter 5, frequent classification rules fall into one of the following two classes: 1) A classification rule (r: $X \rightarrow C$) with negative decision (e.g. denying credit or hiring) is potentially discriminatory (PD) if $X \cap DI_s \neq \emptyset$, otherwise *r* is potentially non-discriminatory (PND). For example, if DI_s = {Foreign worker=Yes}, a classification rule {Foreign worker=Yes; City=NYC} →Hire=No is PD, whereas {Zip=10451, City=NYC} → Hire=No, or {Experience=Low; City=NYC} → Hire=No are PND.

The word "potentially" means that a PD rule could probably lead to discriminatory decisions, hence some measures are needed to quantify the direct discrimination potential. Also, a PND rule could lead to discriminatory decisions in combination with some background knowledge; *e.g.*, if the premise of the PND rule contains the zipcode as attribute and one knows that zipcode 10451 is mostly inhabited by foreign people. Hence, measures are needed to quantify the indirect discrimination potential as well.

As mentioned before, Pedreschi *et al.* (2008) and Pedreschi *et al.* (2009a) translated qualitative discrimination statements in existing laws, regulations and legal cases into quantitative formal counterparts over classification rules and they introduced a family of measures over PD rules (for example *elift*) for direct discrimination discovery and over PND rules (for example *elb*) for indirect discrimination discovery. Then, by thresholding *elift* it can be assessed whether the PD rule has direct discrimination potential. Based on this measure (*elift*), a PD rule (r: $X \rightarrow C$) is said to be *discriminatory* if $elift(r) \geq \alpha$[1] or *protective* if $elift(r) < \alpha$. In addition, whether the PND rule has indirect discrimination potential can be assessed by thresholding *elb*. Based on this measure (*elb*), a PND rule (r': $X \rightarrow C$) is said to be *redlining* if $elb(r') \geq \alpha$ or *non-redlining (legitimate)* if $elb(r') < \alpha$. For more detailed information and definitions of these measures, see Chapter 5.

[1] Note that α is a fixed threshold stating an acceptable level of discrimination according to laws and regulations. For example, the four-fifths rule of U.S. Federal Legislation sets α=1.25.

13.3 Taxonomy of Discrimination Prevention Methods

Beyond discrimination discovery, preventing knowledge-based decision support systems from making discriminatory decisions (discrimination prevention) is a more challenging issue. The challenge increases if we want to prevent not only direct discrimination but also indirect discrimination or both at the same time. In this section, we present a taxonomy of discrimination prevention methods after having reviewed a collection of independent works in the area. Figure 13.2 shows this taxonomy. In order to be able to classify the various approaches, we consider two orthogonal dimensions based on which we present the existing approaches. As a first dimension, we consider whether the approach deals with direct discrimination, indirect discrimination, or both at the same time. In this way, we separate the discrimination prevention approaches into three groups: *direct discrimination prevention methods*, *indirect discrimination prevention methods*, and *direct and indirect discrimination prevention methods*. The second dimension in the classification relates to the phase of the data mining process in which discrimination prevention is done.

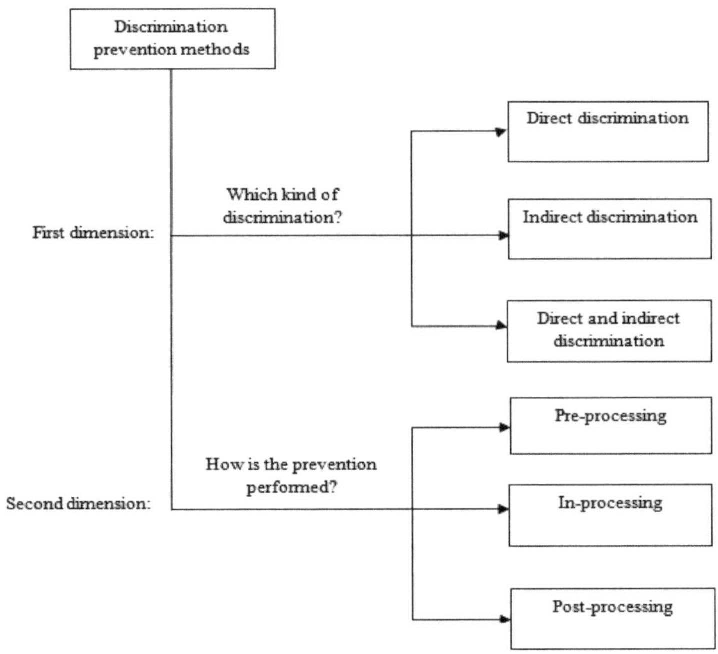

Fig. 13.2 The taxonomy of discrimination prevention methods

Based on this second dimension, discrimination prevention methods fall into three groups (Ruggieri *et al.* 2010): *pre-processing*, *in-processing* and *post-processing* approaches. We next describe these groups:

1.

- Pre-processing. Methods in this group transform the source data in such a way that the discriminatory biases contained in the original data are removed so that no unfair decision rule can be mined from the transformed data; any of the standard data mining algorithms can then be applied. The pre-processing approaches of data transformation and hierarchy-based generalization can be adapted from the privacy preservation literature. Along this line, Kamiran and Calders (2009), Kamiran and Calders (2010), Hajian *et al.* (2011a and 2011b) and Hajian and Domingo-Ferrer (2012) perform a controlled distortion of the training data from which a classifier is learned by making minimally intrusive modifications leading to an unbiased dataset.
- In-processing. Methods in this group change the data mining algorithms in such a way that the resulting models do not contain unfair decision rules (Calders and Verwer 2010, Kamiran *et al.* 2010). For example, an alternative approach to cleaning the discrimination from the original dataset is proposed in Calders and Verwer (2010) whereby the non-discriminatory constraint is embedded into a decision tree learner by changing its splitting criterion and pruning strategy through a novel leaf re-labeling approach. However, it is obvious that in-processing discrimination prevention methods must rely on new special-purpose data mining algorithms; standard data mining algorithms cannot be used because they ought to be adapted to satisfy the non-discrimination requirement.
- Post-processing. These methods modify the resulting data mining models, instead of cleaning the original dataset or changing the data mining algorithms. For example, in Pedreschi *et al.* (2009a), a confidence-altering approach is proposed for classification rules inferred by the rule-based classifier: CPAR (classification based on predictive association rules) algorithm (Yin *et al.* 2003).

13.4 Types of Pre-processing Discrimination Prevention Methods

Although some methods have already been proposed for each of the above mentioned approaches (pre-processing, in-processing, post-processing), discrimination prevention stays a largely unexplored research avenue. In this section, we concentrate on a group of discrimination prevention methods based on pre-processing (first dimension) that could deal with direct or indirect discrimination (second dimension), because pre-processing has the attractive feature of being independent of the data mining algorithms and models. More details, algorithms and experimental results on these methods are presented in Hajian *et al.* (2011a and 2011b) and Hajian and Domingo-Ferrer (2012). The purpose of all these methods is to transform the original data *DB* in such a way as to remove direct or indirect discriminatory biases, with minimum impact on the data and on legitimate decision rules, so that no unfair decision rule can be mined from the transformed data. As part of this effort, the metrics that specify which records should be changed, how many records should be changed and how those records should be changed during data transformation are developed.

There are some assumptions common to all methods in this section. First, we assume the class attribute in the original dataset *DB* to be binary (*e.g.* denying or granting credit). Second, we obtain the database of *discriminatory* and *redlining* rules as output of a discrimination measurement (discovery) phase based on measures proposed in Pedreschi *et al.* (2008) and Pedreschi *et al.* (2009a); discrimination measurement is performed to identify *discriminatory* and *redlining rules* (based on the work in Chapter 5); then a data transformation phase is needed to transform the data in order to remove all evidence of direct or indirect discriminatory biases associated to *discriminatory* or *redlining* rules. Third, we assume the discriminatory itemsets (*i.e. A*) and the non-discriminatory itemsets (*i.e. D*) to be categorical.

13.4.1 Direct Discrimination Prevention Methods

The proposed solution to prevent direct discrimination is based on the fact that the dataset of decision rules would be free of direct discrimination if it only contained PD rules that are *protective* or PD rules that are instances of at least one *non-redlining (legitimate)* PND rule. Therefore, a suitable data transformation with minimum information loss should be applied in such a way that each *discriminatory* rule either becomes *protective* or an instance of a *non-redlining* PND rule. We call the first direct rule protection and the second one rule generalization.

Direct Rule Protection (DRP)

In order to convert each *discriminatory* rule $r': A, B \rightarrow C$, where A is a discriminatory itemset ($A \subseteq DI_s$) and B is non-discriminatory itemset ($B \subseteq nDI_s$)), into a *protective* rule, two data transformation methods (DTM) could be applied. One method (DTM 1) changes the discriminatory itemset in some records (*e.g.* gender changed from male to female in the records with granted credits) and the other method (DTM 2) changes the class item in some records (*e.g.* from grant credit to deny credit in the records with male gender). Table 13.1 shows the operation of these two methods.

Table 13.1 Data transformation methods for direct rule protection

	Direct Rule Protection
DTM 1	$\sim A, B \rightarrow \sim C \Rightarrow A, B \rightarrow \sim C$
DTM 2	$\sim A, B \rightarrow \sim C \Rightarrow \sim A, B \rightarrow C$

Table 13.1 shows that in DTM 1 some records supporting rule $\sim A, B \rightarrow \sim C$ will be changed by modifying the value of the discriminatory itemset from $\sim A$ (Sex=Male) to A (Sex=Female) until *discriminatory* rule $r': A, B \rightarrow C$ *becomes*

protective (i.e. $elift(r') < \alpha$). In order to score better in terms of the utility measures presented in Section 13.5 and 13.6, the changed records should be those among the ones supporting the above rule that have the lowest impact on the other (protective) rules. Similar records are also chosen in DTM 2 with the difference that, instead of changing discriminatory itemsets, the class item is changed from ~*C* (grant credit) into *C* (deny credit) to make *r'* protective.

Rule Generalization

Rule generalization is another data transformation method for direct discrimination prevention. It is based on the fact that if each *discriminatory* rule $r': A, B \rightarrow C$ in the database of decision rules was an instance of at least one *non-redlining (legitimate)* PND rule $r: D, B \rightarrow C$ where *D* is a non-discriminatory itemset ($D \subseteq nDI_s$), the dataset would be free of direct discrimination. To formalize this dependency among rules (i.e. *r'* is an instance of *r*), Pedreschi *et al.* in (Pedreschi *et al.* 2009b) say that a PD classification rule *r'* is an instance of a PND rule *r* if rule *r* holds with the same or higher confidence, namely $conf(r: D,B \rightarrow C) \geq conf(r': A,B \rightarrow C)$, and a case (record) satisfying discriminatory itemset *A* in context *B* satisfies legitimate itemset *D* as well, namely $conf(A, B \rightarrow D) = 1$.

Based on this concept, a data transformation method (*i.e.* rule generalization) could be applied to transform each *discriminatory* rule $r': A, B \rightarrow C$ into an instance of a legitimate rule. Then, rule generalization can be achieved for discriminatory rules *r'* for which there is at least one *non-redlining* PND rule *r* by changing the class item in some records (*e.g.* from "Hire no" to "Hire yes" in the records of foreign and low-experienced people in NYC city). Table 13.2 shows the function of this method.

Table 13.2 Data transformation method for rule generalization

	Rule Generalization
DTM	$A, B, \sim D \rightarrow C \Rightarrow A, B, \sim D \rightarrow \sim C$

Table 13.2 shows that in DTM some records that support the rule $A, B, \sim D \rightarrow C$ will change by modifying the value of class item from *C* (e.g. deny credit) into ~*C* (e.g. grant credit) until *discriminatory* rule $r': A, B \rightarrow C$ *becomes an instance of a non-redlining (legitimate)* PND rule $r: D, B \rightarrow C$. Similar to DRP methods, in order to score better in terms of the utility measures presented in Section 13.5 and 13.6, the changed records should the ones among those supporting the above rule that have the lowest impact on the other (protective) rules.

Direct Rule Protection and Rule Generalization

Since rule generalization might not be applicable to all *discriminatory* rules, rule generalization cannot be used alone for direct discrimination prevention and must be combined with direct rule protection. When applying both rule generalization and direct rule protection, *discriminatory* rules are divided into two groups:

• *Discriminatory* rules *r'* for which there is at least one non-redlining PND rule *r* such that *r'* could be an instance of *r*. For these rules, rule generalization is performed unless direct rule protection requires less data transformation (in which case direct rule protection is used).
• *Discriminatory* rules *r'* such that there is no such PND rule. For these rules, direct rule protection (DTM 1 or DTM 2) is used.

13.4.2 Indirect Discrimination Prevention Methods

The solution proposed in Hajian *et al.* (2011b) to prevent indirect discrimination is based on the fact that the dataset of decision rules would be free of indirect discrimination if it contained no *redlining* rules. To achieve this, a suitable data transformation with minimum information loss should be applied in such a way that *redlining* rules are converted to *non-redlining* rules. We call this procedure indirect rule protection (IRP).

Table 13.3 Data transformation methods for indirect rule protection

	Indirect Rule Protection
DTM 1	$\sim A, B, \sim D \rightarrow \sim C \Rightarrow A, B, \sim D \rightarrow \sim C$
DTM 2	$\sim A, B, \sim D \rightarrow \sim C \Rightarrow \sim A, B, \sim D \rightarrow C$

In order to turn a *redlining* rule $r{:}D, B \rightarrow C$, where D is a non-discriminatory itemset that is highly correlated to the discriminatory itemset A, into a *non-redlining* rule based on the indirect discriminatory measure (*elb*), two data transformation methods could be applied, similar to the ones for direct rule protection. One method (DTM 1) changes the discriminatory itemset in some records (*e.g.* from non-foreign worker to foreign worker in the records of hired people in NYC city with Zip≠10451) and the other method (DTM 2) changes the class item in some records (e.g. from "Hire yes" to "Hire no" in the records of non-foreign worker of people in NYC city with Zip≠10451). Table 13.3 shows the operation of these two methods. Table 13.3 shows that in DTM 1 some records in the original data that support the rule $\sim A, B, \sim D \rightarrow \sim C$ will be changed by modifying the value of the discriminatory itemset from $\sim A$ (Sex=Male) into A (Sex=Female) in these records until the redlining rule $r{:}\ D, B \rightarrow C$ becomes non-redlining (*i.e.* $elb(r) < \alpha$). With the aim of scoring better in terms of the utility measures presented in Section 13.5 and 13.6, among the records supporting the above rule, one should change those with lowest impact on the other (non-redlining) rules. Similar records are also chosen in DTM 2 with the difference that, instead of changing discriminatory itemsets, the class item is changed from $\sim C$ (*e.g.* grant credit) into C (*e.g.* deny credit) in these records to make r non-redlining.

The difference between the DRP and IRP methods shown in Tables 1 and 3 is about the set of records chosen for transformation. As shown in Table 3, in IRP the chosen records should not satisfy the *D* itemset (chosen records are those with $\sim A, B, \sim D \rightarrow \sim C$), whereas DRP does not care about *D* at all (chosen records are those with $\sim A, B \rightarrow \sim C$).

13.5 Measuring Discrimination Removal

Discrimination prevention methods should be evaluated based on two aspects: discrimination removal and data quality. We deal with the first aspect in this section: how successful the method is at removing all evidence of direct and/or indirect discrimination from the original dataset. To measure discrimination removal, four metrics were proposed in Hajian *et al.* (2011a and 2011b) and Hajian and Domingo-Ferrer (2012):

- **Direct Discrimination Prevention Degree (DDPD).** This measure quantifies the percentage of *discriminatory* rules that are no longer *discriminatory* in the transformed dataset.
- **Direct Discrimination Protection Preservation (DDPP)**. This measure quantifies the percentage of the *protective* rules in the original dataset that remain *protective* in the transformed dataset.
- **Indirect Discrimination Prevention Degree (IDPD).** This measure quantifies the percentage of *redlining rules* that are no longer *redlining* in the transformed dataset.
- **Indirect Discrimination Protection Preservation (IDPP).** This measure quantifies the percentage of *non-redlining* rules in the original dataset that remain *non-redlining* in the transformed dataset.

Since the above measures are used to evaluate the success of the proposed methods in direct and indirect discrimination prevention, ideally their value should be 100%.

13.6 Measuring Data Quality

The second aspect to evaluate discrimination prevention methods is how much information loss (*i.e.* data quality loss) they cause. To measure data quality, two metrics are proposed in Verykios and Gkoulalas-Divanis (2008):

- **Misses Cost** (MC). This measure quantifies the percentage of rules among those extractable from the original dataset that cannot be extracted from the transformed dataset (side-effect of the transformation process).
- **Ghost Cost** (GC). This measure quantifies the percentage of the rules among those extractable from the transformed dataset that were not extractable from the original dataset (side-effect of the transformation process).

1.

MC and GC should ideally be 0%. However, MC and GC may not be 0% as a side-effect of the transformation process.

13.7 Experimental Results

This section presents the experimental evaluation of the proposed direct discrimination prevention approaches. We use the German Credit Dataset (Newman *et al.* 1998) in our experiments, since it is a well-known and frequently used dataset in the context of anti-discrimination. This dataset consists of 1,000 records and 20 attributes (without class attribute) of bank account holders. For our experiments with this dataset, we set DI_s= {Foreign worker=Yes, Personal Status=Female and not Single, Age=Old} (cut-off for Age=Old: 50 years old).

Figure 13.3 shows at the left the degree of information loss (as average of MC and GC) and it shows at the right the degree of discrimination removal (as average of DDPD and DDPP) of direct discrimination prevention methods for the German Credit dataset when the value of the discriminatory threshold α varies from 1.2 to 1.7, the minimum support is 5% and the minimum confidence is 10%. The number of direct *discriminatory* rules extracted from the dataset is 991 for α =1.2, 415 for α =1.3, 207 for α =1.4, 120 for α =1.5, 63 for α =1.6 and 30 for α =1.7, respectively.

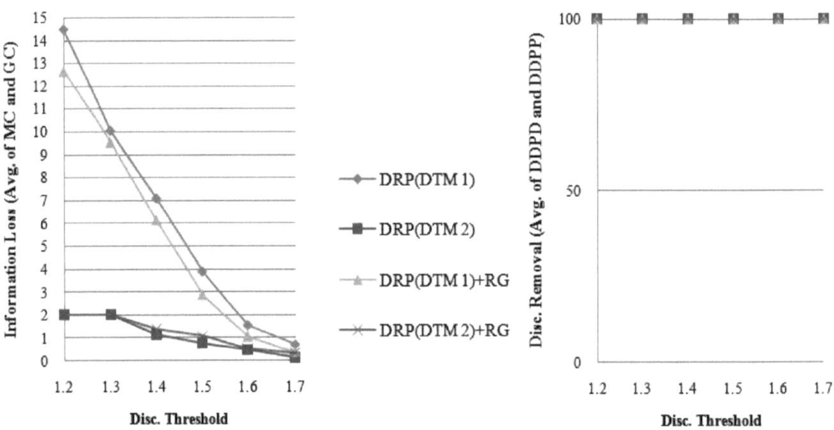

Fig. 13.3 Left: Information loss, Right: Discrimination removal degree for direct discrimination prevention methods for α in [1.2, 1.7]. DRP(DTM *i*): Data transformation method *i* for DRP; RG: Rule Generalization.

As shown in Figure 3, the degree of discrimination removal provided by all methods for different values of α is also 100%. However, the degree of information loss decreases substantially as α increases; the reason is that, as α increases, the number of *discriminatory* rules to be dealt with decreases. In addition, as shown in Figure 2, the lowest information loss for most values of α is obtained by DTM 2 for DRP.

Empirical results on indirect discrimination prevention methods can be found in Hajian *et al.* (2011b).

13.8 Conclusions and Future Work

In sociology, discrimination is the prejudicial treatment of an individual based on their membership in a certain group or category. It involves denying to members of one group opportunities that are available to other groups. Like privacy, discrimination could have negative social impact on acceptance and dissemination of data mining technology. Discrimination prevention in data mining is a new body of research focusing on this issue. One of the research questions here is whether we can adapt and use the pre-processing approaches of data transformation and hierarchy-based generalization from the privacy preservation literature for discrimination prevention. In response to this question, we try to inspire on the data transformation methods for knowledge (rule) hiding in privacy preserving data mining (more discussed in Chapter 11) and we devise new data transformation methods (*i.e.* direct and indirect rule protection, rule generalization) for converting direct and/or indirect discriminatory decision rules to legitimate (non-discriminatory) classification rules; our current results are convincing in terms of discrimination removal and information loss. However, there are many other challenges regarding discrimination prevention that could be considered in the rest of this research. For example, the perception of discrimination, just like the perception of privacy, strongly depends on the legal and cultural conventions of a society. Although we argued that discrimination measures based on *elift* and *elb* are reasonable, if substantially different discrimination definitions and/or measures were to be found, new data transformation methods would need to be designed.

Another challenge is the relationship between discrimination prevention and privacy preservation in data mining. It would be extremely interesting to find synergies between rule hiding for privacy-preserving data mining and rule hiding for discrimination removal. Just as we were able to show that indirect discrimination removal can help direct discrimination removal, it remains to see whether privacy protection can help anti-discrimination or vice versa.

Disclaimer and Acknowledgments

The authors are with the UNESCO Chair in Data Privacy, but the views expressed in this article do not necessarily reflect the position of UNESCO nor commit that organization. This work was partly supported by the Spanish Government through projects TSI2007-65406-C03-01 "E-AEGIS", TIN2011-27076-C03-01 "CO-PRIVACY" and CONSOLIDER INGENIO 2010 CSD2007-00004 "ARES", by the Government of Catalonia under grant 2009 SGR 1135 and by the European Commission under FP7 project "DwB". The second author is partly supported as an ICREA Acadèmia Researcher by the Government of Catalonia.

References

Calders, T., Verwer, S.: Three naive Bayes approaches for discrimination-free classification. Data Mining and Knowledge Discovery 21(2), 277–292 (2010)

Hajian, S., Domingo-Ferrer, J., Martinez-Ballesté, A.: Discrimination prevention in data mining for intrusion and crime detection. In: Proc. of the IEEE Symposium on Computational Intelligence in Cyber Security (CICS 2011), pp. 47–54. IEEE (2011a)

Hajian, S., Domingo-Ferrer, J., Martínez-Ballesté, A.: Rule Protection for Indirect Discrimination Prevention in Data Mining. In: Torra, V., Narakawa, Y., Yin, J., Long, J. (eds.) MDAI 2011. LNCS, vol. 6820, pp. 211–222. Springer, Heidelberg (2011b)

Hajian, S., Domingo-Ferrer, J.: A methodology for direct and indirect discrimination prevention in data mining. IEEE Transactions on Knowledge and Data Engineering (to appear)

Kamiran, F., Calders, T.: Classification without discrimination. In: Proc. of the 2nd IEEE International Conference on Computer, Control and Communication (IC4 2009). IEEE (2009)

Kamiran, F., Calders, T.: Classification with no discrimination by preferential sampling. In: Proc. of the 19th Machine Learning Conference of Belgium and The Netherlands (2010)

Kamiran, F., Calders, T., Pechenizkiy, M.: Discrimination aware decision tree learning. In: Proc. of the IEEE International Conference on Data Mining ICDM 2010, pp. 869–874. ICDM (2010)

Newman, D.J., Hettich, S., Blake, S.L., Merz, C.J.: UCI Repository of Machine Learning Databases (1998), http://archive.ics.uci.edu/ml

Parliament of the United Kingdom. Sex Discrimination Act (1975), http://www.opsi.gov.uk/acts/acts1975/PDF/ukpga19750065en.pdf

Parliament of the United Kingdom. Race Relations Act (1976), http://www.statutelaw.gov.uk/content.aspx?activeTextDocId=2059995

Pedreschi, D., Ruggieri, S., Turini, F.: Discrimination-aware data mining. In: Proc. of the 14th ACM International Conference on Knowledge Discovery and Data Mining (KDD 2008), pp. 560–568. ACM (2008)

Pedreschi, D., Ruggieri, S., Turini, F.: Measuring discrimination in socially-sensitive decision records. In: Proc. of the 9th SIAM Data Mining Conference SDM 2009, pp. 581–592. SIAM (2009a)

Pedreschi, D., Ruggieri, S., Turini, F.: Integrating induction and deduction for finding evidence of discrimination. In: Proc. of the 12th ACM International Conference on Artificial Intelligence and Law (ICAIL 2009), pp. 157–166. ACM (2009b)

Ruggieri, S., Pedreschi, D., Turini, F.: Data mining for discrimination discovery. ACM Transactions on Knowledge Discovery from Data 4(2) Article 9 (2010)

United States Congress. Employment Non-Discrimination Act (1994), http://www.govtrack.us/congress/bill.xpd?bill=h111-3017

United States Congress. US Equal Pay Act (1963), http://archive.eeoc.gov/epa/anniversary/epa-40.html

Verykios, V., Gkoulalas-Divanis, A.: A survey of association rule hiding methods for privacy. In: Aggarwal, C.C., Yu, P.S. (eds.) Privacy- Preserving Data Mining: Models and Algorithms. Springer (2008)

Yin, X., Han, J.: CPAR: Classification based on Predictive Association Rules. In: Proc. of SIAM ICDM 2003. SIAM (2003)

Chapter 14
Introducing Positive Discrimination in Predictive Models

Sicco Verwer and Toon Calders

Abstract. In this chapter we give three solutions for the discrimination-aware classification problem that are based upon Bayesian classifiers. These classifiers model the complete probability distribution by making strong independence assumptions. First we discuss the necessity of having discrimination-free classification for probabilistic models. Then we will show three ways to adapt a Naive Bayes classifier in order to make it discrimination-free. The first technique is based upon setting different thresholds for the different communities. The second technique will learn two different models for both communities, while the third model describes how we can incorporate our belief of how discrimination was added to the decisions in the training data as a latent variable. By explicitly modeling the discrimination, we can reverse engineer decisions. Since all three models can be seen as ways to introduce positive discrimination, we end the chapter with a reflection on positive discrimination.

14.1 Introduction

The topic of discrimination-aware data mining was first introduced in (Calders et al., 2009; Kamiran & Calders, 2009; Pedreschi et al., 2008), and is motivated by the observation that often training data contains unwanted dependencies between the attributes. Given a labeled dataset and a sensitive attribute; e.g., gender, the goal of our research is to learn a classifier for predicting the class label that does

Sicco Verwer
Research and Documentation Centre (WODC) of the Ministry of Security and Justice, The Netherlands
e-mail: siccoverwer@gmail.com

Toon Calders
Eindhoven University of Technology, The Netherlands
e-mail: t.calders@tue.nl

B. Custers et al. (Eds.): Discrimination & Privacy in the Information Society, SAPERE 3, pp. 255–270.
springerlink.com

not discriminate with respect to a given sensitive attribute, e.g., for every sex, the probability of being in the positive class should be roughly the same. For a more detailed description of the problem domain and some algorithmic solutions, see Chapters 3 and 12 of this book. This chapter will discuss different techniques of learning and adapting probabilistic classifiers to make them discrimination-free.

Initially, we concentrate on Naive Bayes classifiers, see, e.g., (Bishop, 2006). These are simple probabilistic models with strong independence assumptions. The main benefit of these assumptions is that they make the problem of learning a Naive Bayes classifier easy. Intuitively, a Naive Bayes classifier can be used to compute the probability that a given combination of attribute values (features or characteristics) obtains a positive class value. If this value is larger than a given threshold (typically 50%), the classifier outputs "yes", otherwise it outputs "no". Consider, e.g., the following example of a spam filter.

Example

Suppose that we have a collection of emails, each of which is marked either as a spam mail or a regular mail. In order to learn a predictive model for spam, we have to transform every message into a vector of values. This is typically done by selecting all words that appear in the emails, order them, and transform every email in a sequence of 0-1 where a 1 in the ith position indicates that the ith word appeared in the mail. Otherwise, the ith position is 0. E.g., suppose that the ordered list of words appearing in the collection of emails is:

(of, a, the, man, $, win, price, task).

Then the vector (1,1,1,0,0,1,0,0) would indicate an email in which the words "of", "a", "the", and "win" appear, but not "man", "$", or "price". Based on the data vectors obtained, a model will be learned that can be used to predict for a new, unlabeled email if it is spam or not. For the Naive Bayes classifier, the model essentially corresponds to assigning a positive or negative score to every word, and setting a threshold. The scores for all words present in the email to be classified are added up, and only if the total score exceeds the threshold, the mail will be classified as spam. The scores of the words and the threshold are the parameters of the model. The Naive Bayes classification algorithm learns optimal values for these parameters based upon the data. For example, suppose that the Naive Bayes algorithm learns the following scores: (-0.5,-0.5,-0.5,0.5,2,1.5,2,-3) and threshold 2, then an email with content "win a price" corresponds to the vector (0,1,0,0,0,1,1,0) and gets a score of -0.5+1.5+2 = 3, which exceeds the threshold. Therefore, the mail is classified as spam.

A more exact definition of Bayesian models will be given in Section 2 of this chapter. The decision of a Naive Bayes classifier is based on a given data set, which is used to fit the parameters of the Naive Bayes model, and the strong class-independence assumption. Although this assumption is often violated in practice, even then good results can be obtained using a Naive Bayes approach (Langley et al., 1992).

Example

In our spam-example above, the class-independence assumption says that the occurrences of the different words in the email are independent of each other, given the class. More specifically, in the spam emails, every word has a probability that it occurs, but all words occur independently; if "a" occurs with 20% probability, and "the" with 50% probability in spam mails, the probability that both words occur in the spam email is 20% times 50% = 10%; the only factor that influences the probability of words occurring is if it is a spam email or not. Obviously this assumption will be violated in real emails. Nevertheless, many spam filters successfully use Naive Bayes classifiers even though they are based upon an unrealistic assumption.

As discussed in much detail in Chapter 3, often there is a need to learn classifiers that do not discriminate with respect to certain sensitive attributes, e.g., gender, even though the labels in the training data themselves may represent a discriminatory situation. In Chapters 12 and 13, preprocessing techniques and an adapted decision tree learner for discrimination-aware classification have already been introduced. In this chapter, we provide three methods to make a Naive Bayes model discrimination-free:

1. Use different decision thresholds for every sensitive attribute value; e.g., females need a lower score than man to get the positive label.
2. Learn a different model for every sensitive attribute value and use different decision thresholds.
3. Add an attribute for the actual non-discriminatory class to a specialized Naive Bayes model and try to learn the actual class values of every row in the data-set using the expectation-maximization algorithm.

Note, however, that all of the above methods can be seen as a type of positive discrimination: they assume an equal treatment of every sensitive attribute value and force the predictor to satisfy this assumption, sacrificing predictive accuracy in the process. Thus, although the off-the-shelf classifier considers it more likely for some people to be assigned a positive class, they are forcibly assigned a negative class in order to reduce discrimination, i.e., they are discriminated positively. Since positive discrimination is considered illegal in several countries, these methods should be applied with care. Applying predictive tools untouched, however, should also be done with care since they are very likely to be discriminating: they make use of any correlation in order to improve accuracy, also the correlation between sensitive and class attributes.

Since it is impossible to identify the true cause of being assigned a positive class using data mining, discrimination in data mining cannot be avoided without introducing positive discrimination. When applying data mining, one thus has to make a choice between positive and negative discrimination. In our opinion, using the assumption of equal treatment in a well-thought-out way is a lesser evil than blindly applying a possibly discriminating data mining procedure.

This chapter is organized as follows. We start with an introduction to the Naive Bayes classifier in Section 2. We then use examples to provide arguments in favor of discrimination-aware data-mining in Section 3. Afterwards, we discuss our discrimination-aware techniques applied to the Naive Bayes classifier in Section 4. In Section 5, we discuss the effects of our techniques on positive discrimination. Section 6 concludes the chapter.

14.2 The Naive Bayes Classifier

We already gave an intuitive introduction to the Naive Bayes classifier in the introduction. In this section we will provide a more in-depth discussion of this classifier introducing the necessary background for understanding the proposed adaptations to the model to make it discrimination-free. The Naive Bayes classifier is a simple probabilistic model that assumes independence between all attributes when given the class attribute, see, e.g., (Bishop, 2006). For example, when predicting whether someone has a high or low income (class attribute), the age of a person correlates with the type of position (s)he occupies. A Naive Bayes classifier assumes that once the income is known, these two attributes are independent. For instance, age no longer correlates with position when considering only people with a high (low) income. Formally, a Naive Bayes model computes the following probability function[1]:

$$P(C,A_1,A_2,\ldots,A_n) \propto P(C)P(A_1|C)P(A_2|C)\ldots P(A_n|C)$$

In this formula, C is the class attribute and $A_1,A_2,\ldots,A_n$ are all other attributes. P(C) is a probability function for the different class values, and P(A|C) is a probability function for A's attribute values given the class value. Due to the independence assumption, the total probability function (or model) $P(C,A_1,A_2,\ldots,A_n)$ can be computed simply by multiplying the individual probabilities of the class and of each attribute given the class. We now show using an example how to estimate these probability functions and use them as a classifier.

Example

Suppose we are given a data-set consisting of 100 people, 40 of which are female and 60 male. We would like to predict whether a new person is likely to have a high or a low income based on this data. In the data-set 20 males and 10 females have a high income. This results in the following probability functions:

P(high income) = 30/100 = 0.3, P(low income) = 0.7
P(male| high income) = 20/30 = 0.67, P(female|high income) = 10/30 = 0.33
P(male| low income) = 40/70 = 0.57, P(female|low income) = 30/70 = 0.43

In addition, suppose we also know the education of these people and that this attribute results in the following probability functions:

P(university|high) = 0.5, P(high school|high) = 0.33, P(none|high) = 0.17
P(university|low) = 0.07, P(high school|low) = 0.57, P(none|low.) = 0.36

---

[1] We disregard normalizing constants. Note that this formulation is consistent with the one used in the introduction, as we can easily move from comparing products to sums via the logarithm.

These functions can all be easily estimated from the data-set by counting how many times each attribute value occurs together with each class attribute value. When we want to determine for instance the probability that a female with high school education receives a high income, we use the total probability function to compute and normalize the probability of these values together with a high and a low income:

P(high income,female,high school) = 0.3•0.33•0.33 = 0.033
P(low income,female,high school) = 0.7•0.43•0.57 = 0.172
P(high income|female,high school) =
P(high income,female,high school)/P(female,high school) =
0.033/(0.033+0.172) = 0.16

Since this is less than 0.5, we estimate that a female with a high school education will not receive a high income. Note that this is estimated based on the assumption that education and gender are independent given the income class.

The above example describes the basic version of a Naive Bayes classifier. Most implementations use Gaussian distributions for continuous attributes and smoothing methods to avoid zero probabilities (Bishop, 2006). In addition, the decision threshold (0.5 in the example) can often be modified. Although using a threshold of 0.5 makes sense intuitively, it is common practice to modify it depending on the situational needs, for instance to increase accuracy, or decrease the number of false positives (Lachiche & Flach, 2003).

14.3 The Problem of Discrimination in Data-Mining

In Chapter 3, it is explained how discrimination may occur, even if the training data is non-discriminatory. In this section we will now show specifically for a Naive Bayes classifier how using an off-the-shelf Naive Bayes classifier can lead to discriminatory results.

We motivate our methods using examples of the discriminatory results that are obtained when using a Naive Bayes classifier[2] on the census income data-set[3]. From this data set we try to learn a Naive Bayes classifier that can be used to decide whether a new individual should be classified as having a high or a low income. Historically, this decision has been biased towards the male sex, as can be seen in the following table:

Table 14.1 The contingency table of the income and gender attributes

	Low income	High income
Female	9592	1179
Male	15128	6662

[2] We use the Naïve Bayes classifier from the e1071 package in the R statistical toolbox (Dimitriadou et al., 2008).

[3] `http://archive.ics.uci.edu/ml/datasets/Census+Income`

This table shows the number of male and female individuals in the high and low income class. About 30% of all male individuals and only about 11% of all female individuals have a high income. Thus, according to the definitions introduced in Chapter 12, the amount of discrimination in the data-set is 30% - 11% = 19%, or 0.19.

Suppose that a bank wants to use such historical information to learn models for predicting the probability that new loan applicants will default their loan. Clearly, the data shows this probability to be dependent on the gender of a person. Nevertheless, from an ethical and legal point of view it is unacceptable to use the gender of a person to deny the loan to him or her, as this would constitute an infringement of the discrimination laws. We now show that this is a serious problem when applying data mining to this type of data.

The problem

If one learns a Naive Bayes classifier from the census income data, the discrimination in the data will be learned as a rule. This can be seen very clearly in the probability tables of the Naive Bayes classifier:

Table 14.2 The P(gender|income) table used in a Naive Bayes classifier

	Low income	High income
Female	0.388	0.150
Male	0.612	0.850

The probabilities in this table denote the probability of being male or female, given the income class of an individual. Thus, if a given person has a high income, the probability that that person is male is 0.85. If the person has a low income, this probability is only 0.61. Since the classifier uses this table in its decision whether someone is more likely to have a high or a low income, the discrimination in its predictions is likely to be worse than 0.19. We test this using the test-set (containing unseen data) included in the census income data folder. The amount of discrimination in the class values of this test-set is approximately equal to the amount of discrimination in the data-set:

Table 14.3 The gender-income contingency table of the test-set

	Low income	High income
Female	4831	590
Male	7604	3256

This changes however when we use the predictions of the Naive Bayes classifier to determine whether someone has a high or a low income:

Table 14.4 The gender-predicted income contingency table for the test-set, assigned by a Naive Bayes classifier

	Low income	High income
Female	5094	327
Male	8731	2129

The amount of discrimination in these predictions is (2129 / (8731+2129)) – (327 / (5094+327)) = 0.20 – 0.06 = 0.14. Thus, surprisingly, the total amount of discrimination has become less. However, notice also that the total positive class probability has dropped from 0.24 to 0.15; I.e., less people get assigned the class label "High income". This drop artificially lowers the discrimination score. We correct for this drop by lowering the decision threshold of the Naive Bayes classifier until the positive class probability reaches 0.24. This results in the following table:

Table 14.5 The gender-predicted income contingency table for the test-set, corrected to maintain positive class (high income) probability

	Low income	High income
Female	4958	463
Male	7416	3444

The positive class probability for females is 0.09, while the positive class probability for males is 0.32, resulting in a total discrimination of 0.32 – 0.09 = 0.23. This is a lot worse than the amount of discrimination in the actual labels of the test-set. One may wonder why this is such a big problem, since the data already told us that females are less likely to have high incomes. Suppose that such a discriminating classifier is used in a decision support system for deciding whether to give a loan to a new applicant. Let us take a look at a part of the decisions made by such a system:

Table 14.6 The corrected gender-predicted income contingency for high income test cases

	Low income	High income
Female	319	271
Male	1051	2205

This table shows the labels assigned by the classifier to people in the test-set that actually have a high income. The ones that get assigned a low income in the table are the false negatives. In the banking example, these are the ones that are falsely denied a loan by the classifier. These false negatives are very important for a decision support system because denying a loan to someone that should actually obtain one can lead to law suits. In fact, when looking at the data, it is obvious that the classifier discriminates females since males have a probability of only 1051 / (1051+2205) = 0.32 to be wrongfully denied a loan, while females have a probability of 319 / (319+271) = 0.54. Using data mining tools unmodified for such decision support systems can thus be considered to be a very dangerous practice.

Removing sensitive information does not help

A commonly used method to avoid potential law suits is to not store any sensitive information such as gender. The idea is that learning a classifier on data without this type of information avoids that the classifier's predictions will be based on the sensitive attribute. This approach, however, does not work. The reason for that is that there may be other attributes that are highly correlated with the sensitive attribute. In such a situation, the classifier will use these correlated attributes and thus discriminate indirectly. This phenomenon was termed the red-lining effect in Chapter 3. In the banking example, e.g., job occupation is correlated with gender. Removing gender will only help a bit, as job occupation can be used as a predictor for this attribute. For example, when we learn a Naive Bayes classifier on the census income data-set without gender information[4] and test it on the test-set with modified threshold, we obtain the following table:

Table 14.7 The gender-predicted income contingency table for the test-set, assigned by a Naive Bayes classifier learned without gender information

	Low income	High income
Female	4900	521
Male	7474	3386

This table shows positive class probabilities of 0.10 and 0.31 for respectively females and males, and thus a discrimination of 0.21. This does not improve a lot over the classifier that used the gender information directly. In fact, the false negatives show the same problem as before:

Table 14.8 The no gender information corrected gender-predicted income contingency table for high income test cases

	Low income	High income
Female	301	289
Male	1079	2177

Thus, even learning a classifier on a data-set without sensitive information can be dangerous. Removing the sensitive information from a data-set actually makes the situation worse because data-mining tools will still discriminate, but in a much more concealed way, and rectifying this situation using discrimination-aware techniques is extremely difficult without sensitive information.

[4] In addition, we replaced "wife" by "husband" in the relationship attribute.

Obviously, one could also decide to remove all of the attributes that correlate with the sensitive ones from the dataset. Although this would resolve the discrimination problem, in this process a lot of useful information will get lost. In fact, the occupation of a person is a very important decision variable when deciding whether to give a loan or not. The occupation attribute can hence, at the same time, reveal information on gender and give useful, non-discriminatory information on loan defaulting. We provide solutions that make use of all the available information, but in a non-discriminatory way.

14.4 Discrimination-Free Naive Bayes Classifiers

In this section, we provide three approaches for removing discrimination from a Naive Bayes classifier.

14.4.1 Using Different Decision Thresholds

The most straightforward method for removing discrimination is to modify the decision thresholds differently for the different sensitive values. For instance, we can decide to give a high income label to females if the high income probability is greater than 0.1, but to males if it is greater than 0.6. This instantly reduces discrimination by favoring females. Note that this is a very direct form of positive discrimination since even though the model considers some males more likely to belong to the positive class than some females; it still predicts a negative class for these males and a positive class for the females.

When using different decision thresholds for different sensitive values, an important question to ask is which ones to use, and why. The answer to this question highly depends on the situation. It is well-known that using a different decision threshold influences the number of positives, false positives, negatives, and false negatives. Since the importance of these values differs per application, several analysis techniques like ROC (receiver operator curve) analysis (Lachiche & Flach, 2003) exist to aid in setting this threshold smartly. By using different decision thresholds for different sensitive attribute values, the threshold settings in addition influence the amount of positive and negative discrimination. Ideally, these should be taken into account when performing such an analysis.

In our work, we assume that the amount of people that are assigned a positive class should remain the same. In many applications, keeping this number close to the number or positive labels in the data-set is highly favorable. For instance, in the setting of banks assigning loans to individuals, the bank does not suddenly want to assign less or more loans. In addition, as explained in Section 3, this assumption makes comparing the different techniques on their discrimination score a lot more fair. We set the decision thresholds using a simple algorithm:

1. Calculate the number of positive class labels P assigned to the data-set.
2. Learn a Naive Bayes classifier on the data-set.
3. Set the decision threshold T_+ and T_- for the favored and discriminated sensitive values to 0.5.
4. Calculate the amount of discrimination in the data-set when using T_+ and T_-.

5. While the discrimination is greater than 0
6. Calculate the number of positive class labels P' assigned to the data-set.
7. If P' is greater than P, raise T_+ by 0.01.
8. If P' is less than or equal to P, lower T_- by 0.01.
9. Iterate
10. Use the resulting decision thresholds to classify the test-set.

The idea of this algorithm is to lower the threshold for females if the classifier assigns less positive class labels than the number of positive class labels in the data-set. Otherwise, we raise the decision threshold for males. In this way, we try to keep the number of positive class labels intact. One may note that since we want to keep this number intact, it is possible to pre-compute the number of males and females that should get a different class label in order to obtain a discrimination score of 0:

$$m_{change} = m_{assigned} - P(\text{positive class}) \bullet m_{total}$$
$$f_{change} = f_{assigned} - P(\text{positive class}) \bullet f_{total}$$

where m_{change}, $m_{assigned}$ and m_{total} (and f) denote the change in the number of males (females) that receive a positive class label, the number of males (females) initially assigned a positive class, and the total number of males (females), respectively. It is straightforward to set the decision thresholds to values that result in these changes. Although this calculation is more efficient, we prefer using our algorithm since it provides an overview of the different threshold settings possible between the original and discrimination-free models. In addition to changing the decision thresholds, we remove the sensitive attribute from the Naive Bayes model.

14.4.2 Two Naive Bayes Models

Using the above method, discrimination can be removed completely from a Naive Bayes classifier. However, it does not actively try to avoid the red-lining effect. Although the resulting classification is discrimination-free, this classification can still depend on the sensitive attribute in an indirect way. In our second approach, we try to avoid this dependence by removing all correlation with the sensitive attribute from the data-set used to train the Naive Bayes classifier.

Removing all correlation with the sensitive attribute from the data set seems difficult, but the solution actually is very simple. We divide the data-set into two sets, each containing people with only one of the sensitive values. Subsequently, we learn two Naive Bayes models from these two data sets. In the banking example, we thus get one model for the male and one for the female population. The model for males still uses attributes correlated to gender for making its decisions, but since it has not been trained using data from females; these decisions are not based on the fact that females are less likely to get positive labels. The predictions made using these models are therefore independent of the sensitive attribute. When classifying new people, we first select the appropriate model, and then use that model to decide on the class label.[5]

---

[5] It has been suggested to swap these models, i.e., use the model learned using males to classify females and vice versa. In our opinion, this makes less sense since this approach

Intuitively, this approach makes a lot of sense since it uses different classifiers to classify data that is known to be differently distributed (males are different from females). Since males are still favored, however, the resulting classification can still contain discrimination. We apply the threshold modification algorithm to remove this discrimination.

14.4.3 A Latent Variable Model

Our third and most sophisticated approach tries to model the discrimination process in order to discover the actual class labels that the data-set should have contained if it would have been discrimination-free. Since they are not observed, these actual class labels are modeled using a latent (or hidden) variable, see, e.g., (Bishop, 2006). Such a latent variable can be seen as an attribute that is known to exist, but its values have not been recorded in the data-set. A well-known example of such a variable is "happiness". It is very difficult to observe if someone is happy, but since we known how being happy influences one's actions, we can infer whether someone is happy by observing his or her actions. In our case, we cannot know who should have gotten a positive class label, but we can make assumptions about how this variable depends on the other variables:

1. The actual discrimination-free class label is independent from the sensitive attribute.
2. The observed class label is determined by discriminating the actual labels based on the sensitive attribute uniformly at random.

These two assumptions might not correspond to how discrimination is being applied in practice. For instance, the females close to the decision boundary could have a higher chance of being discriminated. However, because they result in a simple model, they do allow us to study the problem of discrimination-free classification in detail. The resulting model is given by the following total probability function:

$$P(C,L,S,A_1,A_2,\ldots,A_n) = P(L)P(S)P(C|L,S)P(A_1|L,S)P(A_2|L,S)\ldots P(A_n|L,S),$$

where C is the class attribute after discrimination, L is the latent variable representing the true class before discrimination, S is the sensitive attribute, and A_1, A_2,..., A_n are all other attributes. The formula is similar to the original Naive Bayes formula in the sense that all attributes A_1, A_2,..., A_n are independent from each other given the class label. Except that in this model, we use the actual latent class label L instead of C. In addition, every value except L is conditioned on the sensitive attribute S. The result is identical to the previous approach that used two separate models; for every value of S, a different set of probability functions are used, thus a different model is used for every value of S. The distribution of L however, is modeled to be independent from S, satisfying the first assumption. The probability function P(C|L,S) satisfies the second assumption: for every combination of an actual latent class label value with a sensitive value, a different probability function is used to determine the observed class label. Thus, the

uses classifiers to classify data from different distributions. Also, in our experience it produces worse results.

discrimination depends on both the actual class label, and on the sensitive value, but who is being discriminated is decided at random, i.e., independent of the other attribute values. We now show how to find likely latent class labels, i.e., how to discover who is likely being discriminated.

Finding likely latent values

We need to find good values to assign to the latent attribute in every row from the data-set. Essentially, this is a problem of finding two groups (or clusters) of rows: the ones that should have gotten a positive label, and those that should have gotten a negative label. We now briefly describe the standard approach of expectation maximization (EM) that is commonly used in order to find such clusters. The reader is referred to (Bishop, 2006) for a more detailed description of this algorithm.

Given a model M with a latent attribute L, the goal of the expectation maximization algorithm is to set the parameters of M such that they maximize the likelihood of the data-set, i.e., the probability of the data-set given the model. Unfortunately, since L is unobserved, the parameters involving L can be set in many different ways. Searching all of these settings for the most optimal one is a hopeless task. Instead, expectation maximization optimizes these settings by fitting them to the data-set (the M-step), then calculates the expected values of the latent attribute given those settings (the E-step), incorporates these back into the data-set, and iterates. This is a greedy procedure that converges to a local optimum of the likelihood function. Typically, random restarts are applied (randomizing the initial values of the latent variable) in order to find better latent values.

Using prior information

For the problem of finding the actual discrimination-free class labels we can do a lot better than simply running EM and hoping that the found solution corresponds to discrimination-free labels. For starters, it makes no sense to modify the labels of rows with favored sensitive values and negative class labels. The same holds for rows with discriminated sensitive values and positive class labels. Modifying these can only result in more discrimination, so we fix the latent values of these rows to be identical to the class labels in the data-set and remove them from the E-step of the EM algorithm.

Another improvement over blindly applying EM is to incorporate prior knowledge of the distribution P(C | L, S). In fact, since the ultimate goal is to achieve zero discrimination, we can pre-compute this entire distribution. We show how to do this using an example.

Example

Suppose we have a data-set consisting of 100 rows of people, distributed according to the following occurrence counts:

	Low income	High income
Female	30	20
Male	10	40

Clearly, there is some discrimination: the positive class probability of males (0.8) is much bigger than the positive class probability of females (0.4). Initially, we set the distribution over the latent labels to be equivalent to the distribution over the class labels, keeping the discrimination intact:

	Latent positive		Latent negative	
	Low income	High income	Low income	High income
Female	0	20	30	0
Male	0	40	10	0

Next, we rectify this situation by subtracting occurrence counts from the males with positive latent values, and giving these negative latent values. We do the opposite for females. Since we want the number of rows with actual non-discriminatory positive labels to be equal to the number of rows with positive labels in the data, the amount of such changes we need to make is unique and easy to compute. In the example, it is 10, resulting in the following distribution:

	Latent positive		Latent negative	
	Low income	High income	Low income	High income
Female	10	20	20	0
Male	0	30	10	10

In this table, both males and females have a probability of 0.6 to obtain a positive latent value. The latent values are therefore discrimination-free. We use these counts to determine the probability table P(C / L, S) in the latent variable model.

14.4.4 Comparing the Three Methods

In order to test the three Naive Bayes approaches for discrimination-free classification, we performed tests on both artificial and real-world data (Calders & Verwer, 2010). Here we made use of the latent variable model to generate the artificial datasets. A big advantage of this artificial data is that we can also generate the actual class labels that should have been assigned to the rows when there is no discrimination. These labels are then used to test the accuracy of the classifiers. When using real-world data, we do not have this luxury of a discrimination-free test-set.

When performing such experiments with discrimination-aware methods, one should test at least the following quantities: the loss in accuracy and the amount of remaining discrimination. One always has to make a trade-off between these two values since discrimination can only be decreased by sacrificing accuracy. The main conclusions from experiments in (Calders & Verwer, 2010) are that our second threshold modifying method performs best, achieving zero discrimination with high accuracy. In addition, the expectation maximization algorithm has problems converging to a good quality solution with zero discrimination. In fact, during later iterations, it often finds solutions that are worse both in terms of discrimination and accuracy than solutions found earlier. This strange behavior of the EM algorithm still has to be further investigated. For a more detailed overview and discussion of these results, the reader is referred to (Calders & Verwer, 2010).

14.5 A Note on Positive Discrimination

Although discrimination-aware data-mining is necessary in our opinion, one should be aware that it not only decreases the accuracy of data-mining, it also has a high probability to introduce positive discrimination. For instance, if we repeat the final analysis from Section 3 to results obtained using our first threshold modifying method (until zero discrimination) on the census income data-set, we obtain the following counts on people that should get a high income according to the test-set:

Table 14.9 The gender-predicted income contingency table for high income test cases, assigned by a Naive Bayes classifier with modified decision thresholds

	Low income	High income
Female	101	489
Male	1763	1493

Suddenly, females have a much smaller probability of being falsely denied a high income. This is an example of positive discrimination, and in some countries this type of discrimination is also considered illegal. These numbers, however, are determined using the discriminatory labels in the test-set. The actual difference in false negatives will be smaller using the true non-discriminatory class values. Unfortunately, since we do not know who is being discriminated, we cannot know exactly how to correct these numbers for this discrimination. We can, however, make an estimated guess based on the assumption that discrimination occurs at random, and that the number of positives should remain intact.

Under these assumptions, 690 females with a negative class label in the test-set should actually have a positive label, and 690 males with a positive label should actually have a negative label. The probability that a female is already assigned a positive label is equal to the false positive probability, which is 0.1683 (813 out of 4018). Thus, 690•0.1683=116 discriminated females get a positive label, and 574 discriminated females remain. Since these should get a positive label, these counts are added to the true and false negatives. For the male counts, some positives should actually be negatives. The false positive probability for males is 0.5415 (1763 out of 3256). Thus, 690•0.5415=374 favored males get a negative label, and 316 favored males remain. Since these counts should actually be negative, we subtract them from the counts in the table. This results in the following table:

Table 14.10 The modified threshold gender-predicted income contingency table for discrimination corrected high income test cases

	Low income	High income
Female	675	605
Male	1389	1177

This corresponds to a probability of being denied a loan of 0.53 for females, and 0.54 for males. These probabilities are a lot more reasonable. Although they are based on the not always realistic assumption of equal treatment (and random discrimination), in our opinion, trying to make these false negative probabilities similar for males and females using positive discrimination is a lesser evil than knowingly making them unbalanced by blindly applying a discriminating data mining procedure.

14.6 Concluding Remarks

We introduced the Naive Bayes classifier and argued that naively applying such a classifier to a data-set containing information regarding people automatically introduces discrimination with respect to sensitive attributes such as gender, race, and ethnicity. Using data mining tools in a decision support system based on such data can thus be considered very dangerous since it opens the possibility of law suits. We show using an example that the solution of removing this sensitive information from the data-set does not remove this discrimination. Since data-mining tools use attributes that are correlated with this sensitive information, the decisions made by naively applying data-mining tools will still be discriminating. In fact, removing the sensitive information from a data-set makes the situation worse because data-mining tools will still discriminate, and rectifying this situation without access to sensitive information is extremely difficult. Instead, we introduce three discrimination-aware data-mining methods based on the Naive Bayes classifier that use the sensitive information in order to make non-discriminatory predictions.

In our first method, we use different decision thresholds for different sensitive values. For instance, we can decide to assign a positive class label to females if the positive class probability is greater than 0.1, but to males if it is greater than 0.6. We provide a simple algorithm for making modifications to these thresholds until the resulting classification is discrimination-free.

The second method involves learning two different classifiers; for instance one for all males, and one for all females. This effectively removes all correlation with the sensitive attribute from the data-set used to train the Naive Bayes classifier, thus avoiding that correlated attributes can be used to discriminate. Since there can still be discrimination in the resulting classification, we assign different decision thresholds to them using the algorithm of our first method. Of all three methods, this approach performed best in experiments on artificial and real-world data.

In the third and most involved method we introduced a latent variable reflecting the actual class of a person that should have been assigned if there were no discrimination. This actual non-discriminatory class is assumed to be independent of the sensitive attribute, and the non-discriminatory labels are assumed to be discriminated uniformly at random, resulting in the actual labels in the data-set. The probabilities in this model are learned using the expectation maximization algorithm. We provide ways to incorporate knowledge about the discrimination process into this algorithm. In experiments, this method unfortunately performed poorly due to problems in the behavior of the expectation maximization algorithm.

We ended with a discussion on the positive discrimination introduced by discrimination-aware data-mining and why we believe it is a better option than blindly applying a discriminating off-the-shelf data-mining procedure.

References

Bishop, C.M.: Pattern Recognition and Machine Learning. Springer (2006)

Calders, T., Kamiran, F., Pechenizkiy, M.: Building classifiers with independency constraints. In: IEEE ICDM Workshop on Domain Driven Data Mining, pp. 13–18. IEEE press (2009)

Calders, T., Verwer, S.: Three naive Bayes approaches for discrimination-free classification. Data Mining and Knowledge Discovery 21(2), 277–292 (2010)

Dimitriadou, E., Hornik, K., Leisch, F., Meyer, D., Weingessel, A.: e1071: Misc functions of the Department of Statistics. TU Wien, R package version 1 (2008)

Kamiran, F., Calders, T.: Classifying without discriminating. In: Proc. IEEE International Conference on Computer, Control and Communication (IC4), pp. 1–6. IEEE press (2009)

Lachiche, N., Flach, P.: Improving accuracy and cost of two-class and multi-class probabilistic classifiers using ROC curves. In: Proc. International Conference on Machine Learning (ICML), pp. 416–423. AAAI Press (2003)

Langley, P., Iba, W., Thompson, K.: An analysis of Bayesian classifiers. In: Proc. Conference on Artificial Intelligence (AAAI), pp. 223–228 (1992)

Pedreschi, D., Ruggieri, S., Turini, F.: Discrimination-aware data mining. In: Proc. ACM SIGKDD International Conference on Knowledge Discovery and Data Mining, pp. 560–568 (2008)

Part V

Solutions in Law, Norms and the Market

Chapter 15
From Data Minimization to Data Mini*mum*mization

Bart van der Sloot

Abstract. Data mining and profiling offer great opportunities, but also involve risks related to privacy and discrimination. Both problems are often addressed by implementing data minimization principles, which entail restrictions on gathering, processing and using data. Although data minimization can sometimes help to minimize the scale of damage that may take place in relation to privacy and discrimination, for example when a data leak occurs or when data are being misused, it has several disadvantages as well. Firstly, the dataset loses a rather large part of its value when personal and sensitive data are filtered from it. Secondly, by deleting these data, the context in which the data were gathered and had a certain meaning is lost. This chapter will argue that this loss of contextuality, which is inherent to data mining as such but is aggravated by the use of data minimization principles, gives rise to or aggravates already existing privacy and discrimination problems. Thus, an opposite approach is suggested, namely that of data mini*mum*mization, which requires a minimum set of data being gathered, stored and clustered when used in practice. This chapter argues that if the data mini*mum*mization principle is not realized, this may lead to quite some inconveniences; on the other hand, if the principle is realized, new techniques can be developed that rely on the context of the data, which may provide for innovative solutions. However, this is far from a solved problem and it requires further research.

15.1 Introduction

Gathering, processing and distributing data, distilling patterns, aggregated profiles and statistical or causal relationships from datasets and applying the gathered rules and profiles in practical decisions all have huge opportunities to offer in relation to both the discovery, the application and the dissemination of knowledge. Data

---

Bart van der Sloot
Institute for Information Law, University of Amsterdam, The Netherlands
e-mail: b.vandersloot@uva.nl

B. Custers et al. (Eds.): Discrimination & Privacy in the Information Society, SAPERE 3, pp. 273–287.
springerlink.com

mining and (group) profiling are techniques that have been used since long, but with the emergence of new technical possibilities and processing capacities, these have become the dominant modes of data analyses. Through these techniques, profiles of terrorists are created so as to forestall criminal activities, relationships between specific characteristics and diseases may be discovered so as to prevent them or treat them in an early stage and business profiles are fine tuned to meet consumer interests. However, there are some dangers attached to the use of data mining and profiling. The two major issues regard privacy and discrimination problems.

Privacy might be in danger when personal data of an individual are gathered, used to profile him or used in practical decisions and practices. The discrimination of a particular person or group may occur when personal characteristics, relating to such information as gender, sexual preferences, political and religious believes or ethnicity, are gathered, analyzed and used to bestow upon a person or group a different, disadvantageous treatment. A much used solution in relation to the privacy aspects, but which may also be of use in relation to discriminatory practices, is the implementation of so called privacy enhancing technologies. The technical framework for data processing may be built in such a way that it prevents privacy and discrimination problems, such as by data minimization, which entails a minimum set of sensitive[1] data gathered, stored and used.

Although data minimization sometimes helps to minimize the scale of danger or damage, it has several disadvantages as well. First and most prominently, when valuable data are excluded from the database, it decreases in value and usefulness. Secondly, by deleting these data, the context in which the information was gathered and had a certain meaning is lost. This chapter will argue that from this loss of context, a tendency which is inherent to data mining as such but is aggravated by the use of data minimization principles, problems related to privacy and discrimination arise. Thus, another, opposite approach is suggested, namely that of data mini*mum*mization. This principle requires a minimum set of data being gathered, stored and clustered. Instead of requiring that certain data is not collected, the principle rule of data minimization, the data mini*mum*mization principles requires that the context of the data in the form of metadata is collected along with the data. By requiring and clustering a minimum set of (contextual) information, the value of the dataset is retained or even increased, and the privacy and discrimination problems following from the loss of context might be better addressed than by the data minimization principle.

This chapter will proceed as follows. The first section will shortly distinguish four phases of knowledge discovery in databases. The second and third section will point out some general rules relating to privacy and discrimination, with which these may come into conflict. The fourth section will put forward one of the most prominent solutions for these problems, namely that of privacy enhancing technologies and especially the concept of data minimization. The fifth section will analyze some of the problems relating to this technique. The sixth section will offer an alternative solution: data mini*mum*mization.

[1] In this Chapter, the term 'sensitive data' will refer to both privacy and discriminatory sensitive data, unless where indicated.

15.2 Data Mining and Profiling Techniques

Data mining is commonly used as an umbrella concept for knowledge discovery in databases, though more correct, it is only one of several phases.[2] The first step of knowledge discovery in databases is the gathering of data. Gathering information may be done for example through fieldwork, queries, harvesting the internet and personal observations, but also through interconnecting databases and merging them together. Secondly, storing the data and organizing the material. The latter may be necessary not only in relation to making it computer readable, but also to enable correct analyses of the data and making them comparable. The third phase is that of actual data mining. Data mining refers to the discovery, most commonly with the use of (mathematical) algorithms, of hidden patterns and subtle relationships in data and the inference of rules that allow for the prediction of future results.[3] The patterns and relationships need not to be causal, but may also be statistical. Also, these patterns may be indirect, so that the direct relationship between for example race and solvency is be replaced by the relationship between a racially determined zip code and solvency. This is called redlining or masking.[4] The final stage in the process is applying the knowledge and patterns in real life decisions. This is often done with the assistance of either individual or group profiles.[5] A pattern obtained through data mining will commonly show the probability that characteristic A is combined with characteristic B. For example, it might be discovered that 67% of the people with curly hair use hair products to style their hairdo or that 86% of the people having a certain zip code possess an expensive car. Thus, targeting such groups most commonly entails a certain margin of error.

15.3 Data Protection Legislation

Knowledge discovery in databases may among others come into conflict with two legal values: privacy and equality. To provide for some basic fundaments for assessing the (il)legality of such practices, this section will address the topic of privacy and data protection legislation, the next one will do so with regard to anti-discrimination laws. The main focus will be on European legislation.

Privacy refers to the right to respect for one's private and family life, home and communications, while data protection refers to the right to the protection of personal data concerning a person. The right to privacy is most prominently protected by the European Convention on Human Right and is a moral concept, seen as instrumental in relation to the realisation of autonomy, negative freedom and dignity. If these values are violated or endangered, for example through the use of data mining, then this practice is prohibited unless it is prescribed by law, it is necessary in a democratic society and the infringement is proportional in relation to the goal it serves.

[2] Custers (2004); Skillicorn (2009); Westphal (2009); Larose (2006).

[3] <http://www.gao.gov/new.items/d07293.pdf>.

[4] Squires (2003); Kuhn (1987); LaCour-Little (1999).

[5] Hildebrandt & Gutwirth (2008).

Even more relevant in relation to knowledge discovery in databases is the right to data protection. The European Data Protection Directive, the most important text in this respect, is applicable when personal data[6] are being processed (entailing both the gathering, processing, use and dissemination of data)[7] and spells out several obligations for the so called 'data controller', who determines the purpose and means of processing,[8] in relation to the 'data subject', the one to which the data refer. The directive distinguishes between non-sensitive personal data, with which a person may be identified either directly or indirectly, and sensitive data, relating to information concerning race, ethnicity, political, religious and philosophical believes, trade-union membership and data concerning health and sex life with which a person may be either directly or indirectly identified.[9] The requirements for processing sensitive personal data are stricter then for non-sensitive data.

One of the core doctrines in the directive is that of 'informed consent'. The data controller has certain transparency obligations,[10] correlating with the information rights of the data subject,[11] which relate to information regarding the identity of the data controller, the data processed by him and the purposes for which this is done. Furthermore, the directive requires a legitimate purpose for the data processing, the most prominent possibility being the consent of the data subject;[12] subsequently, the data subject has the ability to object to the processing of his data[13] and to request the erasure or blocking of his personal data.[14] The concept of 'informed consent', relating to the consent or objection to data processing on the basis of adequate and complete information,[15] is instrumental in safeguarding the autonomy of the individual. Besides the doctrine of 'informed consent', two other important principles figure in the directive. Firstly, the so called privacy enhancing principles, regarding the security of processing techniques and data minimization rules, which will be discussed in the fifth section, and secondly, the quality principles, relating to the quality of decision making, the quality of the data themselves and the quality of data processing, which will be elaborated on in the seventh section. Both privacy and data protection problems shall be referred to in this chapter under the umbrella concept 'privacy problems'. First, the general fundaments of anti-discrimination laws will be outlined in the next section.

15.4 Anti-discrimination Legislation

The European legislation regarding discrimination is a bit more scattered. Most importantly, both the Charter of Fundamental Rights and the European Convention on Human Rights contain a general prohibition on the discrimination upon grounds

---

[6] Article 2(a) Data Protection Directive 95/46/EC (hereafter: DPD).
[7] Article 2(b) DPD.
[8] Article 2(d) DPD.
[9] Article 8.1 DPD.
[10] Article 10 DPD.
[11] Article 12 DPD.
[12] Article 7 & 8 DPD.
[13] Article 14 DPD.
[14] Article 12 DPD.
[15] Article 2(h) DPD.

such as gender, race, colour, language, religion, political opinion, nationality, ethnic or social origin, association with a national minority, property, birth genetic features, language, disability, age or sexual orientation. Then, there are also some specific European directives, such as the Employment Equality Directive, prohibiting discrimination on the basis of sexual orientation, religious belief, age and disability in the area of employment, the Racial Equality Directive, among others prohibiting discrimination on the basis of race or ethnicity in the context of employment, the Gender Goods and Services Directive, expanding the scope of sex discrimination regulation to the area of goods and services, and the Gender Social Security Directive, guarantying equal treatment in relation to social security.[16]

Generally, these texts make a distinction between direct discrimination and indirect discrimination. The former is usually described as the situation where one person or group is treated less favorably on one of the above mentioned grounds, while the latter is commonly described as the situation where an apparently neutral provision, criterion or practice would put persons of one group at a particular disadvantage compared with persons of the other group.[17] Two exceptions figure repeatedly in the different legal texts. The first is the case of positive discrimination[18] and the second is the case in which the discrimination on the basis of one of the mentioned grounds is objectively justifiable.[19] Positive discrimination involves specific measures taken with a view to ensuring full equality in practice, that aim to prevent or compensate for disadvantages linked to racial or ethnic origin, sex or any other of the above described characteristics. Proportionate differences in individuals' treatment on the basis of sensitive characteristics may be objectively justifiable if such a characteristic constitutes a genuine and determining requirement or factor, provided that the objective is legitimate and the requirement is proportionate.

15.5 Data Minimization Principles

Knowledge discovery in databases may come into conflict with both privacy and discrimination legislation on several points. These will not be covered extensively, but an example of a privacy violation may be found in the case where personal data are being gathered without a legitimate purpose, where these data are being processed in an 'unsafe' manner, leading to for example data leaks, or where these data are used to undermine the autonomy of the individual. Violations of anti-discrimination laws may for example occur when data regarding gender, religious beliefs, ethnicity and the likes are directly used to bestow on a person or a group a discriminatory treatment or when this is done indirectly, using for example the technique of redlining or masking. Dissemination of such data or knowledge and patterns distilled from them may also lead to a violation of the right to privacy or to stigmatization of individuals and groups. Especially among privacy scholars,

---

[16] <http://www.echr.coe.int/NR/rdonlyres/DACA17B3-921E-4C7C-A2EE-3CDB68B0133E/0/182601_FRA_CASE_LAW_HANDBOOK_EN.pdf>.

[17] Article 2(a) & (b) directives 2000/43/EC & 2004/113/EC.

[18] Article 5 directive 2000/43/EC. Article 6 directive 2004/113/EC.

[19] Article 5.2 directive 2004/113/EC. Article 4 directive 2000/43/EC.

one of the most commonly suggested solutions for such problems is the use of so called privacy enhancing technologies.

(1) Firstly, the Data Protection Directive holds that the controller must implement appropriate technical and organizational measures to protect personal data against accidental or unlawful destruction or accidental loss, alteration, unauthorized disclosure or access, in particular where the processing involves the transmission of data over a network, and against all other unlawful forms of processing.[20] Thus, privacy enhancing technologies may be used to minimize the risk of data security breaches by controlling the access to the data, for example through the use of passwords, by encrypting the data and by protecting databases against cyber-attacks. This way, the risk of privacy violations is minimized.
(2) Secondly, both the danger and the scale of the possible damage are minimized through the use of so called data minimization techniques. Concepts such as privacy by design and privacy preserving data mining are closely aligned to this approach. (2a) The Data Protection Directive holds that personal data may only be processed where they are adequate, relevant and not excessive in relation to the specific purpose for which they are collected.[21] Thus the data controller must specify a specific goal for data processing and the data used should be necessary and proportional in relation to satisfying this objective.
(2b) Another data minimization principle contained in the directive refers to the length of time in which the gathered data may be kept. The directive holds that personal data may be kept in a form which permits identification of data subjects for no longer than is necessary for the specific purpose for which the data were collected.[22] For example, there has been some controversy surrounding Google Street View. Google gathers photographs with cars and people on it. It blurs the faces and the license plates before publishing them on the website. This process takes Google up to a year, but the members of the leading advisory organ of the European Union with regard to data protection (the Article 29 Working Party) have asked Google to limit the period it keeps the non-blurred photographs to six months, since they feel that the period Google maintains is excessive.[23]
(2c) A final data minimization principle embedded in the directive refers to the way in which the data are kept. The principles of the directive do not apply on data rendered anonymous in such a way that the data subject is no longer identifiable. To determine whether a person is identifiable or not, account should be taken of all the means likely reasonably to be used either by the data controller or by any other person to identify the data subject.[24] Thus, anonymous data often refers to data originally able to identify a person, but being stripped of all identifiers, no longer do so. Whether data are able to identify a person must be assessed on a case by case basis. The Article 29 Working Party holds that such assessment '[] should be carried out with particular reference to the extent that the means are likely

[20] Article 17 DPD.
[21] Article 6.1(c) DPD.
[22] Article 6.1(d) DPD.
[23] <http://www.edri.org/edrigram/number8.5/article-29-wp-google-street-view>.
[24] Recital 26 DPD.

reasonably to be used for identification []. This is particularly relevant in the case of statistical information, where despite the fact that the information may be presented as aggregated data, the original sample is not sufficiently large and other pieces of information may enable the identification of individuals.'[25] This refers among others to techniques used in the data mining process.

The data minimization principles are often referred to in technical literature as well. The abovementioned principles are often caught in the phrase 'input privacy data mining'. First, a limitation may be posed on the inclusion in databases of information related to privacy or discrimination sensitive data. Second, limitations may be posed on the use of such data for data mining practices, among others through the use of cell suppression and restricting access to statistical queries that may reveal confidential information.[26] The main goal of 'input privacy data mining' is to minimize the amount of sensitive data, but still allow for an equally valuable data mining process: the so called 'no-outcome-change' property.[27]

Somewhat less well-known and less practiced is the concept of 'output privacy data mining'.[28] This does not refer to the inclusion of data in the database or the use of particular data in data mining processes, but refers to the use of data in the outcome of this process, for example in the rule, pattern or profile distilled from the data.[29] The reason for this additional instrument is that 'input privacy data mining' is not always sufficient to exclude privacy violations or discriminatory results.[30] This may either be caused by masking, indirect discrimination or re-identification, but may also be due to the fact that even although no sensitive data was used in the data mining process, the eventual outcome may still be discriminatory or violate someone's privacy.[31] To address outcome based problems, technical solutions may be implemented to prevent particular data from being used in actual practices and decisions.

15.6 Loss of Contextuality

The principles of data minimization described above help to minimize both the risk and the scale of damage if for example data is misused or a data leak occurs. Also, it may limit the use of particular compromising data in actual practices and decisions. There are however several downsides to using this technique. Firstly, the dataset may lose part of its value through this process. 'From a data mining perspective the primary issue with informational privacy is that by limiting the use of (particular) personal data, we run the risk of reducing the accuracy of the data mining exercise. So while privacy may be protected, the utility of the data mining

[25] Working Party (2007), p. 21.
[26] Ruggieri, Pedreschi & Turini (2010); Pedreschi, Ruggieri & Turini (2008); Custers (2004).
[27] Bu et al. (2007).
[28] Wang & Liu (2008).
[29] Verykios et al. (2004).
[30] Kantarcıoglu, Jin & Clifton (2004).
[31] Porter (2008).

exercise is reduced'.[32] Secondly, knowledge discovery in databases in general and data minimization in particular undermines the context in which data play a role and have a certain meaning, which may create or aggravate (the risk of) privacy violations and discriminatory practices.

Firstly, to retain the value and the meaning of the data, the data itself should be correct and accurate. This may also entail the inclusion of contextual information. However, this principle is often undermined in knowledge discovery in databases, among others since a margin of error is commonly accepted.[33] It also involves a simplification and a decontextualization of reality, since an analysis of few but determining categories is often easier, yields to more direct an concrete correlations and is thus more valuable, then a model which tries to approximate reality's complexity.[34] Last but not least, there are costs involved with accurate and complete data gathering, costs which not all parties involved in data mining are willing to bear because a particular threshold in reliability is often sufficient.

Secondly, the data should be updated so that changed facts or changed contexts are incorporated in the database. Typically however, data mining and profiling are used to predict the behavior of people on the bases of old information. Furthermore, when storing the data, one or more of four weaknesses commonly occurs. 'The data may be incomplete, missing fields or records. It may be incorrect, involving non-standard codes, incorrect calculations, duplication, linkage to the wrong individual or other mistaken inputting; the initial information provided may have been incorrect. It may be incomprehensible, involving (for example) bad formatting or the inclusion of multiple fields in one field. It may be inconsistent, involving overlapping codes or code meanings that change over time. Furthermore, even if data is recorded accurately and properly, different databases may use different formatting standards, making data sharing or the "interoperability" of different databases difficult.'[35]

Thirdly, to retain the value and the meaning of the data, the context of data should be preserved in the process of data analyses and mining. However, harvesting different databases or merging databases together, which is often the case with regard to data mining, may give rise to a problem. '[W]hen data is used in a new context, it may not be interpreted in the same way as previously used, because the new party using the data may not understand how the data was originally classified.'[36] By using data for reasons and purposes not envisaged when gathered, data may be taken and judged out of context. For example, the '[] data which circulate on the web were "issued" by people concerned with a precise objective, or in a particular context. The exchanges of data of all kinds and the possibilities to use search engines with any key words engender the risk that we be judged "out of context". [This also refers to] the question of contextual integrity; the person provides his/her

[32] Schermer (2011), p. 49.
[33] Ramasastry (2006).
[34] Larose (2006), p. 1-2.
[35] Renke (2006), p. 791-792.
[36] Ramasastry (2006), p. 778.

data in a given context and expects reasonably that it will be processed in this same context, at the risk of it being judged "out of context".'[37]

Finally, contextuality is important in assessing the value of the outcomes of the data mining process, either in patterns, profiles or concrete decisions. This may be especially important since, as has been said, automatically processed profiles and decisions usually do not evaluate the outcome and result of the data mining process in specific contexts, effecting specific individuals. Again, there is a tendency in knowledge discovery in databases to disregard the context of data.

The tendency in data mining processes to disregard the context of data are aggravated by the use of data minimization techniques[38] and cannot be addressed if stuck to this principle, since what is needed is gathering a minimum rather than a minimized amount of data, the data must be updated every now and then, which requires a continued search for data, and the context in which the patterns, profiles and rules acquired by data mining are applied must be evaluated after the process is done.[39] Although the principle of data minimization aims at excluding or at least minimizing the risk of privacy and discrimination problems, it may sometimes only aggravate these problems.

For example, if police surveillance mostly takes place in particular neighbourhoods with a lot of immigrants or ethnical minorities, then the gathered data about criminal activities would be heavily tilted towards these groups in society. Incorporation of the methodology of the research in the metadata is thus essential to avoid discrimination and stigmatization towards these minorities.[40] Furthermore, not keeping data accurate and up to date may lead to privacy and discrimination problems. If a person has decided to quit smoking, but a cigarette company keeps on profiling a consumer as a smoker, this might violate his autonomy and privacy.

Subsequently, the data mining and harvesting process must respect the context of the data. First, disregard of the purpose for which the data were gathered, the purpose limitation principle, may not only lead to a loss of the contextuality of data, but may also undermine the autonomy of the individual as his informed consent with regard to data processing for a specific purpose is transgressed.[41] Secondly, data minimization is not always able to exclude privacy violating or discriminatory results[42] given the redlining effect.[43] Data minimization not only offers no adequate solution in this respect, it might also make it difficult to assess whether a rule is indirectly discriminating or privacy violating.[44]

Finally, during the stage in which the acquired patterns and profiles are used in practice it is vital to assess the context in which they are applied. Even although

---

37 Poullet & Rouvroy (2008), p. 10 & 14.

38 Guzik (2009); Müller (2009).

39 The only principle that safeguards the contextuality in data mining that is not in tension with data minimization techniques is the purpose limitation principle, which both limits the use of data and ensures that the context of the data is retained.

40 Custers (2004).

41 Taviani (2004).

42 Calders & Verwer (2010); Ruggieri, Pedreschi & Turini (2010).

43 Calders & Verwer (2010).

44 Pedreschi, Ruggieri & Turini (2008).

rules and profiles are not obtained from analysing sensitive data, they may still have a violating effect in terms of privacy and discrimination. Thus, it would be useful to incorporate background knowledge about the context in which rules and profiles are applied to assess whether such problems or dangers exist.[45] Again, to avoid privacy or discrimination problems, a larger set of data regarding the context in which rules, patterns and profiles are applied is needed rather than a small or a minimal set.

15.7 Data Mini*mum*mization

The loss of contextuality in data mining and profiling leads to privacy and discrimination problems. Implementing the data minimization principle often leads to a further loss of context. A contrary principle might offer a more satisfactory approach. Not minimizing the amount of data gathered, stored and used, but requiring a certain minimum set of (meta)data to be gathered, stored and used when applying the results. In short, the shift from data minimization to data mini*mum*mization.

There are already several legal provisions that safeguard the correct interpretation of data and their context, among others to be found in the Data Protection Directive. These may provide useful building blocks for the data mini*mum*mization principle. The existing safeguards can be summarized as the principles of quality, both of the data themselves, the processing of the data and in the use of the data. These may come in tension with the data minimization principles from the same directive, since the principles of quality may often require additional information, not strictly necessary for the satisfaction of the specific purpose for data processing.

Firstly, the Data Protection Directive spells out that the data must be kept accurately and, where necessary, kept up to date; every reasonable step must be taken to ensure that data which are inaccurate or incomplete, having regard to the purposes for which they were collected or for which they are further processed, are erased or rectified.[46] As data regarding the context of information may be vital for correct interpretations, the first data quality principle may require the collection of such data in the database.

Secondly, the data and the context in which they play a role must be regularly updated, so that a change in facts, their significance and their context will be incorporated in the database. This relates to the second phase in the process of knowledge discovery in databases, as distinguished in section two of this chapter.

Thirdly, the Data Protection Directive spells out that data should be collected for specified, explicit and legitimate purposes and not further processed in a way incompatible with those purposes.[47] This rule entails two separate duties. The purpose for processing data must be explicit and specified. For example, the purpose 'commercial interest' will be insufficiently specific. Secondly, further processing, which means the use of data already gathered by the data controller or by a third party for another purpose then the original one, is prohibited when the purpose for

---

[45] Ruggieri, Pedreschi & Turini (2010).

[46] Article 6.1(d) DPD. Also see article 12 (b) DPD.

[47] Article 6.1(b) DPD.

processing is incompatible with the original purpose. This provision prohibits the so called function creep of data processing, which signifies the tendency to use already collected data, either by governments or by market parties, for all kinds of purposes and functions not originally intended. The third principle of quality restricts the processing of data to one specified sphere, namely the context of and purpose for which the data were originally gathered.

Finally, the Data Protection Directive contains a restriction on the use of personal data and on making of decisions on the basis of such data. The limitation regards decisions which produce legal effects concerning a person or that significantly affect him, which are based solely on the automated processing of data and which are intended to evaluate certain personal aspects relating to him, such as his performance at work, creditworthiness, reliability, conduct, etc. Such automated decision making, which is quite common in data mining processes, entails the danger of reducing a person to a number and so undermines his individuality and his autonomy. This is partially overcome by granting the data subject the right to knowledge of the logic involved in any automatic processing of data concerning him.[48] However, this leaves the problem that automatic, computer based analyses and decisions tend to be viewed by humans as absolute and that the data mining process and the outcome thereof only seldom take into account particular contexts and specific individual characteristics.[49] This risk of contextually detached decision-making is addressed in the directive by granting the individual the right to object to automatic processed decisions, thus granting him the right to be individually judged by another human.[50]

From these existing provisions, a more coherent approach to data mini*mum*mization can be developed. Four data mini*mum*mization principles can be distinguished, relating to the four stages of knowledge discovery in databases distinguished in the section two.

1. Gathering data: firstly, metadata should be registered and conserved about which data was gathered where and when. This makes it easier to assess for example whether databases are tilted towards criminal activities by minorities due to an over analysis of certain neighborhoods. Furthermore, the methodology of the process of obtaining the data, among others what data was gathered, by whom and how, should be incorporated in the metadata as well. Finally, the purpose for the gathering of data must be clear.
2. Storing data: the data gathered in the databases should be both accurate and complete. This means for example that relevant contextual data, which are vital for the correct assessment of gathered data, should be incorporated and clustered in the database. This preserves the context of the data in the further course of the data mining process. Attached to this cluster of information should be the metadata described in the previous point. Furthermore, the gathered data must be kept up to date on a regular basis. Finally, decisions on categorisation and

---

[48] Article 12 DPD.

[49] Com(90) final – syn 287 and 288, Brussels, 13 September 1990. Com(92) 422 final – Syn 287, Brussels, 15 October 1992.

[50] Article 15 DPD.

organisation of gathered material should be clear and metadata about the database itself should be included, for example about who owns it, where it is located, why and when it was build, when the data were included and when they were updated.

3. Analysing data: when analysing data, the previous cluster of data and the metadata about the gathering of information, the database and the organisation and categorization of the material should be preserved. Added should be metadata about the process of analyses, the algorithms used, the databases harvested and the methodology of mining. This may ensure that it can be assessed from hindsight whether patterns, profiles and rules distilled from the data are (indirectly) discriminating or privacy violating. Finally, the context for which the data were gathered, i.e. the purpose limitation, must be respected.
4. Using (aggregated) data: when using the patterns, profiles and rules obtained through data mining, the metadata regarding the gathering of the data, the database, the organisation and categorization of the material and the used analysing techniques as well as the clustered set of data should be accessible. Finally, data must be gathered about in what context the patterns, profiles and rules will be applied and used, so as to assess whether this may lead to privacy violations or discriminatory practices. This may also help to assess whether a discriminatory rule may lead to positive discrimination or is objectively justifiable.

As previously argued, the loss of context may lead to or aggravate privacy and discrimination problems. Inherent to current data mining and profiling practices seems a loss of contextuality, a loss which is not restored, but only aggravated by the data minimization principle. The four data mini*mum*mization principles, on the other hand, may be used and implemented to preserve the contextuality of data in data mining and profiling practices. How this should be done is beyond the scope of this chapter.

15.8 Conclusion

A common definition of autism is context blindness.[51] People suffering from autism treat data, rules and knowledge as isolated facts, as absolute, and thereby disregard the context in which they play a role. Thus, an autistic person may stop at the middle of a zebra-crossing if the traffic light turns red. To him, 'red' signifies 'stop' and nothing else, independent of the given context, while for non-autistic persons, a red traffic light when at the middle of a zebra-crossing signifies 'walk faster', rather than 'stop'. Thus, a set of rules and facts beget a different meaning in different contexts.

Data always signify a certain meaning in a specific context. If this context changes, the information may lose its or beget another meaning. With regard to indexical words such as 'I', 'You', 'Here', 'There', 'This', 'That', 'Now', 'Today', 'Yesterday' and 'Tomorrow', one needs to know where, when and by whom a phrase was uttered to determine the meaning of the phrase. More generally, all

[51] Vermeulen (2009).

data is contextuality determined in time and location, the so called spatio-temporal context. The phrase 'It is cold here' might signify different things in different contexts. If it is uttered after a long trip through the dessert, it might signify a positive feeling, while if it is uttered in a room with an open window, it might signify 'Could you please close the window'. Likewise, the time at which a phrase is uttered is significant.[52] Furthermore, the context may change over time. The phrase 'A bald man living on Abbey Road 4 in London', may originally signify only person A, but over some time could relate to both person A and B, to person B only or to no one at all. Reference can also be made to so called contextual and conversational implicatures. Suppose just after a job interview, the employee would contact one of the persons on the list of references and were to ask that person whether the applicant would be fit for an university job as researcher and the answer would be 'Well, I can tell you for sure that he makes good coffee'. Since the presumption is that a speaker will provide the maximum relevant information and this information is not relevant at all in this specific context, this would presumable mean 'no'.[53] (Again, this changes if uttered when applying for a job in the canteen). Contextuality is essential to understanding and interpreting data and information.

In a way, data mining, profiling and knowledge discovery in data bases give rise to a form of collective autism. Knowledge discovery in databases has the tendency to disregard the contextuality of information. Data are sometimes incorrect, incomplete and out of date, the data set may be tilted towards a certain group of people due to the research methodology, the data may be analyzed and used in a different context and for a different purpose then was originally intended and it's not uncommon that the context in which rules and profiles are put to work in practice are disregarded.

Knowledge discovery in databases may conflict with legal provisions regarding discrimination and privacy. A currently widely propagated solution is that of data minimization, which entails a restriction on the amount of sensitive data gathered, analyzed in the data mining process and used in practical decisions based on the data mining results. The tendency in knowledge discovery in data bases to disregard the context of the data is only aggravated by the data minimization principle.

The loss of contextuality leads to loss of value of the database and the outcome of the data mining process. Moreover, this chapter has argued, the loss of contextuality may give rise to or aggravate already existing privacy and discrimination problems. Thus, sometimes, the data minimization principle may have a counterproductive effect.

Therefore, rather than minimizing the amount of data, this chapter has argued for a minimum amount of data. This replaces the data minimization principle with the data mini*mum*mization principle. The latter principle requires a minimum set of data being gathered, stored and clustered when used in practice. First, with regard to the gathering of data, the methodology with, the context in and the reasons for which the data were gathered should be included. With regard to storing data, the data must be correct, accurate and kept up to date; the decisions on categorization

---

[52] Grice (1975).

[53] This example refers to the maxim of relevance.

and organization of the data should be incorporated. With regard to analyzing data, metadata should be incorporated about the process of analyses, the algorithms used, the databases harvested and the methodology of mining. Finally, with regard to using (aggregated) data, data must be gathered about in what context the patterns, profiles and rules will be applied and used.

By requiring and clustering a minimum set of (contextual) information, the value of the dataset is retained or even increased, and the privacy and discrimination problems following from the loss of context might be better addressed than by the data minimization principle. Nevertheless, it has to be stressed that not all privacy and discrimination problems are caused by a loss of contextuality, nor can all privacy and discrimination problems be solved by the data mini*mum*mization principles. Moreover, the data mini*mum*mization principles are neither totally new to the technical, nor to the juridical doctrine. Finally, no efforts have been made in this chapter to outline how data mini*mum*mization principles may be put into practice or be implemented in data mining rules.

References

Bu, S., et al.: Preservation of Patterns and Input-Output Privacy. In: Proceedings of ICDE 2007, pp. 696–705 (2007)

Calders, T., Verwer, S.: Three Naive Bayes Approaches for Discrimination-Free Classification. Data Mining and Knowledge Discovery 21(2), 277–292 (2010)

Custers, B.H.M.: The Power of Knowledge; Ethical, Legal, and Technological Aspects of Data Mining and Group Profiling in Epidemiology. Wolf Legal Publishers, Tilburg (2004)

Evfimievski, A., Srikant, R., Agrawal, R., Gehrke, J.: Privacy preserving mining of association rules. In: Proceedings of the 8th ACM SIGKDD International Conference on Knowledge Discovery and Data Mining (KDD 2002), pp. 217–228 (2002)

Fulda, J.S.: Data Mining and Privacy. Alb. L.J. Sci. & Tech. (11), 105–113 (2000)

Grice, H.P.: Logic and conversation. In: Cole, P., Morgan, J. (eds.) Syntax and Semantics, vol. (3), pp. 41–58. Academic Press, New York (1975)

Guzik, K.: Discrimination by Design: Data Mining in the United States's 'War on Terrorism'. Surveillance & Society (7), 1–17 (2009)

Hildebrandt, M., Gutwirth, S. (eds.): Profiling the European Citizen Cross-Disciplinary Perspectives. Springer, New York (2008)

Kantarcioglu, M., Jin, J., Clifton, C.: When do data mining results violate privacy? In: Proceedings of the 10th ACM SIGKDD International Conference on Knowledge Discovery and Data mining (KDD 2004), pp. 599–604. ACM, New York (2004)

Kuhn, P.: Sex discrimination in labor markets: The role of statistical evidence. The American Economic Review (77), 567–583 (1987)

LaCour-Little, M.: Discrimination in mortgage lending: A critical review of the literature. Journal of Real Estate Literature (7), 15–50 (1999)

Larose, D.T.: Data mining methods and models. John Wiley & Sons, Inc. All, New Yersey (2006)

Müller, V.C.: Would you mind being watched by machines? Privacy concerns in data mining. AI & Soc. (23), 529–544 (2009)

Pedreschi, D., Ruggieri, S., Turini, F.: Discrimination-aware Data Mining. In: KDD, pp. 560–568 (2008)

Porter, C.C.: De-Identified Data and Third Party Data Mining: The Risk of Re-Identification of Personal Information. Shidler i.L. Com. & Tech. (30) article no. 3 (2008)

Poullet, Y., Rouvroy, A.: General introductory report (2008), http://portal.unesco.org/ci/en/files/27268/12145631033Intro_gen_rapporteur_Y-Poullet_en.pdf/Intro_gen_rapporteur_Y-Poullet_en.pdf

Ramasastry, A.: Lost in translation? Data mining, national security and the "adverse inference" problem. Santa Clara Computer & High Tech. L.J. (22), 757–796 (2006)

Renke, W.N.: Who controls the past now controls the future: counter-terrorism, data mining and privacy. Alta. L. Rev. (43), 779–823 (2006)

Ruggieri, S., Pedreschi, D., Turini, F.,: Data Mining for Discrimination Discovery. Transactions on Knowledge Discovery from Data 4(2), 9:1–9:40 (2010)

Schermer, B.W.: The limits of privacy in automated profiling and data mining. Computer Law & Security Review 2(7), 45–52 (2011)

Skillicorn, D.: Knowledge Discovery for Counterterrorism and Law Enforcement. Taylor & Francis Group, LLC, Boca Raton (2009)

Squires, G.D.: Racial profiling, insurance style: Insurance redlining and the uneven development of metropolitan areas. Journal of Urban Affairs 25(4), 391–410 (2003)

Tavani, H.T.: Genomic research and data-mining technology: Implications for personal privacy and informed consent. Ethics and Information Technology (6), 15–28 (2004)

Vermeulen, P.: Autisme als Context Blindheid. EPO, Berchem (2009)

Verykios, V.S., et al.: State-of-the-art in Privacy Preserving Data Mining. Sigmod Record 33(1), 50–57 (2004)

Wang, T., Liu, L.: Output Privacy in Data Mining. Transactions on Database Systems 36(1), 1–37 (2011)

Westphal, C.: Data mining for Intelligence, Fraud & Criminal Detection. Taylor & Francis Group, LLC, Boca Raton (2009)

Working Party, Opinion 4/2007 on the concept of personal data. WP 136: 01248/07/EN (2007)

Zarsky, T.Z.: Mini your own business!: making the case for the implications of the data mining of personal information in the forum of public opinion. Yale Journal of Law & Technology (5), 1–56 (2003)

Chapter 16
Quality of Information, the Right to Oblivion and Digital Reputation

Giusella Finocchiaro and Annarita Ricci

Abstract. The aim of this chapter is to focus on the quality of information from a legal point of view. The road map will be as follows: the paper will begin by clarifying the definition of quality of information from a legal point of view; it will then move on to draw a link between the quality of information and fundamental rights with particular reference to digital reputation; and finally it will introduce the time dimension and the right to oblivion.

The analysis conducted here will be a scholarly reflection based both on the European Directive and the Italian Law. It introduces an original perspective concerning three different topics: quality of information, right to oblivion and digital reputation.

16.1 Quality of Information

It is well-known and needs no demonstration, that due to its interactive nature the web has become an extraordinary communication system. It is also well-known that to some extent, the web allows for anyone to communicate information without users having any chance to check either its author's identity, or the trustworthiness, accuracy and completeness of the content of information found. This also means that incorrect and false information can easily be introduced. For instance, libellous information affecting an individual's identity[1] can indeed

Giusella Finocchiaro and Annarita Ricci
University of Bologna, Italy
{giusella.finocchiaro,annarita.ricci}@unibo.it

[1] In this chapter the word "identity" means "all personal attributes as a whole". For example, in the Italian legal system the notion of "personal identity" has been for a long time reduced to that of identification, however starting from the '70s several court decisions began to adopt a totally new concept of identity as a complex of spiritual and moral features which are distinctive of individuals, which express their character and autonomy. Such notion of identity is known as the "identity as a projection" of one's moral and intellectual choices. A most recent evolution of such a concept seems to be that of the identity as the right to build oneself, to choose one's moral and spiritual personality rather than merely projecting it on the outside. In such a sense, identity can be defined as an expression of "moral liberty": Zeno-Zencovich (1995), p. 1.

B. Custers et al. (Eds.): Discrimination & Privacy in the Information Society, SAPERE 3, pp. 289–299.
springerlink.com

become more damaging if communicated on the web rather than through conventional channels.

The web is a powerful instrument to spread information but also a potentially powerful instrument to communicate incorrect or inaccurate information. In the words on one scholar: "information can now be regarded as a product which can be measured quantitatively in terms of the time used to process it; it can be stored almost indefinitely, thanks to the possibility of being able to reproduce it constantly; it can be rendered in an exclusive form without others being able to make use of it; and it can be transmitted to many places simultaneously"[2].

Information is personal data. According to the definition of "Directive 95/46/EC of the European Parliament and of the Council of Europe, dated 24 October 1995 on the protection of individuals with regard to the processing of personal data and on the free movement of such data", if information refers to a natural person it is personal data. Therefore, pursuant to art. 3, the above-mentioned directive applies to processing of information.

One of the main principles of this Directive is that of the quality of information stated in art. 6 and in "whereas" 25 and 28.

Article 6 of the Directive lays down the principles relating to the quality of data in order to ensure its accuracy, completeness and relevance and it is not excessive in relation to the purpose for which they are collected[3]. As pointed out by the Economic and Social Commission of the European Union, the said principles are the key criteria of data protection[4]. The legitimacy of the data is subject to compliance with these criteria, representing the most important element for protecting personal identity considered in its entirety. If it is true that only accurate information provides a valuable instrument to protect fundamental rights, it is also true that the requirement of completeness and updated collection of information may be a valuable tool to prevent the creation and subsequent spread of untrue, incomplete or outdated information, likely to create false opinions, to

---

[2] Frosini (1995), p. 12.

[3] "Member States shall provide that personal data must be:

(a) processed fairly and lawfully;

(b) collected for specified, explicit and legitimate purposes and not further processed in a way incompatible with those purposes. Further processing of data for historical, statistical or scientific purposes shall not be considered as incompatible provided that Member States provide appropriate safeguards;

(c) adequate, relevant and not excessive in relation to the purposes for which they are collected and/or further processed;

(d) accurate and, where necessary, kept up to date; every reasonable step must be taken to ensure that data which are inaccurate or incomplete, having regard to the purposes for which they were collected or for which they are further processed, are erased or rectified;

(e) kept in a form which permits identification of data subjects for no longer than necessary for the purposes for which the data were collected or for which they are further processed. Member States shall lay down appropriate safeguards for personal data stored for longer periods for historical, statistical or scientific use.

(...)". Bullesbach et al. (2010).

[4] Kotschy (2010), p. 43.

the point of being discriminatory. The controller must ensure that only accurate data is processed. He must keep the data up to date. These obligations of the controller are independent from the right of the data subject to have their data corrected or deleted. The obligation to only process accurate data may require rectification and sometimes may even require the deletion of data.

It is necessary to take all the measures required to erase or rectify inaccurate or incomplete data[5]. Some useful tools may be: checks at the time of collection, periodic checks (to ensure the update of data) or the use of software to prevent the acquisition of incomplete, irrelevant or inaccurate data. Some countries have been faced with the need to activate a specific mechanism for updating and rectifying the data so as not to alter public or historically valuable records. This was the case, for instance, with the request submitted by a data subject to the Italian data protection Authority to have his data erased from the "Baptism's Register"[6]. The Italian data protection Authority found that it was impossible for a person data to be erased from the "Baptism's Register". "However, it ruled that the applicant could lawfully claim that his religious beliefs should be reflected accurately, and it was thus ordered that a note should be added to the register to specify that the data subject (...) did not intend any longer to be considered a member of that religious confession"[7].

The importance of the quality of information also arises from other provisions of the Directive 95/46/EC.

Firstly, according to art. 12, the data subject has the right to obtain from the controller the access to data and where appropriate, the rectification, deletion or blocking of data the processing of which does not comply with the provisions, "in particular because of the incomplete or inaccurate nature of data".

Moreover, the data subject has the right to control data relating to himself and the controller must implement appropriate measures to guarantee the correct, complete and accurate processing of personal data. According to art. 17 of the Directive, "(...) the controller must implement appropriate technical and organizational measures to protect personal data against accidental or unlawful destruction or accidental loss, alteration, unauthorized disclosure or access, in particular where the processing involves the transmission of data over a network, and against all other unlawful forms of processing".

This provision refers to the risk of data alteration and therefore to the need to ensure the quality of the data.

The principle of quality of information and the data subject's right to access their own data are relevant in all Member State laws on data protection.

For instance, in addition to the rights already recognized by Directive 95/46/EC, the Italian Data Protection Code (Legislative Decree 30 June 2003, no. 196) recognizes the data subject's right to request that the controller integrates and

---

[5] Kuner (2007).

[6] Italian data protection Authority, decision of 13 September 1999, decision of 25 November 2002 and decision of 30 December 2002. These decisions are available on http//www.garanteprivacy, docc. web 1090502, 1067188, 1067171.

[7] Buttarelli (2010), p. 67.

updates their own data[8]. By means of this integration the data subject obtains the adaption of the data to their own personality. In particular, through the addition of new data, the collection of data mirrors the data subject and reflects their identity. Likewise, through the update of information the data subjects obtain the image resulting from the collection of data is exhaustive, so to ensure a faithful representation of their identity.

Through the recognition of these rights, the broader right of data protection is understood as to be a form of informational self-determination and the data subjects can exercise an effective control over their social image[9].

The quality of information will have increasing future importance. The work to amend Directive 95/46/EC seems to be heading in this direction. As stated in the "Communication from the Commission to the European Parliament, the Council, the Economic and Social Committee and the Committee of the Regions. A comprehensive approach on personal data protection in the European Union": "Article 8(2) of the Charter states that 'everyone has the right of access to data which has been collected concerning him or her, and the right to have it rectified'[10]. Individuals should always be able to access, rectify, delete or block their data, unless there are legitimate reasons, provided by law, for preventing this. These rights already exist in the current legal framework. However, the way in which these rights can be exercised is not harmonized, and therefore exercising them is currently easier in some Member States than in others. Moreover, this has become particularly challenging in the online environment, where data are often retained without the consent of the person concerned.

The example of online social networking is particularly relevant here, as it presents significant challenges to the individual's effective control over his/her personal data. The Commission has received various queries from individuals who have not always been able to retrieve personal data from online service providers, such as their pictures, and who have therefore been impeded in exercising their rights of access, rectification and deletion. Such rights should therefore be made more explicit, clarified and possibly strengthened"[11].

---

[8] Art. 7 of the Legislative Degree no. 196/2003: see Finocchiaro (2005), p. 285.

[9] According to art. 1: "Everyone has the right to protection of the personal data concerning them". By protecting personal data we protect the individuals and their self-determination. Under this perspective, privacy should not be seen –negatively– as a mere form of preclusion, of isolation, but –positively– as a choice on the modalities of participation, a mix of intimacy and involvement, left to individual decision. The rationale here seems to be the following: given that personal data "tell" the connections between the person and the outer world, and in particular reveal the wide range of feelings, activities, personal choices of an individual, it can be assumed that to have control over one's engagement in the world requires having control over one's own personal data.

[10] Charter of Fundamental Rights of the European Union (2000/C 364/01). http://www.europarl.europa.eu/charter/default_en.htm.

[11] Communication from the Commission to the European Parliament, the Council, the Economic and Social Committee and the Committee of the Regions. A comprehensive approach on personal data protection in the European Union, 4 November 2010. http://eur-lex.europa.eu/smartapi/cgi/sga_doc?smartapi!celexplus!prod!DocNumber&lg=EN&type_doc=COMfinal&an_doc=2010&nu_doc=0609.

16.2 The Quality of Information as an Instrument to Guarantee Certain Fundamental Rights

As stated above, the creation of a particular social identity may result from data collection[12]. For this reason it is necessary to ensure the quality of information and in this way to guarantee the right of self-determination.

Therefore the quality of information is the focal point of identity protection, when accounting for all of its components. Dignity, reputation, privacy and data protection are influenced by the quality of information. Only information which is qualitatively correct provides a faithful representation of the data subject. And only information having the above-mentioned features, can be defined as true. Information is true when it accurately reflects the image of the data subject. Information is correct when it is complete and up to date, as will be discussed in the next paragraph.

The notion of personal identity is closely linked to the protection of personal data. It is a fact that the interactive nature of the net inevitably affects the quality of the information found on it, with unavoidable repercussions on everyone's right to protect their own personal data. Requiring that the processed information is correct, complete and truthful constitutes a specific right of the person to whom the information refers, also protected by Directive 95/46/EC. Incomplete or simply incorrect information can have negative repercussions on the social image of the entire individual.

It is therefore essential that the related databases be complete, updated and correct. Thus, only the collection of information having these characteristics, in addition to guarantee the fundamental rights of individuals whose information is being processed, can also avoid false representation of their social image.

16.2.1 A New Fundamental Right: The Digital Reputation

If the quality of information guarantees certain fundamental rights, it is particularly true with regard to a specific new fundamental right; namely the right to a digital reputation. What is a digital reputation?

A reputation is traditionally understood as social esteem enjoyed by persons in the community where they live or work. Reputation is generally only considered in a conflict with freedom of expression in defamation cases[13]. Reputation however has the same importance as the other components of an individual's personality. All individuals have a reputation, whether it is good or bad, and have the right not to be subjected to unlawful attacks on their reputation[14]. "Reputation can be a key dimension of our self, something that affects the very core of our identity"[15].

On the net, this essential component of our individual personality acquires features other than those traditionally assigned to it in the physical world. This is due to the particular characteristics that the creation and circulation of information

---

[12] Davis (2009).

[13] Solove and Rotenberg (2003), p. 136.

[14] Zeno-Zencovich (1995), 2, p. 90.

[15] Solove (2007), p. 30.

acquires in the digital environment[16]. In the digital environment, anyone can create information: the Internet is by its very nature interactive, participatory. Anybody can both receive and transmit information. The ease, speed and often free services offered to users have facilitated the creation of a market where even the "buzz conversation" among Internet users can, under certain circumstances, acquire the value of an instrument of knowledge.

We know that inaccurate information is sufficient to damage the reputation of someone, as an individual or as a member of an institution or social group. If the same information is circulated on the web, it may become even more dangerous for an individual's reputation due to the net's particular characteristics which are absence of territorial limits, speed of transmission and problematic identification of the author of the defamatory message[17].

In the digital environment the processes of social knowledge have particular characteristics which may have a significant influence on the various elements of the individual's personality. On the one hand, an individual may be easily damaged by circulation of defamatory information on the web.

On the other hand, a reputation enjoyed on the web is an asset that when positive, may grant an economic benefit to the user, especially if the user is a professional operator in the digital environment.

Reputation can become an instrument of social goodwill which can be used like any other form of advertising and is essential for anyone wishing to operate in the field of business. This statement is valid with regard to reputation in the physical world. However, this statement takes on a very different meaning in the digital environment, due to the particular characteristics of the network as a means of social communication.

We can say that the concept of the reputation in the digital environment takes on a different legal dimension: from a set of opinions expressed about an individual which the individual cannot directly affect except to a small extent it becomes an instrument of personal advantage that the individuals can use for their own personal hands[18].

A digital reputation can be successfully protected also through quality of information.

16.2.2 Quality of Information and Automated Individual Decisions

The risk of distortion to an individual's image increases dramatically when any decision regarding that person is based on automated processing of personal data. Indeed, automatic processing information formulated through the interconnection of different databases, does not necessarily guarantee the accuracy or completeness of data. One of the dangers for an individual is to suffer damage and be discriminated against because of decisions based on incorrect or incomplete data. In this chapter

---

[16] (Solove 2007, 32).
[17] (Solove 2007, 32).
[18] (Ricci 2010, 1297).

we don't refer to the specific concept of "direct discrimination" or "indirect discrimination"[19], but to a more general notion of discrimination as "decision about a person based on a prejudice". It is therefore clear what kind of impact incorrect or simply outdated information can have on the creation of an opinion about an individual. Adopt a decision about a person based solely on an automatic processing of information implies risks due to the use of software. The software may be based on incorrect or outdated information; the software may be bugged; the specific case may not have been considered by the software design.

The processing of automated profiles entails risks which cannot be ignored. The collection, separation and processing of information run the risk of ending up in a huge catalogue. All this raises the need to establish definite rules on data collection, inspired by value and quality.

Moreover, we cannot ignore the fact that individuals can be discriminated against not only when information relating them is inaccurate, out of date or incomplete, but also when, due to the incompleteness of the information, they have been excluded from a certain profile and thus not taken into consideration.

The risk of possible discrimination is even higher if a decision is taken without the data subject being able to modify the data on which it is based.

Therefore the European Directive is aware of the risk of distortions and attacks on the identity of individuals, which may occur as a result of operations of de-contextualization of one or more items of data from their original context[20]. In this way, according to art. 15 of the Directive: "Member States shall grant the right to every person not to be subject to a decision which produces legal effects concerning him or significantly affects him and which is based solely on automated processing of data intended to evaluate certain personal aspects relating to him, such as his performance at work, creditworthiness, reliability, conduct, etc. (...)".

Although it has a specific aim, this provision of the Directive confirms the view expressed in this chapter which is that the quality of information is a value to which all personal data processing should aspire.

16.3 Quality of Information and Time and the Right to Oblivion

Quality of information must also be guaranteed in a time dimension. Information which is qualitatively correct at a specific moment may become inaccurate some time later.

Time brings new events and it is possible for individuals to build themselves new identities. Thus, if information is introduced to a different context, it might take a negative meaning and be a source of possible discrimination.

A distorted image of an individual can in fact be created by referring to out of context data which no longer bear any relationship to their original context of reference. We can all too easily imagine the case in which a decision is taken about a person on the basis of a collection of outdated information.

---

[19] Art. 2 of "Directive 2000/78/EC establishing a general framework for equal treatment in employment and occupation".

[20] (Nissenbaum 2009, 163).

In the same way, outdated information, which does not represent current reality, may well give an impression of the individual which is untruthful or out of context.

By "right to oblivion", according to the historical definition formulated by Italian jurisprudence, we refer to a person's right to prevent the re-publication of information contained in newspaper articles, even though lawfully published in the past. This right comes to a new life on the Internet. On the net, a story or photograph remains forever. However, the right to oblivion both on and off the net, requires balancing with other requirements such as those of the freedom press.

The recognition of the "right to oblivion" (or "right to be forgotten"), i.e. the deletion of information which no longer corresponds to the individual's identity or which is inaccurate, could constitute an adequate form of protection. The recognition of this right, in fact, is not subject to prior evaluation of the information as unlawful or defamatory.

The right to oblivion would not have a different connotation on the web. This right could not be transformed into a right to delete information unconditionally. For example, information published on the web would remain lawfully and circulate in accordance with law.

However, it should be clear that the right to oblivion would apply under certain conditions specified by regulations and by case law: there would be no general "right to delete", according to the wishes of the data subject. From this point of view, the right to oblivion should be generally balanced with other interests or rights. It is well-known that no right exists to construct a subjective identity of oneself, either on or off the Internet, but the identity is, as stated in this chapter, always the result of a social mediation between the one's subjective and a set of objective factors[21].

---

[21] An important decision was rendered by the Italian data protection Authority in connection with a complaint lodged in 2004. The case concerned the retrieval on the Internet of a decision issued by the Italian Antitrust Authority (which is not a judicial authority) against a company, based on misleading advertising; the said decision had been issued in 1996, and was subsequently posted on the Authority's web site. The plaintiff alleged that the fact of the decision being still available on the Internet whenever information concerning his current activities was being retrieved, was in breach of his right to oblivion.

In this decision, the Authority stated that the publication by the Antitrust Authority was lawful. However, in order to ensure that the processing on the Internet was not in breach of the legislation on data protection, two measures were to be taken:

a) the creation of a restricted-access section in the Antitrust Authority's website to post decisions such as the one in question (dating back to 1996), which must not be retrievable by means of the standard external search engines;

b) defining a period by the Antitrust Authority during which posting and free retrieval of a decision on the Authority's website can be regarded as proportionate in view of achieving the purposes sought by the decision in question.

On the issues related to search engines and the right to oblivion, the Italian data protection Authority adopted another decision in November 2005 dealing, in particular, with the retention and availability on the Internet of newspaper articles dating several years back. The articles in question were no longer available on the website of the specific newspaper that had published them, however they could still be retrieved via Google, which showed the parallel processing carried out by Google by means of cache copies and the respective abstracts. These decisions are available on http//www.garanteprivacy, docc. web 1200127 e 1116068.

In this respect two cases, presented by Mayer Schonberger[22], may serve as an example.

In the first one, known as the "drunk pirate", a young trainee teacher had placed on her "My Space page" a photograph of herself wearing a pirate hat while drinking from a plastic cup, with the words "drunk pirate". The photograph was seen by someone from the school where she was a trainee and despite her successfully removing it from the site, it had already been circulated on the web and had already been indexed by search engines. Because of the appearance of this photograph, she was not taken on by the school and did not pursue a career in teaching.

In the second case, a 70 year elder doctor was stopped on the United States-Canada border by a customs officer, who had typed his name into a search engine and found an article written in 2001 in which the doctor mentioned that he had made use of LSD during the sixties. The elderly doctor was stopped at the border and denied permanent access to the United States.

In legal terms, the two cases seem different. In the first one, the processing of personal data by the social network appears legitimate although the circulation of the picture on the network without the data subject's consent is unlawful. In the second case, the doctor should have been able to exercise his right to oblivion.

Both cases reflect the value of information and its influence on the creation of third party opinions. As a consequence, the above examples lead to consider the need to guarantee that every process involving the treatment of personal data fulfils specific requirements, driven by the rule of fairness and by both the need for accuracy and the protection of0 an individual's reputation. In the first case, the information was out of context. In the second case, it was outdated. In both cases the inaccurate information caused a harm to the reputation of the two persons involved.

The above considerations about the principles regarding to data quality take a different perspective on the web, where memory seems to have no limits. As a general rule, the Internet does not forget. It is not common practice to remove data from a website. Data are replicated on other websites and in the cache, in order to make it more readily at the time of request. Therefore, the data published on the net is subsequently traced and rarely deleted. No act of cancellation is commonly performed and would be in any case technically difficult. The net is therefore a repository of global dimensions. There are no fundamental criteria of archiving related to the quality of information, the contextualization in a part of a process or setting up relationships between information (metadata).

This raises the issue of removing data from the network, which does not naturally forget or select information, as well as the well-known problem of quality of information and sources, which are not always reliable or at least recognizable. However, this problem cannot only be solved by the law, but also through technology. For instance, Mayer-Schönberger proposes assigning a deadline to information[23]. Whatever the solution or solutions, the contribution of technology is also essential to allow individuals to exercise their right to

[22] Mayer-Schönberger (2009), p. 1.

[23] Mayer-Schönberger (2009), p. 152.

oblivion[24]. One challenge for operators and scientists working in the world of information, should also be to promote the effectiveness, by using the support of technology, of the individuals' ability to exercise the right to delete data. This right will relate to information which in time appears to be superfluous and with no value or interest for the community, or simply which no longer corresponds to the data subject's identity.

16.4 Conclusions

A qualitative and quantitative change in effects resulting from the collection of personal data arises from information technologies. Therefore, it becomes essential to balance the ease of collecting and using data with the need to protect individual's fundamental rights, which take on a particular aspect on the Internet. The protection of identity, considered in its various aspects (privacy, data protection, reputation, dignity and freedom) becomes necessary to safeguard the fundamental interests of every individual[25].

However, we feel it would be necessary to make changes to the approach to the issue of protection of fundamental rights. In the world of information it has no to make a clear distinction between the right to protection of personal data and other rights, such as the right to reputation, because an individual is the result of a whole which makes no sense to distinguish sharply. On the contrary, we feel it would be necessary to adopt an integrated approach focusing on identity, which would be understood to include several individual components. One of the objectives of protection of social identity should be to ensure the quality of the information as outlined in this chapter. In this respect, we hope that the reform of the Directive

---

[24] At the beginning of 2011 the Agencia Española de Protección de Datos (AEPD) ordered Google to remove certain links to pages hosting personal information regarding Spanish citizens from its results. These are a certain number of pages, most of which are newspaper articles, containing news which can be interpreted as damaging to the social identity and reputation of the subjects involved. One particular case stands out: that of Doctor Hugo Guidotti Russo, a plastic surgeon who in 1991 was involved in a case of medical malpractice and who is now asking Google to remove the related articles from search results connected with his name. In January the controversy between Google and the Spanish Authority ended up in a Madrid Court, where both parties asked the judge to find in favour of the protection of important rights: the Authority asked for the protection of the right to privacy and the right to oblivion whereas Google asked for the protection of the right to inform and freedom of speech. As reported in the Wall Street Journal, during the trial a lawyer representing Google stated that Spain is the only country where a company is obliged to remove links to web pages even if these do not contain illegal content of any description. The Spanish Authority replied that the only way to block access to content is through search engines. This is because newspapers online have the right to refuse to remove legally published news from their archives. So, the Madrid Court asked the European Court of Justice for its opinion on the matter. This Court will now have to establish whether the Spanish Authority's requests are compatible with Community legislation. The European Court's decision is awaited with growing interest in Europe in that it may establish a decisive precedent for the future of the availability of archive information on the Internet: Daley (2011).

[25] Rodotà (2004).

will strengthen the importance of this principle, while broadening its scope and reinforcing the right of the data subject.

References

Buttarelli, G.: Art. 12 – Directive 95/46/CE. In: Bullesbach, A., Poullet, Y., Prins, C. (eds.), Concise European IT Law. Kluwer Law International (2010)

Daley, S.: On Its Own, Europe Backs Web Privacy Fights. Nytimes (August 10, 2011), http://www.nytimes.com/2011/08/10/world/europe/10spain.html?_r=1 (accessed December 2011)

Davis, S.: A Conceptual Analysis of Identity. In: Kerr, I., Steeves, V., Lucock, C. (eds.), Lessons from the Identity Trail. Anonymity, Privacy and Identity in a Networked Society. Oxford University Press (2009)

Finocchiaro, G.: Personal Data Protection in the Workplace in Italy. In: Nouwt, S., de Vries, B.R., Prins, C. (eds.), Reasonable Expectations of Privacy. TMC Asser Press (2005)

Frosini, V.: Law and Liberty in the Computer Age. The Harvard Papers, Tano (1995)

Kotschy, W.: Art. 6 – Directive 95/46/CE. In: Bullesbach, A., Poullet, Y., Prins, C. (eds.), Concise European IT Law. Kluwer Law International (2010)

Kuner, C.: European Data Protection Law. Corporate Compliance and Regulation, Oxford (2007)

Mayer-Schönberger, V.: Delete. The Virtue of Forgetting in the Digital Age. Princeton University Press (2009)

Nissenbaum, H.: Privacy as Contextual Integrity. Washington Law Review 79 (2004)

Ricci, A.: Il valore economico della reputazione digitale. Prime Considerazioni. Contratto e Impresa 6 (2010)

Rodotà, S.: Privacy, Freedom and Dignity. In: Conclusive remarks at the 26th International Conference on Privacy and Personal Data Protection (2004), http://www.garanteprivacy.it/garante/doc.jsp?ID=1049293#eng (accessed December 2011)

Solove, D.J., Rotenberg, M.: Information Privacy Law. Aspen Publishers, New York (2003)

Solove, D.J.: The future of reputation. Gossip, rumor and privacy on the internet. Yale University Press, New Haven (2007)

Zeno-Zencovich, V.: Identità personale. In: Digesto/civ., IX, Torino [1] (1995)

Zeno-Zencovich, V.: Onore e reputazione. In: Digesto/civ., XIII, Torino [2] (1995)

Chapter 17
Transparency in Data Mining: From Theory to Practice

Tal Zarsky

Abstract. A broad variety of governmental initiatives are striving to use advanced computerized processes to predict human behavior. This is especially true when the behavioral trends sought generate substantial risks or are difficult to enforce. Data mining applications are the technological tools which make governmental prediction possible. The growing use of predictive practices premised upon the analysis of personal information and powered by data mining, has generated a flurry of negative reactions and responses. A central concern often voiced in this context is the lack of transparency these processes entail. Although echoed across the policy, legal and academic debate, the nature of transparency in this context is unclear and calls for a rigorous analysis. Transparency might pertain to different segments of the data mining and prediction process. This chapter makes initial steps in illuminating the true meaning of transparency in this specific context and provides tools for further examining this issue.

This chapter begins by briefly describing and explaining the practices of data mining, when used to predict future human conduct on the basis of previously collected personal information. It then moves to address the flow of information generated in the prediction process. In doing so, it introduces a helpful taxonomy regarding four distinct segments within the prediction process. Each segment presents unique transparency-related challenges. Thereafter, the chapter provides a brief theoretical analysis seeking the foundations for transparency requirements. The analysis addresses transparency as a tool to enhance government efficiency, facilitate crowdsourcing and promote autonomy. Finally, the chapter concludes by bringing the findings of the two previous sections together. It explains at which contexts the arguments for transparency are strongest, and draws out the implications of these conclusions.

Tal Zarsky
Faculty of Law, University of Haifa, Israel
e-mail: tzarsky@gmail.com

B. Custers et al. (Eds.): Discrimination & Privacy in the Information Society, SAPERE 3, pp. 301–324.
springerlink.com

17.1 Introduction: Transparency, Technology and Prediction

Can human behavior be predicted? A broad variety of governmental initiatives are using computerized processes to try. Recent advances in mathematics, artificial intelligence and computer science might render this futuristic scenario possible. Vast datasets of personal information, available to commercial and governmental entities, enhance the ability to engage in these ventures, and the appetite to push them forward.

Governments have a distinct interest in automated individualized predictions to foresee unlawful actions. This is especially true when such behavior generates substantial risks or is difficult to enforce. Data mining applications are the technological tools which make governmental prediction possible. They are essential to overcome the vast amounts of personal information at the government's disposal, and the need to analyze the information in real time. These computer programs automatically work through vast datasets to uncover trends in personal data. They then apply the novel trends and patterns revealed to other individuals and actions, while sorting the latter accordingly. In doing so, they try to figure out what the individuals' next steps would be – who of us has a higher chance of being a tax evader, criminal, or even terrorist.

The growing use of predictive practices premised upon the analysis of personal information and powered by data mining, has generated a flurry of negative reactions and responses. An overall concern is the lack of **transparency** these processes entail. A call for transparency emerges from the public, press and even from the US legislator.[1] A need for transparency is commonly cited when calling for changes in these initiatives (TAPAC Report, 2004; Cate, 2008; Solove, 2008).

Although echoed across the policy, legal and academic debate, the nature of transparency in this context is unclear and calls for a rigorous analysis. Transparency might pertain to different segments of the data mining and prediction process. In addition, it flows from different, even competing, rationales, as well as a variety of legal and philosophical backgrounds. When viewed in concert, they lead to different, at times contradicting, conclusions and practical recommendations. This chapter makes initial steps in illuminating the true meaning of transparency in this specific context and provides tools for further examining this issue.

This chapter begins by briefly describing and explaining the practices of data mining, when used to predict future human conduct on the basis of previously collected personal information (Part 2). Part 3 moves to address the flow of information generated in the prediction process. In doing so, it introduces a helpful taxonomy regarding four distinct segments within the prediction process. Each segment presents unique transparency-related challenges. This part also provides for initial strategies as to how transparency could be achieved at every juncture.

Part 4 commences a brief theoretical analysis seeking the foundations for transparency requirements in this context. The analysis addresses transparency as a tool to enhance government efficiency, facilitate crowdsourcing and promote

[1] Federal Agency Data Mining Reporting Act 42 U.S.C. § 2000ee-3(c)(2).

autonomy – a notion that itself carries various meanings in this context. Within this discussion, the chapter explains how the relevance and strength of these theories varies in accordance to the segments of the process which were previously drawn out. Part 5 concludes by bringing the findings of the two previous sections together. It explains at which contexts the arguments for transparency are strongest, and draws out the implications of these conclusions. Finally, Part 6 sets forth a brief coda, which acknowledges that transparency still leaves many important issues unanswered.

17.2 Predictions, Data Mining, Personal Information and Information Flows

Governmental predictions call for the use of sophisticated computer programs and extensive datasets, as well as a role for professional experts and data analysts. The process relies upon the success of specific technological processes. It is also premised upon assumptions; some statistical and some pertaining to society and human nature. To understand these points, the following analysis begins by quickly examining a famous example of governmental predictive tasks. Thereafter, the chapter examines the *technology* enabling these projects, the role of the *human* analyst in what seems to be an automated process and the *policy decisions* underlying many of the steps of these processes. Understanding the intricacies of these processes is crucial for establishing the importance of transparency, and how it could be applied in practice.

17.2.1 Example: Data Mining and Security[2]

Since 9/11 and subsequent attacks around the world, governments are working extremely hard to preempt such events. Among various initiatives, it is reported that governments are employing predictive data mining to study trends in the actions of attackers and attacks.[3] With predictive models in hand, individuals identified as higher risks are contacted or set aside for further questioning or scrutiny.

The public is learning of these practices directly from the government[4] or when they are subsequently leaked to the press (Cate, 2008). In a very famous incident, the public reacted with awe to the Total (and later "Terrorism") Information Awareness ("TIA") project. Parts of this project called for predictive data mining premised upon both public and personal information. It is fair to

---

[2] For a general overview, *see* CRS Report for Congress, JEFFREY W. SEIFERT, DATA MINING AND HOMELAND SECURITY: AN OVERVIEW, Order Code RL31798 (Updated April 3, 2008) available at: http://www.fas.org/sgp/crs/homesec/RL31798.pdf (Hereinafter *CRS Report*).

[3] The Homeland Security Act (HSA) of 2002 (Pub.L. 107-296, 116 Stat. 745, enacted November 25, 2002), 116 Stat. 2135, specifically authorizes DHS to make use of data mining to achieve its objectives. *See* 6 U.S.C. § 121(d)(13).

[4] DHS PRIVACY OFFICE, 2009 DATA MINING REPORT TO CONGRESS (2009), available at: http://www.dhs.gov/xlibrary/assets/privacy/privacy_rpt_datamining_2009_12.pdf.

assume that the lack of transparency in these projects contributed to this negative response. This project was famously halted by the U.S. Congress. However, the development of similar projects continues, under other names and acronyms (Cate, 2008). Beyond TIA, more limited ventures using similar techniques were set in place. The most famous and salient examples pertain to airports and international travel.

At US airports, the DHS is currently using (and further developing) data mining technology to secure the exit and entry of individuals to and from the country. The recent DHS Data Mining Reports address the development of the ATS-P (Automated Target System-Persons) module. The reports explain how various governmental databases (also those which include personal information) are analyzed to generate predictions for achieving these objectives.[5] In other words, personal information within these datasets will be analyzed and used in assessing future risks of specific individuals. DHS has even more ambitious plans in store. It is currently testing new systems which will rely on predictions premised on biological and behavioral information. Such information would be collected in "neutral settings" to establish a baseline, and thereafter at other crucial settings such as airports and sporting events. It should be noted that these systems might not require the collection of personally identifiable information, and thus generate a different set and form of privacy concerns (EPIC, 2011; McCullagh, 2011).

The predictions, which are premised upon data mining analyses, have real world implications. They lead to the fact that some travelers will be engaged with greater security examinations, while others breeze across borders. In most cases, this would result in inconveniencing some individuals for a few minutes, or even seconds. In the rarest of occasions (that are often publicized) it might lead to denial of travel or even incarceration.

17.2.2 Prediction and Data Mining: Technology, Human Discretion and Policy Decisions

The data mining of personal information includes several crucial elements – of **technology**, **human discretion** and **policy**. The key *technological* elements which can enable this process on the one hand, and generate the most difficult normative concerns on the other are data mining tools and protocols. For this discussion, I revert to a somewhat technical definition of data mining; the "nontrivial process of identifying valid, novel, potentially useful and ultimately understandable patterns in data" (Fayyad et al., 1996)[6] (yet as I will explain, the final segment of this definition is probably open for debate). This chapter focuses on "pattern based" searches (also referred to as "event-based" data mining) (Cate

---

[5] The DHS Privacy Office, 2010 Data Mining Report (December 2010), noted that data mining is not yet used, but might subsequently be applied for this objective (at 19).

[6] A somewhat different definition from a Congressional report is: "the use of sophisticated data analysis tools to discover previously unknown, valid patterns and relationships in large data sets." (CRS Report).

2008; Slobogin, 2008). These are searches which are not driven by a specific individual whom generates interest or suspicion. Rather, they focus on events, which lead to identifying patterns of behavior describing them. These patterns are later used to lead back to individuals whom pose greater risks, based on previous occurrences. Data mining methods require analysts to define specific parameters, and thereafter the software itself sifts through data and points out trends within relevant datasets.

While the automated nature of this process generates great public interest, *human discretion* still plays an important role. Analysts carry out extensive tasks at all stages of the analyses process (Zarsky, 2012). The dataset must be actively constructed, at times by bringing together data from many sources (Ramasastry, 2004). This task requires various decisions, such as which databases should be used and how specific attributes are to be matched. Other decisions are more subtle, such as how to define a parameter, and what counts as an "event" which will trigger further analysis. Next, the analysts play an active role in defining the parameters of the actual data mining analysis and the creation of clusters, links and decision trees which are later applied (Zarsky, 2002-3). This is done both in advance, and after the fact, by "weeding" out results the analyst might consider as random, wrong or insignificant. Thus, while the process is seemingly computerized and automated, analysts have ample opportunity to leave an ideological (and potentially, hidden) impression on the process (Friedman and Nissenbaum, 1997).

In addition, applying data mining models calls for several subtle yet important *policy decisions* which can impact the entire process. These decisions are rarely made public. For instance, note the setting of the acceptable level of false negatives in the predictive process. False negatives refer to the inability of the data mining analysis to correctly reveal instances in which the sought event transpires. They result from a broad and diverse mix of factors and are very difficult to establish.

Another, more subtle, policy decision focuses on interpretation. Thus far, we have described data mining as a process which reveals mere correlations. Data mining might point to individuals and events, indicating elevated risk, without telling us *why* they were selected. However, the definition quoted above describes data mining, among others, as a process that is "ultimately understandable." The level of understanding the data mining process provides relates to whether this process is *interpretable* or *non-interpretable*. Data mining can enable non-interpretable processes. In such a case, the reasons for the decisions the algorithm leads to are not explainable in human language. The software makes its decisions based upon multiple variables. Here, the role of the analyst is minimized. The lack of interpretation not only reflects on the role of the analysts, but also on the possible feedback available to individuals affected by the data mining process. It would be difficult for the government to provide a detailed response when asked why an individual received differentiated treatment. The government might be forced to merely state that the individual was singled out based on the algorithm, which was structured on the basis of previous findings.[7]

A policy decision mandating interpretable results calls upon analysts to work through the statistical outputs received, understand their meaning and articulate them clearly. In doing so, analysts note the correlations between higher risks and personal factors (such as height, age, specific credit or purchasing history). With this information, the analyst sets up profiles based on these findings, while defining their parameters, and applies them to future events. When seeking correlations, analysts might choose to ignore findings which seem ridiculous or cannot be explained by an intuitive causation model. Thus, interpretability could be considered as an important step to assure quality and precision, and that the results are not merely anecdotal. The analyst could also provide a response to external inquiries as to what initiated special treatment of an event or individual. The flip side is that interpretability calls for models which are less complex and therefore less accurate (Martens & Provost, 2011).

Interpretability also allows the analyst to go beyond correlation and search for a theory that could uncover causation. For instance, one way, cash-only airline tickets could (in theory) be casually linked to terrorists planning to ignite explosives on an aircraft. Constructing a theory of causation linking these two dynamics is relatively simple (although not necessarily true). Other correlations might call for more elaborate theories of causations. Validating such theories will call for additional study both of fact patterns and possibly in the field – all in an attempt to reveal the forms of causation in play. Therefore, requiring a theory of causation to be set in place prior to taking action based on correlations would further assure the precision of the process. On the other hand, requiring causation theories might potentially slow down and encumber the efficiency of the entire process (and might even be an impossible task). In summary, policy decisions mandating interpretability and causation are subtle, but will have a substantial impact on the prospect of transparency throughout the process.

17.3 The Nature of Transparency in Predictive Modeling: Working through the Information Flow

A call for transparency evolves when considering predictive data mining and its outcomes. Yet transparency can refer to a variety of segments throughout the prediction modeling process. Assuring transparency at every segment generates specific forms of costs and balances, and is derived from a different set of laws and justifications. In some instances, transparency might merely require uploading

---

[7] This is mostly the case when more advanced tools of data mining are applied, such as decision tree learning. Since these tools generate specific concerns of their own, they will not be further addressed here. For a discussion of such instances that at times involved tens of thousands of factors, see David Martens & Foster Provost, *Explaining Documents' Classification*, NYU – Stern School of Business, Working Paper CeDER-11-01, http://archive.nyu.edu/handle/2451/29918.

information and disseminating it. In others, it calls for the creation of guidelines and protocols. In the most extreme cases, transparency might call for proactive research on behalf of the government, which will provide additional insights as to the processes it carries out and their outcomes. To enable full transparency, the conclusions drawn out in these studies must be shared with the public.

Therefore, to properly understand the meaning of transparency in this unique context, the predictive process must be broken down into several segments. To effectively illustrate this point, this part identifies *four* distinct segments of the prediction process. Each such segment generates different transparency requirements and needs on both the technological and administrative level. Current scholarship has failed to properly distinguish among these segments. Yet understanding the different challenges of every segment are the key to resolving the apparent tension between transparency and the will to implement successful and acceptable prediction schemes. The next few paragraphs map out these segments. In addition, they briefly demonstrate the very different meaning of transparency in every context, and how it might be achieved. In doing so the analysis emphasizes the three foundations of the process articulated above: technology, human decisions and overall policy.

Transparency concerns already arise at the first steps of the predictive modeling process – (a) *the collection of data and aggregation of datasets*. At this stage, transparency refers to providing information regarding the kinds and forms of data and datasets used in the analysis. On its face, such disclosures generate limited social risks. When these exist, specific secretive governmental datasets could be excluded. An additional layer of transparency pertains to the human decisions made during the aggregation and collation stage. Human discretion plays out in a broad array of crucial stages. For instance, in the way similar records in different datasets are matched into one source.[8] Transparency at this juncture could be achieved by providing the working protocols analysts use for these tasks. This latter task is easier said than done. Clear protocols on the human role in data aggregation might not exist. Therefore, transparency will call for their creation, updating and enforcement.

Finally, transparency in this early stage has an additional, more extensive meaning. It might call for providing access to the data used in the analysis process. In some contexts, such a right of access already exists, yet to only a limited segment of the population.

Transparency considerations play a role in the next segment of the analysis process as well – (b) *data analysis*. This stage includes both technical and human-related aspects. The "technical" aspect relates to the technology used in this process. It could be rendered transparent by disclosing the names of the software applications used (if they are in commercial use). If these are custom-made, transparency could be achieved by releasing the source code of these programs.

In the realm of human decision and public policy, transparency requirements pertain to a variety of elements. It can relate to the acceptable rates of errors in the

[8] Studies indicate that this stage is a "major contributor to inaccuracies in data mining." Cate, 2008, at 470.

analysis process (such as false positives) and, for instance, the level of esoteric correlations which would be found acceptable for future usage.

Intuitively, however, when addressing transparency, public opinion focuses on stage (c); the actual strategies and practices for using the data that government applies. In other words, these are the *predictive models* formulated through the data mining process. For instance, they are the actual "profiles," according to which DHS or other entities single out individuals or events. Governments are reluctant to provide transparency at this juncture, and expose the relevant information to public scrutiny. Such reluctance is mirrored in the existing legal rules. For instance, in the US, the Internal Review Service does not share the details of the audit profiling algorithm it applies (Schauer, 2003).

Formulating a theoretical framework to achieve transparency at this juncture is challenging. Accounting for the way predictive modeling truly transpires quickly leads to a conclusion that simple solutions previously contemplated are outdated. Regulation cannot merely call for disclosing the factors used in a profiling scheme. With advanced prediction, there is no static "profile" to reveal. There is merely a dynamic learning process. Rather than a set profile, the government uses an algorithm that singles out higher risk events. But such an algorithm cannot be disclosed in a simplified format. The algorithm might be revealing a complex association rule which includes a multitude of factors, as well as the interaction among them. In other instances, the algorithm might be revealing clusters of factors and attributes with blurry and constantly changing borders which are used to identify higher risks. Conveying information about these practices to the public in an understandable way calls for setting new regulatory paradigms in place. Obviously, whether the process is interpretable or non-interpretable will impact the ability to achieve transparency at this juncture.

Moreover, achieving meaningful transparency at this stage calls for an additional set of disclosures rules. The government might not only be required to present the factors correlated with the events it strive to predict, but also establish a causation theory that stands behind the selection of these factors. Furthermore, the government might be required to assure that the prediction schemes do not involve the use of factors (either directly, or by proxy) society finds discriminatory and unethical. For achieving this objective, government would be required to conduct studies examining the impact of the prediction scheme. Only with such information could the process be considered as transparent. In other words, these measures will call upon government to produce new information, rather than provide access to information it already has (Weil, et al., 2011).

Finally, unique transparency requirements relate to the last segment of predictive analysis (d) *the feedback process* following the use of the model. Examining the use of predictive models can lead to important insights. It reveals how many of those indicated as a higher risk turn out to be of no risk at all (false positives). It could further indicate how many of those considered as lower risks should have been indicated as a high risk, yet were "missed" (false negatives) by this analysis. In addition, the analysis of the ongoing process will provide information as to whether the practices facilitated de facto illegal or unethical forms of discrimination.[9] These are extremely important factors which must be produced.

I am currently unaware of concrete policy requirements addressing these transparency elements. However, the foundations for meeting these needs are already in place. In the US, The Office of Civil Rights and Civil Liberties within the DHS is at times called to issue Civil Liberties Impact Assessments ("CLIAs"). These reports directly addressed many of the concerns mentioned above.[10] According to the existing template for such reports,[11] they must examine how new programs and policies will (among others) affect minorities. They also must examine what alternatives routes could be taken to meet the same objectives while limiting harm to civil liberties.[12] In other contexts, however, new rules would be required to generate and later publish feedback studies.

Transparency refers to a broad array of additional factors as well. Transparency requirements pertain to the steps taken to assure data security, retention, and tools for access control. They further might address measures for providing data accuracy, lack of errors and redress for harmed citizens. I choose to set all of these issues aside. These issues are important, and indeed pertain to any general analysis of personal or important information. Yet they probably exist in other digital settings. Predictive modeling calls for additional and even unique dimensions of disclosure which I chose to emphasize here.

17.4 Why Transparency?

17.4.1 General

After understanding *where* transparency would be needed and (very generally) *what* it might entail, we now turn to the foundational normative question – *why* should transparency be mandated. A call for transparency is echoed throughout the debate concerning the implementation of predictive data mining tools for the analysis of personal information. The need for transparency is motivated by a variety of reasons and arguments. Every one of these theories could lead to a different solution. To provide an overall taxonomy of transparency concerns and

---

[9] For instance, *see* results of study concerning NYPD policy for stopping individuals, which turned out to be extremely biased. Floyd, et al. v. City of New York, et al, o8 Civ. 01034 (SAS), REPORT OF JEFFERY FAGAN (October 15, 2010), available at: http://ccrjustice.org/files/CCR_Stop_and_Frisk_Fact_Sheet.pdf.

[10] This is done either by law or within the agency. *See* REPORT TO CONGRESS ON THE DEPARTMENT OF HOMELAND SECURITY OFFICE FOR CIVIL RIGHTS AND CIVIL LIBERTIES 2008, at 20, available at: http://www.dhs.gov/xlibrary/assets/crcl_annual_report_FY_2008.pdf.

[11] DHS Office for Civil Rights and Civil Liberties, CLIA TEMPLATE, available at: http://www.it.ojp.gov/documents/Civil_Liberties_Impact.pdf.

[12] CLIA have yet to examine all the aspects here addressed, but it is possible that such efforts are on their way. *See* Impact Assessments Underway, http://www.dhs.gov/xabout/structure/gc_1273849042853.shtm.

set forth solutions, an overall mapping of these theories is required. This section takes preliminary steps to provide such a theory. In doing so, it will often return to the taxonomy drawn out above. It will explain how the theories vary in relevance and strength when shifting from one segment to another.

Prior to delving into a discussion of detailed theories of transparency and disclosure, we must address the simplest and perhaps most intuitive theoretical explanation. The acts of a liberal and democratic government must, categorically, be as transparent as possible.[13] Indeed, a basic right of transparency could be derived from the notion of democracy. Scholars note that transparency is essential for democracy to function. In doing so, they make reference to an abundance of sources, such as Locke, Mill, Kant, Rousseau, Bentham, and James Madison (Fenster, 2006). They further explain that transparency enables an informed public debate, generates trust in and legitimacy for government. It also informs individual decision on Election Day. A similar notion is reflected by accepting transparency as a basic human right.[14]

Accepting such a categorical argument, on its face, shortens our analysis – as it provides a clear response to the question as to why transparency is important and required. Yet an instrumental analysis of the benefits and outcomes of transparency as it pertains to specific segments of society and the steps of the process is still called for. Any pro-transparency argument is quickly rebutted by powerful and convincing counter arguments (which will be addressed in future work). Central counterarguments note that transparency generates substantial costs, undermines governmental objectives, promotes crime and generates stigma. Without a deeper understanding of the interests in play, correctly balancing transparency against these counterarguments would prove impossible. In addition, a categorical right of transparency will fail to provide many important distinctions between levels of transparency throughout the information flow addressed here. Only a broad and elaborate theoretical foundation will provide specific responses at every juncture.

However, an overarching theory of transparency rights which is premised on democracy still has an important implication. The "default" of governmental actions should be transparency. Yet the precise formation and extent of disclosure would be derived from the specific theories the next paragraphs draw out. In Section 4.2 below, I map out four theories of transparency. These are premised upon the following theories and mechanisms broad array of justification (1) Transparency as a tool for assuring fair outcomes via shaming; (2) ransparency as a measure for engaging the broader public through "Crowdsourcing;" (3) Transparency as a measure for promoting the autonomy of data subjects, or (4) the autonomy of those impacted by the predictive process (the subtle

---

[13] The Obama administration has accepted this notion, as stated on the White House website: "**Government should be transparent.** Transparency promotes accountability and provides information for citizens about what their Government is doing."

[14] For a discussion of this right, and a comparative study, *see* TOBY MENDEL, FREEDOM OF INFORMATION: A COMPARATIVE SURVEY (2nd Ed., UNESCO, 2008), Section 1:30, available at:
http://portal.unesco.org/ci/en/files/26159/12054862803freedom_information_en.pdf/freedom_information_en.pdf.

differences between these two latter arguments will be made apparent in the paragraphs below).

17.4.2 Transparency – From Theory to Policy

Transparency as an Incentive for Fair and Efficient Policy (Or, Transparency and the Role of Shame)

A basic (and intuitive) justification for transparency is that it facilitates a check on governmental actions. Generally, society constantly fears that the acts of its government might be flawed, biased, ineffective or inefficient. The relevant officials might be improperly balancing rights and interests, led by their own bigotry, or are over-influenced by private interests. This outcome might result from the relevant governmental agencies incompetence, corruption, negligence, mere error or perhaps unacceptable point of view. Officials might also try and expand their authority to meet other objectives. They might try to apply tools developed for battling terrorism towards the war on drugs or finding deadbeat parents. "Project Creep" and "Function Creep" are central concerns stemming from the adopting of data mining tools by government (TAPAC Report, 2004). Transparency is a key measure to counter all these concerns.

When discussing this objective, the term "accountability" quickly comes to mind. Transparency is at times considered synonymous to "accountability." Yet these concepts clearly are not the same. Accountability refers to the ethical obligation of individuals (in this case, governmental officials) to answer for their actions, possible failings and wrongdoings. Transparency is an essential tool for facilitating accountability, by subjecting politicians and bureaucrats to the public spotlight. Yet, it is merely one of the strategies that could be applied to achieve this objective. Accountability might be achievable with more limited means. Applying full transparency to achieve this accountability calls for specific justification.

A call for transparency requires the expansion of information sharing schemes beyond internal government review, possibly even to the broadest realm of the entire public. The assumption that broadening the scope of information sharing in this way will promote fairness and efficiency should not be taken for granted (especially in view of transparency's detriments). A constructive way to theoretically approach the benefits of transparency in this context is to return to the work of Louis Brandeis. Brandeis famously advocated the use of transparency to promote fairness. In a recent article, Lessig drew out two basic theories as the foundation of Brandeis' call for transparency which relate to this issue and justification – (1) shaming, and (2) the effects of market or democratic forces (Lessig, 2009). These two theories prove helpful in examining whether transparency is indeed effective in the context at hand, and its proper extent. For this discussion, let us focus on "shaming" and draw out prerequisites for its success. An analysis of the "market forces" element generates similar outcomes, and will be drawn out in future work.

In some instances, transparency will indeed facilitate "shaming."[15] In these cases, the fear that a broad segment of the public will learn of the bureaucrats' missteps will deter them from initially engaging in problematic conduct. Presumably, for effective "shaming," the government must disseminate information to the greatest extent possible. This statement, however, relies on two hidden assumptions: (1) the public takes interest in the relevant workings of government (here, facilitating the predictive modeling dynamic), and (2) the officials and bureaucrats engaging in these practices will react to the public's knowledge and "shaming." Yet when considering predictive data mining schemes, both assumptions could be questioned.

First, there must be public interest. Establishing whether a "shaming" dynamic will transpire calls for examining whether the broad disclosure might generate an interesting story which could be conveyed to the public. The public (directly or through proxies) might shy away from technical, complicated and obscure matters. In such contexts, shaming might not occur. It should be noted that "public interest" does not necessarily call for a direct, active and ongoing interest by a broad segment of the population (Kreimer, 2009). Information flows in "cascades." The limited interest of few experts can generate much greater interest by broader segments of the population at a late stage. The experts encounter information, comment on it and distribute it to the public, which picks up on these dynamics.

With these insights in mind, let us examine the prediction process's segments which were drawn out above. Every one of them includes some issues with broader appeal, and other that are technical and complex. The various segments will be addressed at this juncture in general so to grasp the basic intuitions, and summed up in section 5, below.

Segment (a) includes seemingly salient steps such as the selection of factors for the prediction process. These are decisions which will generate interest and uproar if deemed problematic. Therefore mandating transparency under this theory and at this specific juncture is relatively simple. However, decisions as to what forms of analyses should be applied (which dominate segment (b)) will probably generate far less interest and traction given their technical nature. Segment (c) which includes the models and profiles government will use in the prediction process is a difficult case. Some of the broader issues this segment brings to mind – such as the forms of "discrimination" these models facilitate – will be sure to generate public interest. However, understanding the internal workings of this segment calls for grasping the workings of a predictive modeling process. As mentioned above, novel data mining applications rely upon technical terms. These might be too subtle and complex to generate shame. A similar point could be made

[15] Scholars have recently examined the return of "shame" based punishment in criminal law, while pointing out various benefits and shortcomings. Dan Kahan, for instance, is a famous supporter of shaming, with some limitations, although in a recent paper he expressed some reservations. *See* Dan Kahan, *What's Really Wrong with Shaming Sanctions*, 84 TEX. L. REV. 2075 (2006). I will not address this debate within the confines of this paper, mainly because it does not focus on punishment and the option of imprisonment (which indeed may follow from the actions here described) but that of generating accountability. Thus, only some of the social and psychological dynamics addressed in this literature, apply.

regarding segment (d) – the feedback process following the automated/predictive steps. The public would be interested in the overall success of the project, as well as in systematic errors and failures. Yet it might ignore the more technical aspects of the dynamics. Therefore, for every one of these segments, this transparency justification carries a varying level of persuasive force.

Second, for "shaming" to have an effect, the "shamed" must respond to it (or be deterred by the prospect of such disclosure). This dynamic will fail, for instance, if public opinion does not associate the specific decision maker with the relevant action. In such a case, transparency will not necessarily promote fairness and efficiency. The nature of automated prediction leads to the fact that many important decisions will be made by lower level analysts and IT experts – especially in stages (a), (b) and even (c). Shaming might not have the needed effect on these officials. They might already be at another position at the time of revelation, or not clearly and directly indicated. Again, shaming seems to have a limited effect on the more technical elements of this project, questioning the wisdom of transparency regarding these factors.

Finally, a third underlying assumption which flows from the two already mentioned states that shaming will work well when transparency reveals official conduct that conflicts with well established norms or existing laws. For instance, sloppily constructing the data mining process and operating it, will easily generate backlash. This will counter the accepted norm that governmental work must be carried out with precision and accuracy It can also prove effective if transparency revealed that rather than relying on neutral factors, officials reverted to relying upon "sensitive factors" (either knowingly or unknowingly) such as race and religion – practices which are socially unacceptable. Yet shame might not prove helpful in other important instances, where social norms have yet to formulate. In such cases, disclosure will not lead to "chilling" unwanted governmental conduct. For instance, there has probably yet to emerge a social norm regarding accepted and non-accepted measures of data collation and levels of acceptable in this process.

Returning to the segments of the data mining task drawn out above again shows different outcomes for different segments. Much of the information available in stages (c) and (d) falls within this category. The risks of false positives and the forms of correlations used are currently within "gray areas" of social norms. Transparency could be an important measure to promote a discussion on these issues. However, it is questionable whether this context will generate shaming, which will act as an effective "check" on governmental actions.

To conclude, shaming acts as an effective "check" in instances where decisions are made by high ranking officials and clearly counter social norms held by a broad segment of the population. It is also helpful if the practices at hand are understandable, or at least easily built into a convincing narrative. In all other contexts, a shaming-based transparency theory might be unable to justify the costs and detriments it generates. This distinction will prove helpful in formulating a general blueprint of transparency policy for the data mining context.

Transparency and Crowdsourcing

Transparency might enhance the accuracy and fairness of predictive models in a very different way. Rather than incentivize effective governmental actions, transparency can bring knowledge from *outside* the government to improve the underlying process. Generally, the level of expertise, time and attention available outside the specific agency (and even government in general) are greater than the knowledge available within. Therefore, greater exposure of information regarding the inner works of government to a broader segment of the public will enhance the chances to receive meaningful feedback. This will again lead to a more effective outcome for governmental policy.

These arguments are closely linked to another facet of recent scholarship in IT law. Such scholarship addresses peer-production – the mass participation by individuals from varied walks of life and different skill sets, in joint projects. As Yochai Benkler (Benkler, 2007) and others[16] explain, IT and especially the internet led to the rise of a third collective/industrial force which matches and even surpasses that of the firm and the market. Transparency can enable these powerful dynamics, and thus promote governmental objectives and achieve overall efficiency. In other words, this argument calls for engaging the crowds as a source for achieving social objectives – or "crowdsourcing."

The crowdsourcing argument pertains to almost all stages of the prediction process. Experts and laymen from a variety of disciplines can provided meaningful insights regarding methods of aggregating data, engaging in data mining analyses, examining theories of causations and assessing the feedback. Above all, experts can work through the code of the software operating these schemes, examining its neutrality, and whether it indeed carries out the tasks it purports to doing. Therefore, while this theory can apply to all the process's segments, it is usually linked to stage (b) of the information flow. Here, the disparity between governmental knowledge and freely-available external expertise is the greatest.

A discussion of crowdsourcing and its feasibility quickly leads to the question of motivation – why should the crowd indeed act as a source for these activities, especially when no direct compensation is provided. The incentive structure for external participation in a voluntary venture to assist government in predictive modeling is a complex issue. Indeed, some of the motivations transpiring in other contexts will not play out here.[17] However, several other incentives are extremely

[16] A great deal of popular writing has flourished in these fields – such as Crowdsourcing, Wikinomics. *See* CLAY SHRIKY, HERE COMES EVERYBODY: THE POWER OF ORGANIZING WITHOUT ORGANIZATIONS (Penguin Group, 2008), DON TAPSCOTT & ANTHONY D. WILLIAMS, WIKINOMICS: HOW MASS COLLABORATION CHANGES EVERYTHING (Portfolio, 2006); JAMES SUROWIECKI, THE WISDOM OF CROWDS (First Anchor Books Edition, August 2006).

[17] In other contexts (such as open software source and content projects) scholars indicate that individuals might be motivated by spite (to "get back" at a bad employee or vendor, and in that way inform the public of their wrongdoings). They are also motivated by an aspiration to generate a reputation which will promote the individual within a community or even assist in seeking future employment. Lior J. Strahilevitz, *'How's My Driving?' for Everyone (and Everything?)*, 81 N.Y.U. L. REV. 1699 (2006, Eric Raymond, THE CATHEDRAL AND THE BAZAAR (1997). For a different perspective, which plays down the current level of contribution to open source projects which is altruistically motivated, see Jonathan Barnett, *The Host's Dilemma*, Harv. L. Rev. (*forthcoming*, 2011).

relevant. Individuals will contribute to this project altruistically. Others will do so as a hobby or as a part of their academic research. Yet others might do so as means to contribute to a community which might be emerging (Citron, 2008). Thus, it is fair to assume that a sufficient number of individuals will strive to review and contribute to these policies and governmental initiatives.

Current transparency regulation does not reflect any aspects of this theory. Generally, government does not enable any meaningful feedback of the prediction process. The most practical segment for implementing such policy is where its absence is most noticeable – with regard to the computer code charged with running the analysis. However, rather than allowing experts to review and comment on it, the government provides almost no insights as to the codes inner workings.

While applying this rationale into policy is important, it could be substituted at times by providing information to a selected group of experts These experts will assist the government with feedback on predictive modeling projects without disclosing the information further. Shifting to this limited form of disclosure might be called for given the strong arguments against full disclosure (such as, that disclosing source code will compromise trade secrets and the overall success of the prediction scheme).

Autonomy as Control over Personal Data

In the process of predictive modeling, a requirement for transparency flows directly from the rights of those individuals whose personal information was used throughout the process – the "data subjects."

The basic premise leading to this aspect of transparency is the notion of control individuals have over their personal information (Westin, 1967; Lessig, 1999). This theoretical notion has been broadly accepted in the EU,[18] while only partially recognized in the US. This concept could be understood as an extension of the individual's autonomy. It was translated into several concrete principles that after several transitions formulated the "Fair Information Practices," or FIPs.[19]

"Openness" or "Transparency" was central to FIPs from their earliest stage (Reidenberg, 1995). In FIP's current version, this notion is encapsulated in the principle of "Notice." "Notice" commonly refers to informing individuals that personal information about them is being collected, and its subsequent uses. The analysis of personal information is an essential part of the predictive processes here discussed. Thus, recognizing the principle of "Notice" should lead governments to shed additional light on the data mining processes as far as they pertain to personal information.

---

[18] Directive 95/46/EC on the protection of individuals with regard to the processing of personal data and on the free movement of such data, Official Journal of the European Communities of 23 November 1995 No L. 281 p. 31 (Hereinafter *EU Data Protection Directive*); DANIEL J. SOLOVE & PAUL M. SCHWARTZ, INFORMATION PRIVACY LAW (Aspen Publishers, 2006), 35-8.

[19] ORGANIZATION FOR ECONOMIC CO-OPERATION AND DEVELOPMENT (OECD), GUIDELINES ON THE PROTECTION OF PRIVACY AND TRANSBORDER FLOWS OF PERSONAL DATA (1980).

While the right to "notice" is limited to the data subject, a strong argument could be made to expand this right to the entire public in the predictive modeling context. As prediction methods require the analysis of personal information which pertains to almost the entire population, disclosure according to this theory must also be provided to all. Everyone is a "data subject" one way or the other. Thus, everyone should be provided with information regarding the way personal information is used in this process.

Transparency premised on this specific theory might only have a limited reach. The "notice" requirement usually includes informing individuals as to the way their information is aggregated (segment (a)). Yet can the "notice" principle, which is derived from the notion of data subject autonomy, justify broadening transparency into the latter stages of the prediction process? Under EU Law, the data subject must receive information regarding future purposes of personal information analysis (a right referred to as "purpose specification").[20] These rights might translate into providing some of the information addressed in segment (b) above. Yet reaching farther into the analysis process is quite a theoretical stretch. One can argue that the latter steps of the data flow (segments (c) and (d)) all result from the initial secondary uses of personal information. Therefore, data subjects' autonomy and liberty should be acknowledged by providing them with a full view of subsequent information flow.

Such arguments for the broadening of the autonomy argument to the latter segments of the data mining process will probably be rejected. In the information age, it is difficult to argue for (and surely, enforce) such a broad definition of autonomy and control over personal information. This assertion follows from recognizing the ease with which individuals concede their personal data in commercial settings. It is noted that individuals value their privacy and fear personal data would be subsequently used in a variety of ways. Yet such concerns cannot justify providing individuals with control over events such as those discussed here (in the latter stages of the predictive process). The collection of personal information and its initial analysis are quite removed from the subsequent use. I acknowledge that this analytical position is somewhat different from the existing legal and theoretical setting in EU Data Protection law and theory. I believe it is aligned, however, with actual information flows and market trends. These indicate, in most cases, a loss of interest and control in personal information which travels beyond a specific threshold of proximity to the relevant individual.

Furthermore, this autonomy-based theory is not sufficiently robust to justify the specific forms of transparency the latter stages of the process call for. Autonomy and control might provide a right to understand future analysis of the personal data. Yet the above analysis indicates the need for ancillary information rights as well. For instance, transparency calls for mandating governmental studies of the causation underlying its actions, and the success and failure rates of the project. The connection between such information and the autonomy (in the context of control) rights of data subjects is even beyond incidental.

---

[20] This right is found in Articles 10 and 11 of the Data Protection Directive 95/46/EC.

To conclude, this theory has partial relevance to the issue at hand. It can promote transparency at several early stages of the process. It can lead to providing data subjects with information about the data collection and possibly its analysis. But the subsequent steps of the process are probably beyond its analytic reach.

Autonomy of the Impacted Individual

Autonomy-based theories generate justifications for privacy from a very different perspective as well – that of the individuals adversely affected by the predictive process (as opposed to those whose information is used in the analysis). If individuals are affected by predictive modeling, they have a right to understand why. They should receive an explanation as to the decision criteria and to the logic behind these actions.

This notion can easily be framed in terms of autonomy. An individual has a right to learn the reasons for events which affect him. Such information empowers her, and she senses she is treated with respect, as a human being. This notion is deeply imbedded in European law and specific member states.[21] Beyond autonomy, US scholars strived to embed these specific concerns in the notions of the US Constitution such as the right to Due Process.

For instance, when addressing this issue, Daniel Steinbock finds that measures which resemble "due process" rights should be applied in this context, even though Constitutional protection does not apply under US constitutional doctrine (Steinbock, 2005). These measures are due in view of the individual's dignitary rights. He states that these rights should include some form of notification of the process the individuals were subjected to.

While recognizing the right to this form of transparency is relatively convincing and straightforward in general, applying it in practice raises difficult questions. European laws and US scholarship addressing these requirements in the context of profiling call for informing the affected individuals of the profile they were subjected to (Citron, 2008b). In terms of this chapter's framework (as drawn out is section (3) above), it will call for providing individuals with information regarding stage (c) – while focusing on the actual factors selected for prediction. With such information in hand, the affected individuals would be able to object if they find it to be inaccurate.

However, the general concepts mentioned for implementing transparency on the basis of the abovementioned theory and principles are somewhat outdated. In the age of data mining, conveying information to the impacted individual regarding the profile used is in many instances either meaningless or impossible. The "profile" might prove to be a string of parameters that indicate a problematic

---

[21] In the Netherlands, section 42(4) of DUTCH DATA PROTECTION ACT. In Spain, article 13(3) of the ORGANIC LAW 15/1999 of 13 December on the Protection of Personal Data. In the EU Regulation (EC) No 45/2001 of the European Parliament and of the Council of 18 December 2000, section 13(d) applying the directive to EU government bodies. Note, however, that these rules include exceptions for security and law enforcement.

correlation the affected individual falls within. When provided with such information, the individual could rest assured there was no identification error. She might further understand that the process was not random. Yet without understanding the inner workings leading to this outcome, these results might still appear arbitrary. Therefore, providing limited insights into the governmental actions at stage (c) might be insufficient for restoring autonomy and dignity.

In view of the above and to further empower relevant individuals and provide meaningful feedback, this theory calls for expanding transparency beyond stage (c) of the data mining process. It probably requires that data mining would be an interpretable process. It might even call for assuring a causation theory was found to explain all actions taken. With such additional disclosure, individuals can obtain sufficient insight to the process and how it relates to their lives. This theory could also be understood to call for transparency in other stages of the prediction process. It could call for the measures described above as part of stage (d). In other words, to assure dignity and promote autonomy, the individual should receive assurances as to the precision, effectiveness and lack of discrimination in the process. The information provided through the feedback loop can promote these objectives.

Furthermore, this theory could also justify transparency in stage (b). Information regarding the use of data mining algorithms is essential to allow the affected individual to retain autonomy and dignity. With information regarding these important steps of the process, the mere correlations used (in which an individual was implicated) can be understood as part of a broader picture. This will prove helpful in understanding that targeting was not arbitrary, and perhaps even in challenging its findings. It should be noted, however, that this segment of the argument is relatively weak.

Thus far, this section of the analysis has shown that this autonomy-based rationale provides a powerful argument for transparency in a broad variety of segments along the data mining process. However, this theory also includes a central flaw – it provides a transparency justification for merely a small segment of the population – those adversely impacted by the relevant predictive practices. Only such individuals face the potential of autonomy-based harms and are thus entitled to autonomy-based remedies.

Yet one can argue convincingly that if the government must disclose such information to a limited population segment, it should already provide it to the entire public. This argument flows from acknowledging that the information regarding the data mining practices vested with the few will make its way to the entire population anyway. In today's information age, it is quite common that disclosure to a limited group of disgruntled individuals quickly leads to spreading such knowledge to the entire public. Those adversely affected will provide their information online (and if stigma may attach, will do so anonymously). With time, the pieces of the puzzle will come together and a full picture would emerge in the public realm. For that reason, government should initially go ahead and provide such information to all.

At this point, some might argue that disclosing these governmental practices to the affected few will not lead to a broad understanding of what the government is actually doing. Rather, it will lead to lead to a partial and in many instances biased and wrong overall picture of governmental practices. For that reason, the government would be foolish to disclose its entire array of activities, especially given the various negative impacts of such disclosures on the effectiveness of prediction schemes.

While I agree that indeed a distorted understanding of what government is doing might follow from only selective transparency, I believe the abovementioned assertion is yet another strong argument for broad transparency to the entire public regarding the effects of the prediction schemes (and not an argument against enhanced disclosure). Such broad disclosures are in the governments interests. A distorted public opinion regarding the actual data mining practices might have devastating outcomes. It might lead the public to believe that the government is engaged in unfair or racial discrimination or even acting arbitrarily. If the government is not doing so, it is within its interests to fully reveal its strategies to the public.[22]

17.5 Bringing It All Together: Towards a Policy Blueprint for Transparency

Our short journey through the theoretical justifications to transparency in this unique context is nearing its end. Let us return to the taxonomy of the flow of information throughout the four segments drawn out above, and explain the limited policy implications this study can provide. I do so by summarizing the theoretical findings of the previous section.

Before proceeding, it is important to note the limits of the recommendations to follow. Their main flaw is their general scope. When these issues are to be examined in a specific context, several key elements must be rethought or introduced. First, we must examine the feasibility of transparency in the specific context – what it might entail in terms of costs and technical difficulties. Second, the strength of the "general" pro-transparency arguments which are premised upon democracy and basic human rights will vary as well. In some contexts (for instance, when core democratic rights such as speech might be compromised) this justification holds greater force than others. Third and perhaps most importantly, are the arguments for opacity – a matter of central importance when deciding on the extent of transparency in governmental schemes. Governments are often concerned that transparency will lead to unintended consequences and even allow for the circumvention of its efforts. Clearly these concerns must be examined on a case-by-case basis.

Returning to our summary of transparency recommendations, I begin with *segment (A)*. Strong transparency-based justifications exist for making public the lists of datasets applied at this stage. Much weaker justifications exist for the

[22] For those concerned with security issues, note that if policy considerations allow for revealing the governmental strategies to those indicated as higher risks, it would be quite difficult to argue that such publication of such information to the broader public would harm government interests.

disclosure of the information within them (with the exception of the relevant data subjects). The difficult questions pertain to the nature of transparency regarding the technical measures for collating these datasets. These are probably best kept out of the reach of the public eye. Auditing of this process would be carried out internally, with the help of selected experts.

These conclusions follow from accounting for the elements discussed above. With the exception of the actual datasets used, I doubt whether any disclosures made regarding this segment will generate sufficient public interest to "shame" lower level officials (who will be making most of the technical decisions) into changing their practices. On the other hand, the "crowdsourcing" objective carries merit at this juncture, especially regarding technical decisions. However, they probably could be acheived by engaging in selective disclosure to experts.

The "autonomy" based arguments do not provide substantial insights. Those premised upon the rights of data subjects (addressed in section 4.2.3) might justify additional disclosure of these factors – especially regarding the personal information used in this process. Yet I am skeptical whether this theory (which, as mentioned suffers from several analytical flaws) can justify the disclosure of ancillary information regarding the collation and matching process of the analysis. On the other hand, it is quite a long shot to connect disclosure requirements at this segment to the autonomy rights of those affected by the data mining analysis – concerns arising on the opposite side of the information flow (and addressed in section 4.2.4).

Segment (B) presents more of an analytic challenge. Currently, the public is left almost entirely in the dark at this stage, in which the data is analyzed and patterns formulated. This must change. Additional layers of disclosure should be applied to both technological elements and human and policy decisions.

These arguments can be justified under several theories. Let us separately approach the technology and policy aspects of this segment. In terms of the *technology* used at this juncture, transparency will only serve as a minimal "check" on governmental actions. The public would have limited interest in these technical details. Thus, there is only limited potential for effective shaming. "Crowdsourcing" is the argument which seems to have the greatest force. Both autonomy based arguments are quite a stretch, as "data subjects" and "affected individuals" will have a difficult case linking these rights to the actual computer analysis. In view of the obvious detriments of sharing the technology with external sources, a possible compromise calls for releasing the software to a selected group of experts throughout the industry. These experts will be barred from sharing such code commercially. They, however, would be able to inform the public if hidden agendas are imbedded within the code. This seems to be a reasonable policy strategy for disclosure of technology-related information at this juncture.

Moving to the realm of *policy* decisions, I find that information concerning the decisions regarding the acceptable level of errors within the process (sometimes referred to in the technical jargon as "support" and "confidence" (Zarsky 2002-3)) requires greater transparency. The internal balances between accuracy and security will generate public interest that will prove to be an effective check on government. These decisions also impact the personal autonomy of those affected by the analysis; with such data in hand they can have a better understanding of the connection between their actions, the government's findings and their implications. It also empowers data subjects who understand how their data was used (although this is an overall weak argument). Thus, many of the theories are aligned at this juncture, and lead to the conclusion that transparency at this juncture is crucial, and must be attended to with vigor.

Segment (C) has generated the greatest interest in the context of governmental data mining and proper disclosure. It also raises several interesting related issues. First, we examine the notion of disclosure of the actual patterns used. The arguments for transparency are strong; these are matters within the public interest, and both shaming and political forces will be in place. Autonomy interests will be in place as well – especially in terms of those affected by the process (crowdsourcing arguments, however, are relatively weak). Yet at this juncture, an ounce of realism is called for. In this context, the arguments regarding opacity are of greatest strength; revealing the actual factors used will allow individuals to circumvent the governmental objective. While taking into account the existing legal rules and governmental sentiment, calling for transparency in this element has no chance – and probably with good reason.

Yet this should not be the ending point of a discussion of transparency at this juncture. Transparency could be reflected in other aspects of this segment. One is interpretability – whether we must require that all relevant processes will be understandable to humans even if the process is not disclosed to the entire public. I believe such a duty is crucial. Furthermore, a requirement to set it in place could be derived from transparency justifications.

Applying the various transparency theories to this specific issue easily leads to the conclusion set forth above. Interpretability could promote effectiveness, via fear of shaming. While the information revealed will not be shared with the public, if really ridiculous factors are applied, such information has the risk of leaking (and thus launching a shaming dynamic). Thus, the government will think twice before using problematic correlations. To some extent, this requirement will enhance the autonomy of those affected by the process. Individuals might not be privy to "the logic" behind the decision, but will at least know someone is looking into the matter, and has additional tools to do so. I would thus recommend that all processes be interpretable, even at the cost of lowering overall efficiency.

The same arguments cannot hold, however, when examining a call for causation on the basis of transparency-related considerations. On its face, transparency might call for developing causation theories prior to using predictive proxies in the field. A causation requirement could be derived from the transparency theoretical framework. A causation requirement will promote effectiveness as an important check on governmental actions. Causation will also generate public interest. Developing such models, even internally, will enhance autonomy, as an additional element to assure the process is not arbitrary. These studies might also be built into a "crowdsourcing" dynamic (even when only shared minimally). Experts will examine the strength of the causation theory, or try and come up with an alternative one.

However, causation has its downsides. It will slow down the overall process and render it cumbersome and possibly inefficient. Furthermore, disclosure of such theories might open the door to serious privacy and stigma concerns. Causation (in addition to mere correlation) will strengthen the negative stigma attached to those indicated by the prediction model. This might even follow from developing such theories and examining them internally, in view of potential leaking of such information to the general public. Balancing these concerns leads to recommending that a mandatory causation requirement is unnecessary. In addition, human imagination could probably find causation at almost every point. Thus, its effectiveness as a "check" on governmental actions could be seriously doubted.

Segment (D) is a crucial (yet often overlooked) element in the policy discussion of achieving transparency in a data mining process. Disclosure at this stage must be enhanced by proactive governmental research. Almost all of the theories mentioned above indicate such an outcome. Incentives in accordance to the first theory ("effective policy" or "shaming") are especially strong. The issues addressed in this segment are those that are most likely to gain public and political traction – false negatives in the project, the actual success of the program and studies regarding its inner dynamics and its effects on minorities and weaker segments of the population. Autonomy would also be enhanced if individuals will know the process which impacted them or used their personal data is overall successful, and thus worth their personal sacrifice. Thus, the government must initiate studies examining the impact on minorities. These should join studies as to the level of false negatives, and overall whether the data used is helpful in predicting human behavior.

17.6 Coda: The Limits of Transparency

Transparency is hailed as an important policy tool which could enhance autonomy and forward democracy. Its role in the age of information technology has yet to be firmly established. This chapter takes initial steps in setting forth a comprehensive mapping for meeting the transparency challenge in a specific context – that of predictive data mining of personal information.

While acknowledging the important strengths of transparency, it is crucial to recognize that there is much harm that governmental prediction models could generate, and transparency alone cannot cure. For instance, one must question whether allowing the government to obtain a powerful tool, which can generate such insights, is wise. Additional powerful arguments set forth are that the process is ineffective, ridden with errors, generates chilling effects, leads to unfair discrimination and is prone to facilitate function creep. Transparency provides a partial response to these problems. For instance, enhanced disclosure might chill the government from expanding data mining initiatives into unacceptable realms. Additional work must establish how effective a cure transparency is to the various ills mentioned, and what other steps must be taken. For this reason, the analysis here presented is an essential, yet certainly not a final step. I hope, however, that the blueprint here provided for understanding the role and limits of transparency in this novel context will prove helpful in approaching these difficult problems.

References

Benkler, Y.: The wealth of networks, ch. 4. Yale University Press, New Haven and London (2007), http://www.benkler.org/Benkler_Wealth_Of_Networks.pdf (accessed December 29, 2011)

Cate, F.H.: Government, data mining: The need for a legal framework. Harvard Civil Rights-Civil Liberties Law Review 43(2), 435–489 (2008)

Citron, D.: Open code governance. University of Chicago Legal Forum 2008, 355–387 (2008)

Citron, D.: Technological due process. Washington University Law Review 85, 1249–1313 (2008)

EPIC. Future attribute screening technology (FAST) project, http://epic.org/privacy/fastproject/ (accessed December 30, 2011)

Fayyad, U., Piatetsky-Shapiro, G., Smyth, P.: From data mining to knowledge discovery: An overview. In: Fayyad, U., Piatetsky-Shapiro, G., Smyth, P., Uthurusamy, R. (eds.) Advances in Knowledge Discovery and Data Mining. AAAI/MIT Press, Cambridge, Mass. (1996)

Fenster, M.: The opacity of transparency. Iowa Law Review 91, 885–949, 895–896 (2006)

Friedman, B., Nissenbaum, H.: Software agents and user autonomy. ACM, New York (1997)

Kreimer, S.F.: The freedom of information act and the ecology of transparency. University of Pennsylvania Journal of Constitutional Law 10(5), 1011–1080, 1056–1056 (2008)

Lessig, L.: Code and other laws of cyberspace. Basic Books, New York (1999)

Lessig, L.: Against transparency: The perils of openness in government. The New Republic (October 9, 2009), http://www.tnr.com/article/books-and-arts/against-transparency (accessed December 29, 2011)

Martens, D., Provost, F.: Explaining documents' classification. New York University - Stern School of Business Working Paper No. CeDER-11-01, pp. 1-39 (2011), http://pages.stern.nyu.edu/fprovost/Papers/martens-CeDER-11-01.pdf (accessed December 29, 2011)

Mendel, T.: Freedom of information: a comparative survey, 2nd edn., (UNESCO), Section 1:30 (2008), http://portal.unesco.org/ci/en/files/26159/12054862803freedom_information_en.pdf/freedom_information_en.pdf

McCullagh, D.: Real-life "minority report" program gets a try-out. CBSNews (2011), http://www.cbsnews.com/stories/2011/10/07/tech/cnettechnews/main20117207.shtml (accessed December 29, 2011)

Ramasastry, A.: Lost in translation? data mining, national security and the adverse inference problem. Santa Clara Computer & High Technology Law Journal 22(4), 757–796 (2005)

Reidenberg, J.R.: Setting standards for fair information practice in the U.S. private sector. Iowa Law Review 80(3), 497–551, 515 (1995)

Schauer, F.: Profiles, probabilities and stereotypes, p. 31. Harvard University Press, Cambridge (2006)

Slobogin, C.: Government data mining and the fourth amendment. University of Chicago Law Review 75, 317–341 (2008)

Solove, D.J.: Data mining and the security-liberty debate. University of Chicago Law Review 74, 343–362 (2008)

Steinbock, D.J.: Data matching, data mining, and due process. Georgia Law Review 40, 1–86, 23 (2005)

TAPAC. The report of the technology and privacy advisory committee, safeguarding privacy in the fight against terrorism (2004), http://epic.org/privacy/profiling/tia/tapac_report.pdf (accessed July 12, 2011)

Fung, A., Graham, M., Weil, D.: Full Disclosure: The Perils and Promise of Transparency. Cambridge University Press, New York (2007)

Westin, A.: Privacy and Freedom. Atheneum, New York (1967)

Zarsky, T.Z.: Mine your own business!: Making the case for the implications of the data mining of personal information in the forum of public opinion. Yale Journal of Law & Technology 5, 1–56 (2002-2003)

Zarsky, T.Z.: Governmental data mining and its alternatives. Penn State Law Review 116(2), 285–330 (2012)

Chapter 18
Data Mining as Search: Theoretical Insights and Policy Responses

Tal Zarsky

Abstract. Data mining has captured the imagination as a tool which could potentially close the intelligence gap constantly deepening between governments and their new targets – terrorists and sophisticated criminals. It should therefore come as no surprise that data mining initiatives are popping up throughout the regulatory framework. The visceral feeling of many in response to the growing use of governmental data mining of personal data is that such practices are extremely problematic. Yet, framing the notions behind the visceral response in the form of legal theory is a difficult task.

This chapter strives to advance the theoretical discussion regarding the proper understanding of the problems data mining practices generate. It does so within the confines of privacy law and interests, which many sense are utterly compromised by the governmental data mining practices. Within this broader theoretical realm, the chapter focuses on examining the relevance of a related legal paradigm in privacy law – that of governmental searches. Data mining, the chapter explains, compromises some of same interests compromised by illegal governmental searches. Yet it does so in a unique and novel way. This chapter introduces three analytical paths for extending the well accepted notion of illegal searches into this novel setting. It also points to the important intricacies every path entails and the difficulties of applying the notion of search to this novel setting. Finally, the chapter briefly explains the policy implications of every theory. Indeed, the manner in which data mining practices are conceptualized directly effects the possible solutions which might be set in place to limit related concerns.

18.1 Introduction: Beyond the Visceral Response to Governmental Data Mining

Governments around the world are facing new and serious risks when striving to assure the security and safety of their citizens. Perhaps the greatest concern is the

Tal Zarsky
Faculty of Law, University of Haifa, Israel
e-mail: tzarsky@law.haifa.ac.il

B. Custers et al. (Eds.): Discrimination & Privacy in the Information Society, SAPERE 3, pp. 325–338.
springerlink.com

fear of terrorist attacks. Various technological tools are used or considered as means to meet such challenges and curb these risks. Of the tools discussed in the political and legal sphere, data mining applications for the analysis of personal information have probably generated the greatest interest. The discovery of distinct behavior patterns linking several of the 9/11 terrorists to each other and other known operatives (Taipale, 2004) has led many to ask: What if data mining had been applied in advance? Could the attacks and their devastating outcomes been avoided?

Data mining has captured the imagination as a tool which could potentially close the intelligence gap constantly deepening between governments and their new targets – terrorists and sophisticated criminals. Data mining is also generating interest in other governmental contexts, such as law enforcement and policing. In recent years, law enforcement worldwide has shifted to "Intelligence Led Policing" (ILP) (Cate, 2008). Rather than merely reacting to events and investigating them, law enforcement is trying to preempt crime. It does so by gathering intelligence, which includes personal information, closely analyzing it, and allocating police resources accordingly – all tasks which data mining could enhance. It should therefore come as no surprise that, at least in the United States, data mining initiatives are popping up throughout the regulatory framework (GAO, 2004).

The visceral feeling of many is that the outcome of data mining analyses, which enable the government to differentiate among individuals and groups in novel ways, is extremely problematic. Yet framing the notions behind this strong visceral response in the form of legal theory is a difficult task. Even though governmental data mining is extensively discussed in recent literature, an overall sense of confusion is ever present. Additional thought is still required to properly articulate the concerns these practices generate, and the context in which they arise. While mapping out these issues, scholars as well as policymakers must further establish which paradigms of legal thought are suitable for addressing these matters. Central potential paradigms are constitutional law, privacy law and anti-discrimination, yet other fields will surely prove relevant.

This chapter strives to advance the theoretical discussion regarding the understanding of the problems data mining practices generate. It does so within the confines of privacy law and interests, which many sense are utterly compromised by the governmental data mining practices. Within this broader theoretical realm, the chapter focuses on examining the relevance of a related legal paradigm in privacy law – that of governmental searches. Examining whether an intrusive act is a legal or illegal search is a common analytical query invoked when approaching various governmental actions which might compromise privacy interests. It is analytically helpful – this chapter will explain – to conceptualize the privacy harms data mining might cause by using paradigms of thought arising in "search" related analyses. To some extent and from some perspectives, data mining compromises the same interests affected by illegal governmental searches. Yet it does so in a unique and novel way. This uniqueness renders the discussion of data mining and its detriments difficult and complex. This chapter introduces three analytical paths for extending the well accepted notion of illegal searches into this novel setting. It also points to the important intricacies every path entails and the difficulties of applying the notion of search to this novel setting.

Addressing this interesting comparison need not be a mere theoretical exercise. The theoretical concepts drawn out here will prove important in the future. Regulators will surely strive to move from theory to practice, approach data mining initiatives and establish which practices are to be allowed, and which must be prohibited. Therefore, this chapter would be of interest not only to readers interested in legal theory. It might also prove helpful to regulators and ractitioners seeking ways to ground the novel data mining practices in existing legal concepts.

Before proceeding, several analytical foundations must be set in place. Therefore, in section 18.2, the chapter briefly demonstrates and explains the meaning of data mining initiatives and what they might entail. This is a crucial step, as the term "data mining" has almost taken on a life of its own, and is applied in several - at times contradictory - ways. Data mining also presents specific unique traits, and sets distinct roles for humans and machines. Section 18.3 sets forth the central thesis of this chapter. It first explains why the chapter chose to import theoretical insights from "search" related interests in privacy law. It also explains why specific theories of search were selected for this discussion. It thereafter moves on to map out three ways in which the somewhat abstract notion of "search" could be conceptualized, and applies these notions to the data mining context. In doing so, the analysis addresses specific points where applying the relevant theory to the data mining context might face theoretical and practical obstacles, and discusses ways to overcome them. The chapter concludes in section 18.4, where it briefly explains the policy implications of applying every theory, both in terms of direct and ancillary policy measures which might be called for to minimize privacy related concerns. In these last two sections, the chapter demonstrates the importance of the theoretical analysis presented; indeed, the manner in which data mining practices are conceptualized directly effects the possible solutions which might be set in place to limit related concerns.

The chapter specifically focuses on the data mining practices of government, while purposefully neglecting similar initiatives carried out by commercial entities. This is not to say that the latter practices do not raise privacy concerns in general, and those related to the concepts of unacceptable searches in particular. Indeed, marketers, advertisers and insurers are all crunching away on the vast datasets of personal information at their disposal. In doing so, they open the door to a flurry of policy and legal problems regarding the permitted scope of using personal data and (among others) the form of consent data subjects must provide prior to such uses. This chapter, however, sets these issues aside for now. While the commercial-related issues are severe, governmental data mining leads to concerns of a far greater magnitude. The government has great datasets of personal information at its disposal and almost endless resources and opportunities to obtain many more. It can collect such information without the data subjects' consent (and in many cases without their knowledge). Perhaps most crucially, it can potentially use such information to impact the property, liberty and even life of the data subjects, given the government's almost limitless powers. For these reasons (and others) I choose to focus on governmental data mining and leave a discussion of the actions of the commercial entities for a later day.

In addition, at this point it is useful to point out what this chapter will *not* discuss (even within the realm of governmental data mining) given the chapter's focus on privacy. The analysis here presented will be premised on an underlying assumption that the tools here discussed are effective in achieving their analytical objectives while maintaining an acceptably low level of false positives and negatives. Whether this is indeed true is currently hotly debated (Harper & Jonas, 2006; Schneier, 2006) and notoriously difficult to measure and prove. Those opposing data mining can make a strong case that these predictive automated processes are, in general, inherently flawed and ineffective. In addition, they might argue they are particularly unfair to the individuals they implicate. This position has merit, and is no doubt true in specific contexts. The critiques presented below, however, will be premised upon the contrary assumption (which I believe is true in a variety of other settings), that data mining is effective and operational. Yet even so, such forms of analyses might prove problematic as they clashes with other important interests. In addition, data mining generates concerns related to the lack of transparency this practice entails, as well as discrimination it could generate. These too are important aspects which are addressed elsewhere within this volume (Chapter 17 and 19).

18.2 Governmental Data Mining: Definitions, Participants and Problems

The term "data mining" has recently been used in several contexts by policymakers and legal scholars. For this discussion, I revert to a somewhat technical definition of this term of art. Here, data mining is defined as the "*nontrivial process of identifying valid, novel, potentially useful and ultimately understandable patterns in data*" (Fayyad et. al, 1996). Within this broader topic, the core of this chapter focuses on data mining which enables "pattern based searches" (also referred to as "event-based" data mining). These methods provide for a greater level of automation and the discovery of unintended and previously unknown information. Such methods can potentially generate great utility in the novel scenarios law enforcement and intelligence now face – where a vast amount of data is available, yet there is limited knowledge as to how it can be used and what insights it can provide.

With "pattern based analyses," the analysts engaging in data mining do not predetermine the specific factors the analytical process will apply at the end of the day. They do, however, define the broader datasets which will be part of the analysis. Analysts also define general parameters for the patterns and results which they are seeking and that could be accepted – such as their acceptable level of error. Thereafter, the analysts let the software sift through the data and point out trends within the relevant datasets, or ways in which the data could be effectively sorted (Zarsky, 2002-2003). The data mining process could achieve both descriptive and predictive tasks. In a predictive process (on which this chapter is focused), the analysts use data mining applications to generate rules based on preexisting data. Thereafter, these rules are applied to newer (while partial) data, which is constantly gathered and examined. In doing so, software searches for the patterns and rules it previously established and encountered. Based on new information and previously established patterns, the analysts strive to predict outcomes prior to their occurrence (while assuming that the patterns revealed in the past pertain to the current data and environment as well).

A notion usually mentioned when considering data mining analyses is the level of automation this process facilitates. Data mining analyses indeed provide a higher level of automation than that available with other governmental alternatives; the predictive process somewhat limits the extent of human discretion in the process. Yet the level of automation this process entails might be easily overestimated. Analysts play important, yet at times hidden roles in the online process. Their actions (such as those mentioned in the previous paragraph) directly impact the outcome of the process and therefore affect actual governmental policy.

18.3 Governmental Data Mining and/as (Illegal) Searches?

18.3.1 Finding a Theory

A governmental data mining process inherently calls for automatically reviewing and analyzing profiles filled with personal information regarding many individuals. Such data was previously collected by either government or commercial entities. It is hard to imagine that individuals conceded to the data mining process here described at the time of collection or at any later stage. If the information was collected by the government, citizens might not have even provided consent at the point of collection. Rather, they merely received a basic and vague notice of the collection and future uses provided by the government.

Engaging in personal data analysis without the direct consent of relevant data subjects contradicts to several "privacy" related legal concepts. However, the precise meaning of privacy is elusive, and the privacy concerns arising in this context could be articulated in a variety of ways. In this chapter, I choose the salient paradigm of "searches" to try and illustrate the nature of privacy concerns data mining analyses generate. Of course, other paradigms of privacy might pertain to the data mining context as well. Yet this chapter focuses on a relatively specific privacy notion, which on its face is relevant and can prove insightful.

Applying the search paradigm to this context would imply that given various traits of the data mining process, this form of analysis should not be considered reasonable. Applying "search" related arguments to the data mining context has several implications. On the theoretical level, such a linkage will allow for "importing" well developed concepts of law into a novel context where they can potentially enrich a confused discourse. However, such linkage can have far reaching practical ramifications. In many cases, for a legal search to commence, various forms of ex ante judicial approval and supervision are required. If data mining will be considered as a search, data mining analyses would be considered an illegal search when carried out without sufficient judicial approval – approval which is not currently sought.

The link between data mining practices and the concept of search can be made on several levels – only one of which would be examined in this chapter. It could be carried out on an intuitive level. It could also be carried out on a doctrinal level. Finally, it could be carried out on a theoretical level. This chapter merely focuses on the latter aspect. Yet before doing so, I hereby provide a few explanations about the former two realms, and explain why I chose to set them aside for now.

On an intuitive level, data mining seems to invoke the notion of "searching" and perhaps therefore, the legal implications of such terminology. The data mining process calls for the substantial analyses of personal information pertaining to specific individuals. In this process, computer programs work through a broad array of datasets on their way to developing clusters, links, and other outputs. Thereafter, the programs examine specific sets of personal data in real time in an effort to establish whether they fit the predictive models previously constructed. This is a process which will certainly be referred to as "searches" in laymen's terms (Slobogin, 2007). Yet intuition is a fickle prospect. In many instances it could be plainly wrong, as the public might be ill-informed regarding the true meaning and implications of data mining – including its vast benefits. For that reason, I set this discussion aside. Indeed, not all activities which are "searches" to laymen are or should be considered as searches in the eyes of the law.

Linking data mining and searches will have real world implications and therefore opens the door to an elaborate doctrinal analysis. When the law recognizes searches as such, it moves to regulate them, limits their scope, and sets systematic boundaries to assure the protection of rights. It is however unclear whether under current case law and the existing concept of "search" as articulated by the courts, data mining analyses constitute searches. In the US, for instance, these steps are commonly discussed in the Fourth Amendment context, which protects the people from unreasonable searches (Kerr, 2007). Whether current Fourth Amendment doctrine will find data mining to be a "search" is a difficult doctrinal question, which is beyond the scope of this chapter, but will probably be answered negatively (Cate, 2008). Therefore, the starting point for this discussion is that data mining analyses are not "searches." The analysis set forth assumes that data mining (or other forms of data analysis) is carried out while relying upon data which was initially collected lawfully by either third parties and later passed on to the government or directly (yet lawfully) by the government itself. With this assumption in place, American law regarding searches generally assumes that individuals have a very limited subsequent privacy interest (at least in terms of "searching" and the Fourth Amendment) given the initial lawful collection of data (Kerr, 2010). The point of data collection is where data subjects relinquish control over the data and its future uses. To summarize, the governmental data mining initiatives usually do not amount to breaches of constitutional rights; or, as Daniel Solove succinctly states, "Data mining often falls between the crevices of constitutional doctrine" (Solove, 2008). At least in the US, these initiatives are also probably permitted according to current privacy laws in view of various exceptions and loopholes (Cate, 2008).

As mentioned above, this chapter sets aside the doctrinal analysis and examines the issue at hand from a theoretical and normative perspective. Doing so allows for quickly working through the relevant issues, and leaving room for an in-depth discussion of various perspectives. Yet this discussion might not remain entirely theoretical for long. It should be noted that the doctrinal outcome mentioned is not set in stone. Data mining allows the government to add additional layers of knowledge after further analyzing the data – knowledge previously undiscovered by either side. This novel development might lead to changing the abovementioned assumptions regarding privacy expectation in lawfully-collected datasets. Thus, courts might choose to change the existing doctrine in view of new theoretical understandings (which I strive to promote here), changes in public opinion, or other factors.

18.3.2 Data Mining as "Searches": Introducing Three Perspectives

On a theoretical level, linking data mining concerns to search-related interests in the privacy context can be an illuminating exercise. This is because some of the underlying theories for articulating the interests compromised by illegal searches directly address the elusive privacy interests compromised by data mining initiatives. These nexuses between search interests and data mining practices are indeed the premise of this entire chapter. However, linking data mining and the notion of illegal searches in privacy law must be done with caution. This is due to the lack of consensus among scholars regarding the definition of illegal searches and the rationale behind their prohibition.

This chapter sets forth three normative theories, which are especially helpful in understanding concerns related to governmental data mining. These theories are drawn from the existing literature and case law examining searches in the technological age in general, and in the context of data mining in particular. With these theories in mind, it is easy to see how privacy concerns in the context of data mining could be articulated using the terminology and concepts of illegal searches.

As presented below, not all of these theories are of equal strength. Some (the first) are weaker than others in explaining the privacy concerns arising in this context. Every theory however addresses a different aspect of the harms of privacy. The first focuses on the individuals and their state of mind while the second on the government and its unchecked powerful force. The third theory presents somewhat of a combination of both elements, and calls for limiting the government's ability to engage in "fishing expeditions." I now move to present these theories, how they might apply to the data mining context and what analytical obstacles might arise when doing so.

Searches as Psychological Intrusions

The *first* theory for distinguishing between legal and illegal searches looks to their *intrusive* nature. In other words, the government should meet a higher threshold of scrutiny if its actions are understood to be intrusive.[1] While intrusion is usually understood to be one that is physical, it has a psychological aspect as well (it should be noted, that this theory was *not* yet accepted in US courts).[2] The notion of psychological intrusion can be easily identified when government searches the home and self of citizens. Yet it need not be limited to these instances. Intrusiveness of various forms is the mirror image of key privacy interests, such as the right to solitude and "to be left alone."

Examining whether mere psychological (as opposed to physical) intrusions are afoot can lead this normative theory to the data mining context. It is fair to assume that many will feel intruded when confronted with the existence of data mining practices carried out with regard to their personal data. This aggravated sense of intrusion (as opposed to any other form of review of personal information on file with the authorities) could be derived from two key unique elements of the data mining process (which distinguish it from other governmental practices). First, the process's *automated* nature might generate additional anxiety. Second, data mining's ability to *predict* future behaviors could cause greater worry. These predicted behaviors might be premised upon thoughts and traits that relevant individuals have strived to keep secret or perhaps did not fully grasp. Yet now they are in the hands of the government. Empirical data gathered regarding the public attitude towards searches upholds this theory, while showing indications of anxiety towards these novel and (assumedly) "intrusive" practices (Slobogin, 2007).

The "psychological intrusion" theory provides an interesting perspective for examining the extent of privacy concerns arising from governmental data mining analyses. However, when rigorously applying this theory to the data mining context, it does not provide a conclusive response as to the intrusiveness of these governmental practices. This should come as no surprise, as psychological

---

[1] In the US, the test for the legality of searches is one which is premised upon the "reasonable expectation of privacy." Such expectation has two elements – subjective and objective/normative. Clearly, this discussion pertains to the subjective element – and a search might indeed be found to be subjectively unreasonable if considered intrusive – even merely on a psychological level. Indeed, wiretapping which does not involve a physical intrusion is considered unreasonable as well. However, the test includes an important objective/normative layer. Here, justices decide which form of subjectively unreasonable conduct is objectively unacceptable as well. As mentioned in the text, the courts have yet to find that psychological intrusions in the form of governmental searches throughout legally obtained data are unreasonable. For more on the theoretical analysis of the Fourth Amendment, see Orin Kerr, *Four Models of Fourth Amendment Protection*, 60 STAN. L. REV. 503 (2007) (mapping out four theoretical models to understand and analyze the Fourth Amendment which are used interchangeably by courts). The theory presented in this segment coincides with his first model – the *Probabilistic* Mode – a descriptive model which is premised about expectations based on current social norms. *Id.* at 508-13.

[2] This notion of "psychological intrusion" in computer searches (as a notion that would provide Fourth Amendment protection) was not accepted by the Sixth Circuit Court of Appeals in *United States v. Ellison*, 462 F.3d 557 (6th Cir. 2006). It was, however, noted by the dissent. *Ellison*, 462 F.3d at 568 (Moore, J., dissenting).

intrusion is a complicated notion. Some individuals might be greatly troubled by the automated nature of the data mining process, and the lack of human decision-making and discretion. Yet others might have a very different set of preferences when it comes to governmental analyses of personal data. To properly assess the notion of psychological intrusion at this specific juncture, one must remember the alternative strategy to governmental data mining. This would call for broader roles for experts and field officers in the law enforcement decision making process; in such a non-automated process, actual humans are those sifting and considering the individual's files. For some individuals, data mining generates greater anxiety than this latter options given concerns with automated and computerized decision-making processes. For others, however, the opposite would be true.[3] These persons would not be alarmed by the faceless computer searching their data (Tokson, 2011). They would, however, be gravely concerned with actual individuals looking through their information.

A similar complication will follow when considering the psychological intrusion resulting from fears of powerful revelations made by a computer algorithm. While this is the perspective of some, others might have greater fears of the other practices government might apply if data mining is set aside. When relying on experts and field officers, the process might be ridden with errors and biases which result from the cognitive limitations and opinions of humans (Zarsky, 2012). These are concerns that the computer analysis could avoid with greater success.

The last few paragraphs set out arguments which explain that data mining processes might generate a sense of psychological intrusion for some, yet might be comforting to others. The latter are individuals whom believe that this process is preferable to its inevitable alternatives. Both arguments and points of view seem acceptable, even reasonable. The differences of opinion people will have regarding the intrusiveness of data mining will result from differences in their understanding of the data mining technology, its benefits, and its detriments. A possible measure to overcome the analytical obstacle this theory faces might be through conducting surveys to establish the public's position. Yet administering such surveys would be a very difficult, perhaps near-impossible task (Solove, 2010).

To conclude, the "psychological intrusion" perspective to the law of searches can easily be applied to the context of governmental data mining. It can easily explain why, for some, the governmental actions breach privacy rights. However, this perspective – if ever accepted and applied by law – will face problems when moving from theory to practice. Establishing whether data mining is indeed intrusive will depend on a variety of unpredictable factors. Thus, this theory will probably fail to provide clear-cut policy.

Limiting Searches to Limit the Force of Government

A *second* theory distinguishing between legal and illegal searches which can illuminate the privacy-in-data mining debate looks to the normative reasons (as

---

[3] For instance, see Goldman, *Data Mining and Attention Consumption,* 225, 228, as discussed by SOLOVE, NOTHING TO HIDE, *supra* note 7, at 183.

opposed to visceral feelings) for limiting governmental power. This theory notes that searches are found to be illegal when they are a *powerful tool government should not be entrusted with* (at least without various forms of judicial supervision). Again, this rationale applies naturally to searches of the home and self, as well as wiretapping of communications.

When considering the use of data mining for automated predictive modeling, one can easily argue that government should not be entrusted with such a powerful tool without being closely scrutinized. Data mining can potentially turn even seemingly benign factors into a powerful mapping of an individual's persona and insights. For that reason, ex-ante judicial (or other forms of scrutiny) must be applied.

The challenge of applying this theory to the data mining context and finding that a privacy interest was compromised is that the analysis here discussed uses information which was collected lawfully by government. Therefore, the power of government was already examined and limited when information was collected. Accepting that a search-related interest might have been compromised in the data mining context calls for accepting a non-trivial argument: at times the knowledge provided by the analysis of the sum of the dataset goes beyond the value of the parts of the dataset previously collected, when viewed on their own. If this is indeed true, then the fact that the governmental actions were reviewed by courts at the data collection stage is insufficient. Additional scrutiny is required at the data mining "search" stage. Given the enhanced ability of data mining tools to engage in broad, automated and predictive tasks, this argument seems quite convincing. Data mining transforms small segments of information into an overall "mosaic" of human behavior.

The provocative notion that many, seemingly innocuous, bits of information which were collected lawfully should be treated differently in the aggregate is slowly gaining recognition in US courts which examine the notion of "search" (although it has yet to be accepted into Fourth Amendment doctrine). Most famously, this issue is fiercely debated in the context of location-based data (which is currently easily collected by mobile phone operators and other GPS devices), while questioning whether there is a difference between collecting limited and vast amounts of such data. For instance, in a controversial opinion, the Federal Court of the D.C Circuit chose to restrict governmental collection of location-based data over an extensive time period while promoting the "Mosaic Theory."[4] This finding contradicted previous cases which found that individuals have no privacy in GPS information which pertained to their actions in the open.

[4] *U.S. vs. Maynard,* 615 F.3d 544, 562 (D.C.Cir. 2010), *cert. granted,* 131 S.Ct. 3064 (2011). For a critique, see Orin Kerr, *Applying the Mosaic Theory of the Fourth Amendment to Disclosure of Stored Records*, THE VOLOKH CONSPIRACY (Apr. 5, 2011, 4:54 pm), http://volokh.com/2011/04/05/applying-the-mosaic-theory-of-the-fourthamendment-to-disclosure -of-stored-records. Several courts have taken the opposite position and allowed for these forms of surveillance. *Cf.* United States v. Hernandez, 647 F.3d 216 (5th Cir. 2011) (holding that government's use of hidden GPS to track defendant's movements was not an unconstitutional warrantless search); United States v. Cuevas–Perez, 640 F.3d 272 (7th Cir. 2011) (holding that placement of GPS tracking unit on defendant's vehicle did not violate Fourth Amendment).

The "Mosaic Theory" argues that small bits of innocuous information, when brought together, can provide a full mosaic of an individual's persona. Therefore such practices of aggregation should be further scrutinized. It should be noted, that very recently the US Supreme Court addressed this case on appeal (United States v. Jones, 2012). It unanimously found the governmental search to be unconstitutional, yet the majority relied on other grounds and left the acceptance of the "mosaic theory" into the law for another day.

To conclude, this search-related theory of privacy can explain why data mining must be limited, and when this must be done: in instances in which the tools used by government prove extremely effective! The theory here presented is premised on an interesting insight; data mining's analytical strength is the key to its normative disadvantage. The public has learned to live and accept decision making processes involving experts and field officers with their limited abilities. These existing alternatives strike an acceptable balance between law enforcement needs and civil liberty interests, even though they might compromise overall effectiveness. Data mining presents a challenge which law must now answer to, and a force which the law might find to be excessive if not properly checked. However, this theory has clear limits – if the data mining process is not found to be more powerful and insightful than other acceptable practices, this argument loses its analytical force.

Limiting Searches to Limit "Fishing Expeditions"

A *third theory* which can prove helpful in articulating privacy-related concerns from the "search" perspective in the context of data mining analyses pertains to their very broad scope. Usually, when considering invasive searches, laws and courts find that they must be carried out narrowly, while limiting the gaze of government as much as possible. Searches which fail to do so amount to a "fishing expedition" on behalf of the state – the practice of looking through the files and personal effects of individuals who raise no suspicion while striving to build a case on the basis of information they might recover. Curbing "fishing expedition" by governments is one of the central roles of judicial review (Solove, 2002). Thus, this theory finds a normative flaw with very broad searches, which impact to non-suspects.

Data mining initiatives famously call for actively examining and analyzing datasets pertaining to a very broad realm of individuals, including those whom are substantially removed from the matter at hand. The software does so while striving to formulate patterns, trends, and clusters. Thus, data mining generates a massive "fishing expedition" which resembles the most feared practices of government – searching datasets in mass, while hoping to locate relevant evidence (as opposed to initiating a search based on suspicion). On its face, this paradigm of thought might be extremely helpful in grasping the concerns data mining generates.

Yet again a theoretical obstacle blocks the application of this perspective in the data mining context. If, under existing doctrine (and as explained above) the government may review and analyze information which was lawfully collected in any way it deems fit, data mining cannot be considered a “fishing expedition.”[5] In other words, no “search-related” interest is compromised by the analysis (or, to carry through the metaphor, there is no “expedition” in these actions), as the government is clearly operating within its mandate, rather than intruding on the rights of the innocent. Therefore, this perspective does not prove helpful in mapping the boundaries of legitimate and excessive data mining practices.

This theoretical obstacle follows from today's understanding of “searches” as an almost dichotomous variable; actions are either a search (and thus lead to a harsh legal analysis usually calling for the finding of “probable cause”) or they are not (in which case no constitutional form of protection is called for). The harsh implications (for government) of actions being considered as “searches” have led courts to limit the breadth of this term. Yet this dichotomous perspective of the concept of searches is not set in stone and the problem not beyond repair. Recent scholarship argues that rather than a dichotomy, searches should be viewed on a sliding scale. In other words, the legitimacy of search should be established through a proportionality-based analysis (Slobogin, 2007); different forms of intrusions will be met by different forms of legal thresholds to protect search-related interests. Every such intrusion will call for a proportionate level of protection and standard of review.

The proposed shift to a proportionality based analysis of search interests will force policy makers to address the “fishing expedition” problem data mining practices set forth. Data mining analysis could be considered as a minute intrusion on its own (rather than a process which is not a “search” at all), when examing the impact on a single citizen. Yet when considering the aggregated impact on a broad segment of the population subjected to the data mining analysis, the result might be quite different. Indeed, in cases where the benefits of the data mining analysis are limited or unsure, and the population segment extremely broad, such practices might be found to be a disproportionate measure (Slobogin, 2007). Therefore, this specific theoretical perspective of “searches” can provide a different form of “calculus” for configuring whether a governmental data mining is acceptable – and a balance which is quite different than the one called for under the previously mentioned theories.

18.4 Conclusion: Novel Practices, Classic Concepts and Policy Proposals

This chapter draws out a basic conceptual framework for "importing" theoretical concepts used in the “search” discourse to properly understand concerns associated with governmental data mining practices. Yet the discussion need not stay in the realm of theory. This conceptual framework can also assist policymakers searching for a balance in today's world of global insecurity. These policymakers are now challenged with the structuring of schemes striving to use databases of personal information to promote law enforcement and stability. In

[5] Note that most recently, in *US v. Jones* (1.2012), the count’s concurring opinion questioned the wisdom of the third-party doctrine.

doing so, they must figure out ways in which "search" related interests could be answered within the governmental data mining analysis process.

As explained above, every such theory calls for a different set of balances and findings. However, the implications of these theories – if they are indeed accepted in the data mining contexts – run deeper. Every one of the theories mentioned above might point regulators in a different regulatory direction when considering ancillary privacy rights to overcome the concerns at hand. The *first* theory points to the sense of intrusion data mining generates. If this is the privacy-based theory which generates concerns in the data mining context, then this concern could be partially mitigated by a greater degree of transparency in the data mining process. With additional knowledge as to the process, the public aversion might be limited. Therefore, accepting this theory should promote this ancillary right.

The *second* theory points in a different direction. Addressing this concern should call for various measure for assuring that data mining analysis are only used for the specific tasks they are needed for the most. In other words, steps must be taken to assure that the use of these methods does not "creep" into other realms. This could be achieved by both technological measures (which safeguard the use of these tools) and a regulatory structure which closely supervises these uses. Again, the theoretical perspective can point policymakers (if convinced by this argument in the relevant context) in the direction of relevant (yet different) ancillary rights.

The *third* theory might call for yet a different regulatory trajectory in terms of regulatory steps. The concern it addresses relates to the very nature of the data mining practice – one that finds its broad scope problematic. Therefore, this theory might indeed lead to limiting data mining analysis. Another possible option might call for engaging in the mining of anonymized data – a practice which might somewhat mitigate these concerns (yet raises others) (Zarsky, 2012). Arguably, if the search is of anonymized data, the interests of the many subjected to it are not compromised by the vast net the data mining practice apply. Therefore the use of this measure would be found proportionate.

As there is probably a kernel of truth in every one of the theories, it would be wise to take all these proposals under consideration. However, at some points they might prove contradicting. Therefore, additional analytical work must prioritize among them, while relying on social norms and the balancing of other rights. This analysis of course must be context-specific, as in different contexts the relative force of every theory will vary.

In conclusion, existing risks call for analyzing and using personal information in an effort to preempt possible harms and attacks. Society will be forced to decide among several non-ideal options. At the end of the day, the solution selected will no doubt be a compromise, taking into account some of the elements here set forth. The theoretical analysis here introduced strives to assist in the process of establishing such a compromise, while acknowledging that there is still a great deal of work to be done.

References

Cate, F.H.: Government, data mining: The need for a legal framework. Harvard Civil Rights-Civil Liberties Law Review 43(2), 435–489 (2008)

Fayyad, U., Piatetsky-Shapiro, G., Smyth, P.: From data mining to knowledge discovery: An overview. In: Fayyad, U., Piatetsky-Shapiro, G., Smyth, P., Uthurusamy, R. (eds.) Advances in Knowledge Discovery and Data Mining. AAAI/MIT Press, Cambridge, Mass. (1996)

Henderson, S.: Nothing new under the sun?: A technologically rational doctrine of fourth amendment search. Mercer Law Review 56, 507–563, 544 (2005)

Jonas, J., Harper, J.: Effective counterterrorism and the limited role of predictive data mining. CATO Institute Policy Analysis 584, 1–12 (2006)

Kerr, O.: Four models of fourth amendment protection. Stanford Law Review 60, 503 (2007)

Kerr, O.: Applying the fourth amendment to the internet: a general approach. Stanford Law Review 62, 1005 (2010)

Schneier, B.: Why data mining won't stop terror. Wired (September 3, 2006), `http://www.wired.com/politics/security/commentary/securitymatters/2006/03/70357?currentPage=all` (accessed December 30, 2011)

Slobogin, C., Schumacher, J.E.: Reasonable expectations of privacy and autonomy in fourth amendment cases: An empirical look at "understandings recognized and permitted by society." Duke Law Journal 42(4), 727–775, 743–751 (1993)

Slobogin, C.: Privacy at risk: The new government surveillance and the fourth amendment, p. 194. The University of Chicago Press, Chicago and London (2007)

Solove, D.J.: Digital dossiers and the dissipation of fourth amendment privacy. Southern California Law Review 75, 1083–1167, 1106–1107 (2002)

Solove, D.J.: Data mining and the security-liberty debate. University of Chicago Law Review 74, 343–362 (2008)

Solove, D.J.: Fourth amendment pragmatism. Boston College Law Review 51, 1511–1538, 1522–1524 (2010)

Taipale, K.: Technology, security and privacy: The fear of frankenstein, the mythology of privacy, and the lessons of king ludd. Yale Journal of Law and Technology 7, 123–221, 134 (2004)

Tokson, M.: Automation and the fourth amendment. Iowa Law Review 96, 581–647, 602–609 (2011)

United States General Accounting Office. Data Mining: Federal Efforts over a Wide Range of Uses GAO-04-548, pp. 1–64 (2004), `http://www.gao.gov/new.items/d04548.pdf` (accessed December 30, 2011)

Zarsky, T.Z.: Mine your own business!: Making the case for the implications of the data mining of personal information in the forum of public opinion. Yale Journal of Law & Technology 5, 1–56 (2002-2003)

Zarsky, T.Z.: Governmental data mining and its alternatives. Penn State Law Review 116(2), 285–330 (2012)

Part VI
Concise Conclusions

Chapter 19
The Way Forward

Bart Custers, Toon Calders, Tal Zarsky, and Bart Schermer

Abstract. The growing use of data mining practices by both government and commercial entities leads to both great promises and challenges. They hold the promise of facilitating an information environment which is fair, accurate and efficient. At the same time, they might lead to practices which are both invasive and discriminatory, yet in ways the law has yet to grasp. This point is demonstrated by showing how the common measures for mitigating privacy concerns, such as a priori limiting measures (particularly access controls, anonymity and purpose specification) are mechanisms that are increasingly failing solutions against privacy and discrimination issues in this novel context.

Instead, a focus on (a posteriori) accountability and transparency may be more useful. This requires improved detection of discrimination and privacy violations as well as designing and implementing techniques that are discrimination-free and privacy-preserving. This requires further (technological) research.

But even with further technological research, there may be new situations and new mechanisms through which privacy violations or discrimination may take place. Novel predictive models can prove to be no more than sophisticated tools to mask the "classic" forms of discrimination, by hiding discrimination behind new proxies. Also, discrimination might be transferred to new forms of population segments, dispersed throughout society and only connected by some attributes they have in common. Such groups will lack political force to defend their interests. They might not even know what is happening.

With regard to privacy, the adequacy of the envisaged European legal framework is discussed in the light of data mining and profiling. The European

---

Bart Custers · Bart Schermer
eLaw, Institute for Law in the Information Society, Leiden University, The Netherlands
e-mail: bartcusters@planet.nl, schermer@considerati.com

Toon Calders
Eindhoven University of Technology, The Netherlands
e-mail: t.calders@tue.nl

Tal Zarsky
Faculty of Law, University of Haifa, Israel
e-mail: tzarsky@law.haifa.ac.il

B. Custers et al. (Eds.): Discrimination & Privacy in the Information Society, SAPERE 3, pp. 341–357.
springerlink.com

Union is currently revising the data protection legislation. The question whether these new proposals will adequately address the issues raised in this book is dealt with.

19.1 Concise Conclusion: Shifting Paradigms

In this book, we discussed issues regarding privacy and discrimination due to data mining and profiling techniques and provided technological and non-technological (directions for) solutions. In this chapter, we draw some general conclusions and discuss the way forward.

Throughout the book, we have shown that a powerful paradigm shift is transpiring. The growing use of data mining practices by both government and commercial entities leads to both great promises and challenges. They hold the promise of facilitating an information environment which is fair, accurate and efficient. At the same time, they might lead to practices which are both invasive and discriminatory, yet in ways the law has yet to grasp. Approaching these new risks call for joint work of experts from both the computer science and legal realm. This book is a first step in this important direction.

To demonstrate this point, in this section, we will show how the common measures for mitigating privacy concerns, such as a priori limiting measures (particularly access controls, anonymity and purpose specification) are mechanisms that are increasingly failing solutions against privacy and discrimination issues in this novel context. We argue that a focus on (a posteriori) accountability and transparency may be more useful. This requires improved detection of discrimination and privacy violations as well as designing and implementing techniques that are discrimination-free and privacy-preserving. In Section 19.2 we will focus on further research (particularly technological research) in this field. In the future, there may be new situations and mechanisms through which discrimination and privacy violations may take place. Therefore we discuss the future of discrimination in Section 19.3 and the future of privacy in Section 19.4. In Section 19.3, we discuss two very different forms of discrimination-based problems which might arise in the future. First, novel predictive models can prove to be no more than sophisticated tools to mask the "classic" forms of discrimination, by hiding discrimination behind new proxies. Second, discrimination might be transferred to new forms of population segments, dispersed throughout society and only connected by some attributes they have in common. Such groups will lack political force to defend their interests. They might not even know what is happening. In Section 19.4 we will discuss the future of privacy and data protection and the adequacy of the current legal framework regarding these new technological developments. The European Union is currently revising the data protection legislation. The question whether these new proposals will address these issues raised in this book effectively will be dealt with.

19.1.1 The Failure of Access Controls

While privacy and antidiscrimination concerns are derived from different legal sources, they are commonly cured by a similar remedy – the limitation of data collection. Discrimination concerns usually focus on distinguishing among individuals on the basis of particular sensitive attributes (such as gender, ethnic background, religion, et cetera). Privacy concerns usually focus on the use, exposure or analysis of identifying attributes (such as name, address, etc.) in combination with sensitive attributes.[1] Usually the advice to citizens who want to protect their privacy is not to disclose their personal data. The advice to citizens who want to protect themselves against discrimination is the same. Data subjects, i.e., the people the data in databases relate to, may have good reasons not to provide particular data. For instance, people may consider such information not to be someone else's business, they may consider disclosure as not improving their reputation, or they may fear disadvantageous judgments of others about themselves. Some information may not be considered appropriate for disclosure to anyone, but more often information may not be considered appropriate for disclosure to particular people or institutions. For instance, people may want to share medical information with their physician and their hospital, but not with their car insurance company, employer or supermarket. People may want to discuss their sexual preferences with friends, but not with their parents. Such a partitioning of social spheres is referred to as audience segregation.[2] In short, people may prefer that others who collect, process, and analyze data have some blanks in their databases.

Let's focus this argument on privacy issues. From a legal perspective, people have, to some extent, a right to refuse disclosure of their personal information.[3] Everyone has a right to privacy, according to Article 12 of the Universal Declaration of Human Rights. What this right to privacy exactly means and encompasses, is not entirely clear. When it comes to informational privacy (contrary to, for instance, spatial privacy) a commonly used definition (particularly in the United States) is that of Alan Westin, who refers to privacy in terms of control over information.[4] Privacy is a person's right to determine for himself when, how, and to what extent information about him is communicated to others. In other words: who has access and who does not. This definition is sometimes referred to as *informational self-determination* and has a strong focus on the autonomy of the individual.[5] Based on this perspective, people were equipped (through data protection regulation) with access controls. Such access controls focus on limiting the collection and distribution of personal data. The concept of informational self-determination is an example of this. Other examples are concepts

---

[1] See Chapter 4 and also Custers B.H.M. (2010).

[2] Van den Berg, B. and Leenes, R. (2010).

[3] See Chapter 7.

[4] Westin, A. (1967).

[5] Other common definitions of the right to privacy are the right to be let alone, see Warren and Brandeis (1890) and the right to respect for one's private and family life (Article 8 of the European Convention on Human Rights and Fundamental Freedoms).

like 'need to know', 'select before you collect', and many of the OECD privacy principles[6] (including the purpose specification principle, see below).

However, these mechanisms for limiting the collection and distribution of information are failing, for several reasons. First, from a practical perspective, informational self-determination is complicated because people often do not know who collects and processes their personal data. This is mainly due to the fact that most personal data collecting no longer takes place directly, i.e., by asking data subjects for the data, but indirectly, for instance, by sharing or buying datasets or coupling databases. When collecting data indirectly, it is far more difficult for data subjects to know who is processing their personal data and to exercise any form of control over their data.

Second, already in 1948, it was shown that the dissemination of information follows the rules of entropy.[7] Basically this means that it is easy to spread information, but very difficult to withdraw information from the public sphere. In the information society this is more obvious than ever. Everyone knows that it only takes two mouse clicks to copy and send information to dozens of people (or many more). Since the spreading of information always proceeds in one direction (towards a larger entropy), principles focusing on access controls are increasingly inadequate in a world of automated and interlinked databases and information networks, in which individuals are rapidly losing grip on who is using their information and for what purposes. Due to the movement towards larger entropy, it may be difficult for people to know where their information will end up. This is, in fact, an argument for greater control over information. However, according to the rules of entropy, the extent of this control is limited to stopping or slowing down the increase of entropy. According to the rules of entropy, it is impossible to reverse the increase of entropy.[8]

Third, throughout this book, it was shown that data mining technologies are useful tools for profiling, i.e., ascribing characteristics to individuals or groups of people. Most data mining technologies are very good at dealing with datasets that are incomplete or incorrect. Missing data generally do not constitute a problem when searching for patterns, as long as the total amount of missing data is not too large compared to the amount of data available. Hence, with the help of data mining predictions, the blanks (missing data) can easily be completed in datasets.

---

[6] See the 1980 principles for fair information processing developed by the Organization for Economic Co-operation and Development (OECD).
See http://www1.oecd.org/dsti/sti/it/secur/prod/PRIV-EN.HTM.

[7] Shannon, C.E. (1948) and Shannon, C.E. (1949). The entropy of data item X is expressed as $H(X) = -\sum_{i=1}^{n} p(X_i)\log_2 p(X_i)$ with $X_i, \ldots X_n$ the n possible values with probabilities $p(X_1), \ldots, p(X_n)$, where $\sum p(X_i) = 1$; see also Denning, D.E. (1983), p. 17.

[8] Deleting data from databases does in fact decrease the entropy, but when copies of particular data remain, sooner or later the entropy may increase again.

In short, even if people refuse to disclose their personal data, these characteristics can easily be predicted with data mining tools.

19.1.2 The Failure of Anonymity

The arguments above underlining the failure of access controls are particularly applicable to hiding two types of information: identifying information and discrimination-sensitive information. Starting with the first, identifying information is important in establishing whether information is anonymous or not. Current European legislation protects and limits collecting and processing personal data, but not the collecting and processing of anonymous data. For this reason, data controllers may prefer to process anonymous data, which allows profiling on an aggregate (group) level. Despite false negatives and false positives, such profiles may be sufficiently accurate for decision-making.[9] The characteristics may be valid for the group members even though they may not be valid for the individual group members as such.[10] For instance, predicting that people driving white cars cause less traffic accidents on average or predicting that people who refrain from eating peanut butter live longer on average may be (hypothetical) data mining results based on anonymous databases. Ascribing an anonymous profile to a data subject (if John drives a white car, then he is likely to be a careful driver, or if Sue regularly eats peanut butter, then she is likely to live long), implies ascribing personal data to individuals. This process creates new personal data. Compared to a situation in which a data subject voluntarily provided personal data to a data controller, it is much more difficult for a data subject to know about the existence and the contents of such ascribed personal data. In fact, characteristics may be attributed to people that they did not know about themselves (such as life expectancies or credit default risks). People may be grouped with other individuals unknown to them (such as being on flight KL611 to Chicago together).

This process may seem harmless, but may be considered less harmless to the individuals involved when information is combined and used to predict or deduce, with slight margins of error, particular sensitive data. Furthermore, predicting or deducing missing values and subsequently ascribing them to individuals may cause friction with informed consent from those individuals. In Europe, in many cases (though not always), data subjects have a right to consent to the use of their data. When people do not know the ways in which their personal data are processed, which characteristics are ascribed to them, and what are the consequences of this, it is very difficult for them to object.

The mechanisms involved in anonymity are also applicable to a certain extent to discrimination-sensitive information. Under discrimination laws, several characteristics are considered unacceptable for decision-making. For instance, ethnic background or gender should not be used to select job applicants. However, everyone knows that a trivial attribute like a name can often predict the ethnicity

[9] Zarsky, T. Z. (2003).
[10] Vedder, A. H. (1999).

or gender of a person. The same may be true for attributes like profession (there are still very few female airline pilots or males working as an obstetrician) or zip code (some neighborhoods are predominantly 'black' whereas others are predominantly 'white').

The use of data mining may further increase the possibilities of predicting sensitive characteristics. From a legal perspective, no employer looking for a new employee is allowed to ask for these characteristics and no job applicant must provide them, but it is obvious that anti-discrimination legislation is extremely difficult to enforce nonetheless. The point here is that hiding particular characteristics is not sufficient. In fact, research has shown that leaving sensitive data like ethnic background and gender out of a database may still yield discriminatory data mining results.[11,12]

In summary, using what is considered today to be anonymous data does not properly resolve concerns related to both privacy and anti-discrimination, as data may sooner or later be ascribed to individuals again.[13] In fact, when particularly identifying characteristics, such as name, address, and social security number, are missing, data mining technologies and database coupling may also be used to predict the missing characteristics. Deleting sensitive data from databases does not work either, as these sensitive characteristics may also be predicted. Prohibiting data mining (a radical measure) is not realistic given the enormous amounts of data we are facing in our information society, as it would imply less insight in and overview of the data available. Thus, relying on anonymity as a solution to both privacy and discrimination concerns is problematic. It is difficult to achieve given technological advances and even if achieved many of the concerns will still manifest. Note, though, that anonymity can be an objective of its own. Therefore, anonymity can be very important, for instance, in a context without data mining and profiling, but may be insufficient in other contexts, particularly when data mining and profiling are used.

19.1.3 The Failure of Purpose Specification

In order to protect data subjects from collecting and using personal data very broadly, there is a strong focus on purpose specification in European data protection legislation. The purpose specification principle states that the purposes for which personal data are collected should be specified and that the data may only be used for these purposes.[14] This principle is included in the Treaty of Strasbourg (art. 5b and 5e) and the EU Data Protection Directive[15] (art. 6.1b, 10

---

[11] Verwer and Calders. (2010).

[12] Pedreschi, D., Ruggieri, S., and Turini F. (2008).

[13] Ohm, P. (2010).

[14] See the 1980 principles for fair information processing developed by the Organization for Economic Co-operation and Development (OECD).
See http://www1.oecd.org/dsti/sti/it/secur/prod/PRIV-EN.HTM.

[15] Directive 95/46/EC of the European Parliament and of the Council on the protection of individuals with regard to the processing of personal data and on the free movement of such data, passed 2 February 1995.

and 11). The idea behind this principle is closely related to informed consent: a person is only able to provide informed consent if he or she knows for which purposes his personal data will be used. This requires a clear description of these purposes. If the purposes are not clear, consent would imply *carte blanche*. In fact, it is arguable whether such consent can be considered *informed* consent. The purpose specification principle is thus closely related to the notions of autonomy and control discussed in the previous subsections.

However, in the information society and particularly with the rise of data mining techniques, the purpose specification principle is gradually losing meaning for several reasons. First of all, many organizations have a great need for information. The basic idea is that in an information society, it is necessary to base all decisions on as much information as possible. Organizations want to know who their clients are, how they behave, et cetera. This creates a drive to collect large amounts of data and to analyze these data for useful patterns. The purpose specification principle limits the collection and analysis of personal data, but organizations may not like to be limited in this respect. This does not mean that organizations plainly ignore the purpose specification principle. Many organizations nowadays simply formulate their purposes rather broadly, so that concrete purposes do not necessarily have to be known at the time of collection. Common phrases include: "we use your information to fulfill your requests", "to personalize your experience with us", "to keep you updated", "to better understand your needs", etc. This development implies that the purpose specification principle rapidly loses its meaning.

The second reason why the purpose specification principle is failing is that purposes of data collectors and processors may change. Obviously, when this happens, organizations may change the text in their privacy statements regarding their purposes accordingly, but this does not necessarily mean that they will delete personal data collected for the previous purposes. As indicated in the previous subsections, once a piece of information has been disclosed (or discovered by data mining), it is practically impossible to withdraw it. The information may easily spread through computer systems by copying and distribution. It may be difficult to trace every copy and delete it. Furthermore, it may be very difficult to retrieve which part of the data were collected for which purpose (present or past). Moreover, when a data subject becomes 'attached' to a certain service, for instance because all his peers use the service, or a great deal of data is stored in the service in a proprietary format, withdrawing from the service becomes more difficult. In many cases data subjects therefore simply accept changes in the privacy policies, as leaving the service because of these changes is often an unattractive option.

A third reason is related to the very nature of data mining: data mining aims at discovering patterns and relations that were previously unknown. Data mining is not a theory-driven approach, starting with reasonable hypotheses, but a bottom-up approach, starting without any hypothesis at all.[16] This is the core of the innovative nature of data mining technologies, which may result in very useful,

[16] See Chapter 1. Note that there are different ways of generating hypotheses, see Chapter 2.

previously unknown and sometimes completely unexpected patterns and relations. Specifying purposes before the start of any data mining exercise is basically impossible, because it is not yet known what will be discovered.

Note that it may be argued that the purpose in such cases could be specified as something like: "we plan to use your data for data mining purposes", but obviously it is impossible for data subjects to know what the outcome of such an exercise will be and may raise difficulties for them assessing the positive and negative effects of disclosing information. Consent in such cases is unlikely to be *informed*. Furthermore, the knowledge discovered with data mining is likely to be used for further decision-making. For data subjects it is even more difficult to overview how data mining results will be used and may affect them at a stage where these data mining results are not yet known.

19.1.4 Focus on Transparency and Accountability

In the previous subsections, we have shown that a priori limiting measures (particularly access controls, anonymity and purpose specification) are failing solutions for privacy and discrimination issues. In this subsection, instead, we argue that a focus on (a posteriori) accountability and transparency may be more useful.[17] Instead of limiting access to data, which (as shown above) is increasingly hard to enforce in a world of automated and interlinked databases and information networks, we must stress the question as to how data can and may be used. Whereas a more traditional approach focuses on the concepts of access controls, anonymity, 'need to know' and 'select before you collect', our approach focuses on other legal concepts, such as those used in tort law: accountability, liability, redress, etc. Another option is that of considering data mining as (legal or illegal) search.[18] The use of such concepts is familiar to legal experts and, as such, may help them understand what data mining and profiling are about.

From a technological perspective, transparency and accountability can be improved in several of the technological measures suggested in this book. For instance, the architecture of data mining technologies can be adjusted ('solutions in code')[19] to create a value-sensitive design, that incorporates legal, ethical and social aspects in the early stages of development of these technologies. This is exactly what privacy preserving data mining techniques strive to achieve.[20] These may aim at protecting identity disclosure or attribute disclosure, but also at prevention or protection of the inferred data mining results. Similarly, discrimination-free data mining techniques have been developed, by integrating legal and ethical concerns and interests in data mining algorithms.[21] Such

[17] Weitzner, D.J., Abelson, H. et al. (2006).

[18] See Chapter 18.

[19] See Chapters 11-14. For more information, see also: http://wwwis.win.tue.nl/~tcalders/dadm/doku.php

[20] See Chapter 11.

[21] See Chapters 12-14 and Calders, T., and Verwer, S. (2010).

technological measures and transparency about their design may prevent data mining results that may easily lead to discrimination or privacy infringements.

The use of discrimination-free and privacy preserving data mining techniques may prevent many problems, but may not be sufficient. Transparency regarding the use of these techniques is required to create more awareness and understanding among data subjects on how their data is used.[22] In the end, more transparency may create more trust, provided that the data mining and profiling methods used are not discriminating or violating privacy. Even if they are, disclosing these facts will bring these issues to the forefront of the political discussion. Thus, political forces might assure that the data mining processes carried out by the state or private firms are acceptable by the broader public.

Apart from more transparency and generating greater trust, accountability is a key element in a call for greater transparency in data mining and profiling. To enforce proper use of personal data, it is crucial that cases of discrimination and infringements of privacy are easily and quickly detected, even internally. For this, technology may be useful again.[23] With proper detection systems, a rapid and adequate response can be given to situations where discrimination or privacy violations take place. The next section will discuss this in more detail.

19.2 Further Research

In today's society we are continuously profiled, the profiles used may be intrinsically discriminatory and our privacy may be violated. Discovering discrimination and privacy violations, however, is difficult, since they can be hidden in very specific niches. We can use data mining, i.e., the use of automated data analysis techniques to uncover previously undetected relationships among data items, for discovering hidden discriminative contexts. Data mining does not only come to our help, however: more and more data mining is becoming a crucial tool when designing decision procedures. Many decision procedures are, at least partially, being automated and in this automation, unfortunately, often little attention is paid to anti-discrimination and privacy preservation. We will argue that in many circumstances the use of data mining will lead to the construction of discriminating models, which is particularly dangerous as these techniques offer a false comfort of providing unbiased solutions based upon solid statistical evidence, not affected by subjective human interpretation. Ethical and legal implications regarding anti-discrimination legislation have been underestimated and neglected for far too long by the data mining and machine learning communities. In order to breach this gap between legislation and technology the following directions need to be explored in future:

- Using existing and newly developed data mining tools for automatically detecting and assessing discriminative profiles used in a decision process.

[22] See Chapter 3 and Chapter 17.
[23] See Chapter 5.

- The design, implementation and testing of classification and clustering techniques that produce non-discriminatory models of the data by design; i.e., the techniques will be constrained in such a way that they can only produce models that are non-discriminatory. These techniques will offer a safe alternative to the current techniques which only offer a false comfort of providing unbiased solutions.
- Based upon research in the previous two topics, we may discover situations and new mechanisms through which discrimination can take place. These discoveries can be profitably used to update current legislation.

In this book the first two orthogonal directions are discussed. The first direction concerns the *detection* of discrimination in a dataset, whereas the second concerns *avoiding* discrimination. For both directions, the development of adequate technological solutions is a necessity to implement discrimination control in practice. As detailed in Chapter 3, the application of data mining techniques may lead to subtle forms of indirect discrimination, even if there was no direct intention to discriminate. As data mining is a very active research domain continuously being further developed, there will be an increasing need for sophisticated discrimination detection techniques. The situation can be compared to that of spam email; without proper spam filters and techniques to investigate the originator of spam, legislation outlawing spam has little power. But even with spam filters, spam detection remains a moving target. As soon as new detection techniques are developed, spammers change their strategy to fool the filters. A similar race can be expected in discrimination detection; legislation alone will not suffice to stop discriminatory practices in large scale profiling by companies or governmental institutions. As discrimination detection techniques will improve, profiling software will exploit increasingly more subtle and ingenious ways to circumvent restrictions imposed by technological solutions. This will transfer, not for the purpose of discriminating per se, but for reasons of predictive accuracy or efficiency; as long as sensitive attributes such as ethnicity serve as a proxy and indirectly provide otherwise inaccessible information relevant for the profiling task it will be interesting to use sensitive information either directly or indirectly. In Chapter 5 several techniques to detect discrimination in decision-making records were proposed. The chapter sketches the idea of a discrimination "audit", aimed at post-factum identification of discriminative context. Such audits will be important in discrimination law enforcement. As explained in Chapter 8, however, in practical applications it will be very hard to assess which forms of discrimination results from an acceptable use of informative attributes, and what part represents unjustified or illegal discrimination, i.e., discrimination that cannot be justified by objective arguments or supported by a legal basis.

In addition to the discrimination detecting technology in post-factum data, tools have to be developed so to allow for learning unbiased models. Several of such techniques are detailed in this book; for instance, in Chapters 12, 13, and 14. None of these techniques, however, can guarantee that the model built will stand the test of judicial trial. Unlike some of the technological solutions in anonymization and privacy prevention surveyed in Chapter 11 that guarantee anonymity *by design*, we are quite still far from *discrimination-freeness by design*.

19.3 The Future of Discrimination

As the previous chapters of this book indicate, there is a growing appetite in both the private and public sector to try and predict what specific individuals will do in the future based on what happened to them in the past. Entities with vast datasets of personal information at their disposal try and utilize the information they have, by using advanced analytical tools. The outcomes of these processes are individualized forecasts and predictions: what the specific individual will consume, where she will travel, will he be ill, or will she break the law or default on a loan. The government is already using similar mechanisms to establish who is most likely to lie on her tax return, or become a security risk at the border.[24] Yet the reach of predictive modeling will not stop with these examples.

Stepping beyond privacy law and the way it might limit the concerns these practices generate, law will also deal with these practices using other existing doctrines. In doing so, lawmakers will be required to establish the legitimate borders of this growing practice, while determining which predictive tasks are acceptable and which go too far. One of the key doctrines which will surely be called into play in this context is that of *unfair discrimination*.[25]

Before proceeding, it is first worth mentioning that a shift to automated predictive modeling as means of decision making and resource allocation might prove to be an important step towards a discrimination-free society – at least in terms of the salient features of discrimination as we understand them today. A computerized decision-making process could be monitored in real time and reviewed after the fact with ease. Therefore, discriminatory practices carried out by officials and employees, that counter governmental or business policies, could be limited effectively. Furthermore, the physical interaction between the decider and the subject are usually non-existent. Thus, the sensory cues which usually trigger discrimination – a different skin color, accent or demeanor – are removed from the process, thus limiting additional opportunities for discriminatory conduct. While these arguments might all prove true, it is possible that automated practices substitute well accepted discrimination concerns with newer ones we have yet to fully understand.

Connecting the broad notion of unfair discrimination to the novel practices of automated prediction will prove to be an elaborate task for the next generation of jurists. As discussed in Chapter 5 of this book, discrimination is a broad term, which generated a breadth of legal thought and case law. It is also a charged term, which quickly triggers visceral reactions and responses. Discrimination seems intuitively relevant to the issue at hand. It usually refers to instances in which

---

[24] For a discussion of the deployment several systems, see Cate, F.H. (2008) at 447, and referenced sources.

[25] See, for instance, Solove, D. (2011).

seemingly equal individuals are treated differently. Here, however, we are dealing with a more nuanced situation – individuals are found to be different based on various statistical analyses. If we assume for the sake of this analysis (and this is no trivial assumption) that the data used is correct and the statistical models valid, we are *not* addressing situations in which equals are treated differently (as the analysis itself indicates that the individuals are, in fact, different and not equal). Rather, we are referring to situations in which individuals are distinguished from each other on the basis of factors, which might be mathematically correct, yet are rendered normatively *irrelevant* by society. Or, we are concerned with various negative social outcomes which tend to follow from discriminatory practices, such as stigma, stereotyping and the social seclusion of the group subjected to discrimination.

To unpack these difficult questions and bring them into the context of data mining and automated prediction, we must note two very different forms of discrimination-based problems which might arise. First, a discussion of discrimination in this context quickly leads to considering the relevance of racial discrimination and other repugnant practices of the past. In other words, novel predictive models can prove to be no more than sophisticated tools to mask the "classic" forms of discrimination of the past, by hiding it behind scientific findings and faceless processes. It is possible that data mining will inadvertently use proxies for factors which society finds socially unacceptable, such as race, gender, nationality or religion. This might result from various reasons: a faulty learning dataset, problematic motivations (or subconscious biases) plaguing those operating the system, improper assumptions regarding the data and the population, and other reasons society is only beginning to explore.

In the next few years, law must map out the ways in which discrimination is to be defined in this context, and how it should be measured and distinguished from acceptable data practices. Some of the technological measures to do so are already being discussed in Chapter 12 of this book. Yet we do hope future analyses of these matters will follow. In addition, many policy decisions are still unanswered: at what point should these discriminatory outcomes be measured – after the fact, or before running the analyses (by testing the data on a sample data set)? Who would be responsible for establishing whether a process is discriminatory and what data sources will be required for doing so? For instance, it is possible that to establish whether a seemingly innocuous analysis process is in fact a proxy for unacceptable discriminatory practices, a vast amount sensitive information is required; information indicating whether individuals within the dataset are members of protected groups. Clearly, obtaining and using such sensitive datasets leads to complicated privacy problems which require additional thought.

Yet data mining might lead to an additional, second set of discrimination-based concerns. These novel concerns result from the fact that the data mining processes might systematically single out individuals and groups. In these cases, the process could potentially lead to a flurry of novel ethical and legal concerns which society has yet to consider – concerns that these groups are treated unfairly, or will be subjected to the detriments of stereotype and seclusion which plagued the weaker segments of society in the past and now might be transferred to new forms of population segments. Furthermore, the condition of these groups might potentially be far worse than even those commonly discriminated against in the past and today. These new groups might be dispersed throughout society. Thus, they will lack the minimal political force to bring the issues of their misfortune to the forefront of the legal discussion. Even worse, given the inherent obscurity of the data mining practices (features this book tries to somewhat mitigate in the discussion set out in Chapter 17) those adversely impacted by these processes might not even know this is happening!

Dealing with these new sets of concerns calls for substantially altering the current understanding of anti-discrimination theory; for many years now the focus of discrimination law has been on generation of bigotry towards "protected groups" which were (and some still are) systematically discriminated against. Even with the bigotry gone, much of the structural discrimination is set in place. Therefore, much of the existing law is tuned towards discrimination which is motivated by discriminatory intent – even if such intent cannot be proven. Or, it is focused upon reintegrating insular groups into the general society. Yet the novel forms of discrimination data mining might be setting forth do not feature these elements. Therefore, they call upon academics and policymakers to rethink the theory and practice of discrimination in this unique context.

As these last few paragraphs indicate, data mining practices might lead to new and serious fears of growing discrimination. Therefore, should this understanding not lead to an overall recommendation to limit or even ban these forms of analyses? While this recommendation might have merit in some limited contexts, we generally find it should be treated with caution. The issue of discrimination in the context of data mining must be approached with an open mind. While the potential detriments must be acknowledged, we must also consider a very different option – public intuition is wrong, and data mining does *not* pose serious or unique discrimination-based concerns.

Furthermore, we must consider whether the discrimination-based concerns have resulted from an irrational, Luddite-like fear of these advanced models.[26] Or, it is also possible that a much greater and sinister force is in play. The seeming intuition that data mining leads to unacceptable discrimination is merely a manipulation of the powerful trying to influence the weak. While automated practices might finally lead to equal treatment, they might compromise the elite's dominance and subject them to the same level as scrutiny as everyone else (something they are not used to). Therefore, the elite might forcefully advocate against these practices, pointing us back in the direction of human discretion which has ruled in their favor time and again (and in that way hiding its self-interest). For that reason, legal and other scholars must exercise extreme caution when pointing to the discrimination-based flaws of data mining practices – as they might be merely pawns in a much greater game. These last few arguments might seem unnecessarily paranoid and probably are. Yet they still demonstrate the importance of seeking out a sound analytical foundation to any regulatory step taken to battle discrimination in the novel context of data mining.

[26] See discussion in Taipale, K.A. (2004). For a very different perspective, see Solove, D. (2011).

19.4 The Future of Privacy and Data Protection

From a legal perspective, the right to privacy and personal data protection are the main bulwarks against risks associated with data mining and profiling. Technological developments influence our perception and notion of privacy. New applications may limit the privacy of the individual and the processing of personal data may entail risks for the individual. In a sense, the right to privacy and the right to data protection try to re-erect barriers of access to the private sphere that have been removed by the possibilities of the technology. An important function of the right to privacy is therefore to regulate the use of technologies that can be used to encroach upon the private sphere. Since technological developments raise new questions on how to interpret the right to privacy, the legal framework for privacy protection is in a constant state of flux.

Originally, the private sphere was made up of the home, the family life, and correspondence. Mainly as a result of digitization, the private sphere has grown to include personal data. As described in the introduction of this chapter, by incorporating personal data into the private sphere, a new type of privacy emerged: informational privacy.[27]

An important aspect of informational privacy is personal data protection. Given the growing importance of personal data processing in modern society, the OECD set forth principles for the protection of personal data in 1980.[28] The goal of these principles was not only to protect personal privacy but also to ensure that disparities in national privacy laws would not lead to interruptions in the trans-border flows of data. These OECD principles, together with the Council of Europe Convention for the Protection of Individuals with regard to Automatic Processing of Personal Data of 1981, formed the basis for the EU Data Protection Directive (95/46/EC), which was adopted in 1995.

Since 1995 a lot has happened. In particular the advent of the Internet has led to the massive proliferation and processing of personal data. It is estimated that the average person is now registered in several hundred databases.[29] Apart from the digital traces we leave behind on the Internet, we also leave more and more digital traces in the physical world, through technologies like smartphones and RFID-chips that enable geo-location and interaction with the Internet. As such, the physical world and the 'virtual world' merge to an increasing extent. Augmented reality, which enables us to enhance the physical world with a layer of digital information, is a good example of this development.[30]

The question is thus whether the current legal framework for data protection is adequately suited to deal with these new technological developments and applications of (personal) data. Because the Data Protection Directive is technology

---

[27] See the Section 19.1 for Westin's definition of privacy.

[28] OECD Recommendation of the Council concerning guidelines governing the protection of privacy and transborder flows of personal data (23 September 1980)

[29] Schermer & Wagemans (2009).

[30] See Chapter 11.

neutral, it has withstood the test of time remarkably well. Nevertheless, the Data Protection Directive is starting to show its age. The ever-expanding scope of the Data Protection Directive has led to a patchwork of case law and interpretations by the Data Protection Authorities. Furthermore, there is a lack of awareness on data protection, costs of compliance and administrative burdens are high, and enforcement seems ineffective.

Therefore, the EU is currently in the process of rethinking the legal framework for data protection. To further strengthen and harmonize data protection in Europe, a proposal for a General Regulation on Data Protection was published by the European Commission on January 25th 2012. With regard to the topic of this book, the most relevant development is that more strict rules on profiling and automated decision making are introduced. Furthermore, many ideas in this book are also considered in the new Regulation (e.g., 'privacy by design' and the 'right to be forgotten').

In particular the notion of privacy by design is relevant when we look towards the future of privacy in relation to profiling. As the technology advances, the necessity to build limitations, restraints and protective measures into the technology itself becomes apparent. Regulation through the technology itself is oftentimes more efficient and effective than traditional modes of regulation and enforcement. However, we must also be cautious not to overestimate the power and possibilities of privacy by design. Furthermore, we must not take too narrow an approach when it comes to privacy by design. As the legitimacy of personal data processing is always context-dependent and the right to privacy is not absolute, simply prohibiting the use of certain types of data, or limiting the use of personal data to or within a particular application, is most likely not a viable option.

Furthermore, it is likely that as the technological capabilities for gathering and processing (personal) data continue to grow, the borders of the personal sphere will recede further. As a result of this the *communis opinio* on what is considered a reasonable expectation of privacy may also change. To ensure the protection of the interests of the data subjects we must look beyond the protection mechanism itself (i.e., privacy) and more towards the underlying goals (e.g., equal treatment, prevention of harm). From a technical perspective this may even mean that we need to process data in a way that is currently at odds with the current system of data protection.

Thus, in the future, privacy might be less about erecting barriers when it comes to processing personal data, but more about defining boundaries in terms of the ethical use of IT in general and personal data in particular. In the context of data mining and profiling, avoiding discrimination will likely be one the most important aspects of ethical IT design. When we look at this possible future of privacy, we may conclude that although the new Regulation on Data Protection would be a significant step towards strengthening data protection within the EU, questions remain. Given the fact that the Data Protection does not significantly change the current system of data protection, it is questionable whether it will address the issues raised in this book effectively. The new Regulation will most likely not remedy the failure of access control, the failure of anonymity and the failure of purpose specification. It is also questionable whether it will facilitate technical solutions such as those proposed in this book. While we have focused on the legal framework for privacy and data protection in Europe, similar conclusions may likely be drawn for other legal systems in the world. We hope that this book will contribute to the ongoing discussion on how the protect privacy and avoid discrimination using both law and technology.

References

Calders, T., Verwer, S.: Three Naive Bayes Approaches for Discrimination-Free Classification. Special issue of ECML/PKDD (2010)

Cate, F.H.: Government, Data Mining: The Need for a Legal Framework. Harvard Civil Rights-Civil Liberties Law Review 43, 436 (2008)

Custers, B.H.M.: Data Mining with Discrimination Sensitive and Privacy Sensitive Attributes. In: Proceedings of ISP 2010, International Conference on Information Security and Privacy, Orlando, Florida, July 12-14 (2010)

Denning, D.E.: Cryptography and Data Security, p. 17. Addison-Wesley, Amsterdam (1983)

Ohm, P.: Broken Promises of Privacy: Responding to the Surprising Failure of Anonymization. UCLA Law Review 57, 1701–1765 (2010)

Pedreschi, D., Ruggieri, S., Turini, F.: Discrimination-aware Data Mining. In: 14th ACM International Conference on Knowledge Discovery and Data Mining (KDD 2008), pp. 560–568. ACM (August 2008)

Schermer, Wagemans: Onze digitale schaduw, een verkennend onderzoek naar het aantal databases waarin de gemiddelde Nederlander geregistreerd staat. College Bescherming Persoonsgegevens, Den Haag (2009) (in Dutch)

Shannon, C.E.: The mathematical theory of communication. Bell Systems Technology Journal 27, 379–423, 623–656 (1948)

Shannon, C.E.: Communications theory of secrecy systems. Bell Systems Technology Journal 28, 656–715 (1949)

Solove, D.: Nothing to Hide: The False Tradeoff Between Privacy and Security. Yale University Press (2011)

Taipale, K.A.: Technology, Security and Privacy: The Fear of Frankenstein, the Mythology of Privacy and the Lessons of King Ludd. Yale Journal of Law and Technology 7(123) (December 2004)

Van den Berg, B., Leenes, R.: Audience Segregation in Social Network Sites. In: Proceedings for SocialCom2010/PASSAT2010 (Second IEEE International Conference on Social Computing/Second IEEE International Conference on Privacy, Security, Risk and Trust), pp. 1111–1117. IEEE, Minneapolis (2010)

Vedder, A.H.: KDD: The Challenge to Individualism. Ethics and Information Technology (1), 275–281 (1999)

Verwer, Calders: Three Naive Bayes Approaches for Discrimination-Free Classification. In: Data Mining: Special Issue with Selected Papers from ECML-PKDD 2010, Springer (2010)

Warren, S.D., Brandeis, L.D.: The right to privacy; the implicit made explicit. Harvard Law Review, 193–220 (1890)
Weitzner, D.J., Abelson, H., et al.: Transparent Accountable Data Mining: New Strategies for Privacy Protection. MIT Technical Report. MIT, Cambridge (2006)
Westin, A.: Privacy and Freedom. Bodley Head, London (1967)
Zarsky, T.Z.: "Mine Your Own Business!": Making the Case for the Implications of the Data Mining of Personal Information in the Forum of Public Opinion. Yale Journal of Law & Technology 5, 56 (2003)

Author Index

Subject Index

E

Q

R

MIX
Papier aus verantwortungsvollen Quellen
Paper from responsible sources
FSC® C105338

If you have any concerns about our products,
you can contact us on
ProductSafety@springernature.com

In case Publisher is established outside the EU,
the EU authorized representative is:
Springer Nature Customer Service Center GmbH
Europaplatz 3, 69115 Heidelberg, Germany

Printed by Libri Plureos GmbH
in Hamburg, Germany